ENGLISH-CHINESE MARINE MANAGEMENT DICTIONARY

英汉海洋管理词典

Edited by LIU Dahai　　YANG Hong

刘大海　杨　红　主编

海洋出版社

2017 年·北京

图书在版编目(CIP)数据

英汉海洋管理词典/刘大海,杨红主编. —北京:海洋出版社,2017.4
ISBN 978-7-5027-9741-6

Ⅰ.①英… Ⅱ.①刘… ②杨… Ⅲ.①海洋-管理-词典-英、汉 Ⅳ.①P7-61

中国版本图书馆 CIP 数据核字(2017)第 050686 号

责任编辑:方 菁
责任印制:赵麟苏

海洋出版社 出版发行

http://www.oceanpress.com.cn
北京市海淀区大慧寺路 8 号　邮编:100081
北京朝阳印刷厂有限责任公司　　新华书店发行所经销
2017 年 4 月第 1 版　2017 年 4 月北京第 1 次印刷
开本:880mm×1230mm　1/32　印张:11.125
字数:500 千字　定价:68.00 元
发行部:62147016　邮购部:68038093　总编室:62114335
海洋版图书印、装错误可随时退换

ENGLISH-CHINESE MARINE MANAGEMENT DICTIONARY
Editorial Committee

Consultants: DING Dewen　LI Tiegang　MA Deyi　WANG Dianchang

　　　　　　　SHEN Jun　SHI Xuefa　WANG Fang

　　　　　　　SHAO Guilan　HU Guobin

　　　　　　　XU Wei　GUAN Lijuan

Editor-in-chief: LIU Dahai　YANG Hong

Editorial members: LIU Dahai　YU Ying　XING Wenxiu

　　　　　　　　　LI Xiaoxuan　WANG Chunjuan　LIU Yang

　　　　　　　　　MA Yunrui　MA Xuejian　ZHANG Jinxuan

　　　　　　　　　WANG Jing　CHEN Ye　GONG Wei

Translation members: YANG Hong　YU Ying

　　　　　　　　　　XIN Haiyan　ZOU Shanshan　SUN Guangfeng

　　　　　　　　　　LIU Yan　LIU Huiran　XING Wenxiu

　　　　　　　　　　LI Xiaoxuan　XU Meng

《英汉海洋管理词典》
编撰委员会

顾　问：丁德文　李铁刚　马德毅　王殿昌　沈　君
　　　　石学法　王　芳　邵桂兰　胡国斌　徐　伟
　　　　关丽娟

主　编：刘大海　杨　红

编写组：刘大海　于　莹　邢文秀　李晓璇　王春娟
　　　　刘　洋　马云瑞　马雪健　张金轩　王　晶
　　　　陈　烨　宫　伟

翻译组：杨　红　于　莹　辛海燕　邹姗姗　孙广峰
　　　　刘　艳　刘会然　邢文秀　李晓璇　徐　孟

Editors' Note

In 2012, the report of the 18th National Congress of the CPC stated definitely that we should enhance our capacity for exploiting marine resources, develop marine economy, protect marine ecological environment, resolutely safeguard China's maritime rights and interests, and build China a maritime power. Recently, President Xi Jinping pointed out that we should give more concern to ocean, learn about ocean, exploit and develop ocean and put forth effort to promote "Four Shifts". President Xi's speech offered a clear direction and crystallized a goal for the development of marine industries of our country, marked that China has entered a new stage in ocean development and management.

With the rising status of marine development and exploitation, the increasing awareness of ocean from society and the deepening of international exchanges in the ocean field, the non-standard related concepts have made problems such as communication difficulties and research ambiguities, increasingly prominent. Therefore, for the purpose of truth-seeking and on the basis of the current situation of the development of marine economy and management at home and abroad, this book compiling group selects up-to-date terms from relevant standards, documents, monographs and papers, standardizes their definitions, scopes and translations, and endows them with the latest ideas in the hope that it can become a popular and practical reference book helpful both for marine staff to conduct research and practice, and for the increasing popularity of knowledge on marine economy and management.

Nothing can be accomplished without norms or standards. At present, with a rapid and vigorous development of marine economy and a successive introduction of marine planning policies, scientific and standard terms are urgently needed to ensure the accurate and smooth implementation of policies. Meanwhile, as international marine exchanges become more frequent, there are higher requirements for the English-Chinese translation of terms on marine economy and management. In view of the above, this research team has spent two years, consulting a large number of materials including laws and regulations, policy planning, terms, etc. to ensure the scientific authority of terms and the standardization and practicality in editing.

As this is the first version, there may be limitations and omissions in this Marine Management Dictionary (English-Chinese Version). Therefore, readers' comments and opinions would be highly appreciated. E-mail : mpc@ fio.org.cn.

Special thanks are given to "Non-profit Marine Science Research Project" which funds the publishing of this book.

30th June, 2015

编者的话

2012年党的十八大报告明确提出要"提高海洋资源开发能力,发展海洋经济,保护海洋生态环境,坚决维护国家海洋权益,建设海洋强国",习近平总书记指出,要进一步关心海洋、认识海洋、经略海洋,着力推动"四个转变",这为我国海洋事业发展指明了方向,明确了目标,标志着我国海洋开发与管理进入一个全新的发展阶段。

随着海洋开发地位的不断提升,社会各界对其认识不断提高以及海洋领域国际交流的日益深化,但因相关概念不规范,造成交流困难、研究歧义等问题。针对此问题,本词典编写组本着求真求实的宗旨,根据国内外海洋开发与管理发展现状,选取相关标准、文件、专著和论文中的最新术语,规范其定义、范围与译名、译文,并赋予最新理念。希望其不但有助于海洋工作人员开展研究与实践,更能向大众普及相关的海洋开发与管理常识,最终成为一本深受广大读者欢迎的实用工具书。

没有"规矩"不成方圆,当前我国海洋事业蓬勃快速发展,各类海洋规划政策纷纷出台,迫切需要科学规范的术语以保证政策的准确贯彻和顺利实施;同时,国际海洋交流的日益频繁也对海洋开发与管理术语的中英文对接提出了更高的要求。结合以上,课题组耗时两年,查阅了大量法律法规、政策规划、术语标准等资料,力求做到术语上的科学权威,编排上的规范实用。

作为《英汉海洋管理词典》第一版,书中可能存在缺欠疏漏之处,诚望读者批评指正!邮箱地址:mpc@fio.org.cn。

2015 年 6 月 30 日

Guide to the Use of the Dictionary

Headword

1. This dictionary compiles a total of 2,151 terms which cover the new progress in the activities and ideas of marine economy and management in recent years, reflecting the advanced achievements in the field. The dictionary unifies and standardizes the expressions of marine economic terms.

2. Each term is accompanied by its corresponding Chinese term. A standard term generally corresponds to only one Chinese expression.

3. The first letter of the English word is all in lowercase when it can be either uppercase or lowercase.

Page layout

1. Standard terms are preferred in compiling format. Other terms such as "also called", "synonym", "permitted", "alias", are listed in annotations in the way of "also called".

2. The English abbreviation follows the English term.

3. When the term has different definitions in different professional fields, its professional field is marked in parenthesis before the definition.

4. When the term has a wider range of applications, such as "interference" and "slope", it is defined based on the ocean and related fields in this dictionary.

Annotations

1. Standard modern Chinese is used in annotations, all in simplified Chinese characters.

2. Terminology annotations in this dictionary are completed by referring to the national and industry standards in the areas of marine economy and management. The reference materials mainly include:

①Authoritative literature of laws, regulations, standards, etc.

②Literature which is widely acknowledged by academic organizations such as textbooks, scientific papers, scientific journals, etc.
③Terminology database.
④Glossary of terms set, dictionaries, encyclopedia and thesaurus.

3. Appropriate English is selected based on Chinese annotations in translation, which is not entirely consistent with the Chinese annotations.

Others

1. The compilation of this dictionary is led by the Marine Policy Research Center of the First Institute of Oceanography of State Oceanic Administration (SOA).

2. The scientific terms used in this book are subject to the various scientific terms published by the China National Committee for Natural Scientific Terms. For those which are not verified or have not been unified, idioms are followed.

3. The legal units of measurement of China are adopted for all the related units of measurement in this dictionary and the statistical figures are written with Arabic numerals.

4. The related terms of marine economy and management are so abundant that the number of them may far exceed what is covered in this dictionary. The terms selected in this book is based on the results of previous studies and current situations.

5. This dictionary is to be revised and edited every two or three years so as to add new concepts and terms needed for the development and management of marine economy, facilitating better development of marine causes.

凡　例

词目

1. 本词典共收录术语 2 151 个，约计 500 千字，充分吸收海洋管理相关领域近年来在活动、理念方面的新进展，力求反映海洋开发与管理领域的前沿成果，形成海洋管理术语的统一规范表述。

2. 每个词都附有对应的中文，一个规范术语，一般只对应一个中文词。

3. 凡英文词的首字母大小写均可时，一律小写。

编排

1. 编写格式为"规范术语"，即首推选用的术语；其他"又称"、"同义词"、"许用"、"别名"等术语，全部以"又称"方式放到释文中。

2. 英文缩略排在英文之后。

3. 若一个术语在不同的专业领域有不同的概念，则在释文前的角括号中标明所属的专业领域。

4. 若一个术语适用范围较广，如"干涉"、"斜坡"等，本词典皆立足海洋及其相关领域，对其进行解释。

释文

1. 释文使用规范的现代汉语，一律使用通用简化字。

2. 本词典的术语释文是在参考海洋管理相关领域的国家标准、行业标准的基础上完成的，其中查阅资料的主要类型包括：

① 法律、法规、标准等权威性文献；

② 教科书、科学论文、科技期刊等学术团体普遍公认的文献；

③ 术语数据库；

④ 术语词汇集、辞典、百科全书、叙词表。

3. 由于中英文语法和表达习惯不同，汉英释义并非完全一致。

其他

1. 本词典由国家海洋局第一海洋研究所海洋政策研究中心牵头组织编写。

2. 本词典所用的科学名词,以全国自然科学名词审定委员会公布的各学科名词为准,未经审核或尚未统一的,遵从习惯。

3. 本词典有关计量单位采用中华人民共和国法定计量单位,统计数字用阿拉伯数字记写。

4. 海洋管理相关术语数量繁多,实际上远不止此数,本词典立足前人研究成果,根据现阶段的实际情况进行甄录。

5. 出版以后,本词典拟2~3年修编一次,以期补充海洋经济与管理发展所需的新概念、新术语,助力海洋事业更好地发展。

目 次

正文 …………………………………………………………（1）
参考文献 ……………………………………………………（338）

A

abrupt change of climate 气候突变

It is a phenomenon occurring when a climate system suddenly changes from one stable condition to another stable condition.

气候从一种稳定状态跳跃到或转变为另一种稳定状态的现象。

abundance of natural resources 自然资源丰度

It indicates the total amount of certain natural resources owned by a geographical unit and the state of the resources compared with a comparable region. It may also indicate the grade of utilisation of certain natural resources or the proportion of high-grade resources in a geographical unit.

表明一个地域单元所拥有的某种自然资源的总量及其与可比地域相比较的状况，或一个地域单元所拥有的某种自然资源中可利用品位或高品位资源所占的比例。

abysmal area 深海区

It refers to a sea area with a water depth of 2-10 kilometres.

深海区指水深在 2~10 千米的海域。

abyssal clay 深海黏土

It refers to the clay in brown red or marron color in the deep sea and the sediment distributed in the deep sea, whose particle size is no more than 0.0039 millimetres.

又称"远洋黏土"、"褐黏土"、"红黏土"，深海呈褐红色、棕红色分布的黏土，分布于深海，粒径不大于 0.0039 毫米的沉积物。

abyssal hill 深海丘陵

It refers to a small uplift zone at the bottom of the ocean, composed of lava flows and laccolith, which is hundreds of metres long with small slopes in the shape of circles or ovals. Some abyssal hills are long and appear in rows, such as parallel ridges with the width of 1-10 kilometres.

大洋底小面积隆起的地带。由熔岩流和岩盖组成，起伏数十米至数百米，坡度较小，外形多呈圆形、椭圆形，也有长条状分布，趋向于成列出现。如平行的山脊，宽 1~10 千米。

abyssal plain 深海平原

It refers to a flat area at the bottom of an ocean basin, generally located between continental slopes and submarine hills, whose water depth is between 3,000 - 5,000 metres.

大洋盆地底部的平坦区域，通常位于陆坡(陆隆)和海底丘陵之间，水深一般为 3 000~5 000 米。

abyssal sediment 深海沉积

It refers to the sediment existing in the deep sea which is below 2,000 metres, including both biogenic sediments and non biogenic sediments.

大洋中 2 000 米以深的深海范围内的沉积，既有生源沉积物，又有非生源沉积物。

abyssal zone 深渊带

It refers to a concave area with a clear outline in an abyssal sea with a depth of more than 6,000 metres.

深海中轮廓清楚的"凹"形地区,深度大于 6 000 米。

accumulation of mud 淤积

It is a process in which sediment particles from a current sink down to a river bed, causing the uplift of bottom elevation or river bank.

水流挟带的泥沙颗粒沉落到河床上致使河底高程抬升或河岸淤涨的过程。

accumulated island 堆积岛

It refers to an island formed due to the accumulation or deposition of rivers, tides and waves in an estuary area or coastal zones.

在河口区及海岸带中由于河流、潮流和波浪的堆积作用而形成的岛屿。

actual amount of foreign investments 实际使用外资金额

They refer to overseas remittance, equipment, technology, etc. financed by governments at all levels, departments, enterprises and other economic organizations through overseas borrowing, Foreign Direct Investments (FDI) absorption, etc. The unit of measurement is one hundred million yuan.

我国各级政府、部门、企业和其他经济组织通过对外借款、吸收外商直接投资以及用其他方式筹措的境外现汇、设备、技术等。计量单位:亿元。

actual expenditure 实际支出

It is the amount of money spent on products and services by families, enterprises and governments.

家庭、企业和政府花在产品和服务上的数额。

added value of primary industry 第一产业增加值

It refers to the added value created in the production process by resident units and the transferring value of fixed assets of the primary industry (including agriculture, forestry, husbandry and fishery). The unit of measurement is ten thousand yuan.

第一产业(农、林、牧、渔业)常住单位生产过程创造的新增价值和固定资产的转移价值。计量单位:万元。

added value of secondary industry 第二产业增加值

It refers to the added value created in the production process by resident units and the transferring value of fixed assets of a secondary industry (such as mining industry, manufacturing industry, electricity production and supply industry, gas and water industry, or construction industry). The unit of measurement is ten thousand yuan.

第二产业(采矿业,制造业,电力、煤气及水的生产和供应业,建筑业)常住单位生产过程创造的新增价值和固定资产的转移价值。计量单位:万元。

added value of tertiary industry 第三产业增加值

It refers to the added value created in

the production process by resident units and the transferring value of fixed assets of a tertiary industry (other industries except for primary industries and secondary industries). The unit of measurement is ten thousand yuan.

第三产业(除第一、二产业以外的其他行业)常住单位生产过程创造的新增价值和固定资产的转移价值。计量单位:万元。

adjacent coasts 相邻海岸

They are coasts on both sides of the land borders between two neighboring countries.

两个相邻国家之间的陆地边界两侧的海岸。

administration management of resources 资源行政管理

It is a term that indicates that national resource management agencies use administrative means to manage development, utilization and protection of resources.

国家资源管理机构采用行政手段对资源开发利用与保护进行的管理。

advance coast 前进海岸

It refers to a coast extending into the sea, which varies in location and appearance and is generally distributed in an estuary of a delta region.

向海推进的海岸。海岸的位置、形态经常发生变化,多分布于大江河口三角洲地区。

advance of coast 海岸主体走向

It refers to the overall direction of coastal extension passing through a starting point.

通过起始点的海岸整体延伸方向。

advance of sea 海进

It refers to the process during which seawater slowly transgresses into land from the coast due to the rising of sea level or sink of crustal structures.

又称"海侵",由海平面上升或地壳构造下沉等引起的海水缓慢地从海岸入侵陆地的过程。

advection 平流

It refers to the horizontal movement of a large group of air.

大块空气的水平运动。

aeolation 风化作用

It refers to the process in which the physico-chemical properties of minerals and rocks at or near the Earth's surface have undergone changes under the action of solar radiation, atmosphere, water, and living things. During the process, the rocks and materials break down into grains or soils, and their mineral compositions change into new substances.

地表岩石与矿物在太阳辐射、大气、水和生物参与下理化性质发生变化,颗粒细化,矿物成分改变,从而形成新物质的过程。

affluent 支流

It is a river that flows into a main stream directly or indirectly.

直接或间接流入干流的河流。

Agenda of 21st Century《21世纪议程》

It is one of the important documents passed at the *United Nations Conference on Environment and Development*, convened in

Rio de Janeiro, Brazil, from June 3rd to 14th in 1992 and it is a worldwide sustainable development program. It is a comprehensive action blueprint of governments, UN organizations, development agencies, non-governmental organizations, and independent groups on a worldwide scale in various aspects of environmental impacts caused by human activities from the time it was passed to the 21th century.

1992年6月3日至14日在巴西里约热内卢召开的"联合国环境与发展大会"通过的重要文件之一,是"世界范围内可持续发展行动计划",它是从缔约期至21世纪在全球范围内各国政府、联合国组织、发展机构、非政府组织和独立团体在人类活动对环境产生影响的各个方面的综合行动蓝图。

agricultural and fishery zone 农渔业区

It refers to a sea area which is exploited to expand agricultural development spaces, develop marine biological resources, construct fisheries infrastructure for agricultural reclamation, fishing pot and breeding grounds, enhance marine culture and fishery resources, protect and maintain precious fish species, which covers agricultural reclamation area, fisheries infrastructure areas, aquaculture areas, mariculture and fishery resources enhancement areas, fishing areas and aquatic species protection reserves.

农渔业区是指适于拓展农业发展空间和开发海洋生物资源,可供农业围垦、渔港和育苗场等渔业基础设施建设、海水增养殖和捕捞生产,以及重要渔业品种养护的海域,包括农业围垦区、渔业基础设施区、养殖区、增殖区、捕捞区和水产种质资源保护区。

agricultural modernization 农村现代化

The ultimate goal of modernization, namely, the ultimate goal of Chinese countryside modernization is as follows: people possess higher ideological and ethical standards, cultural and scientific qualities and healthy qualities; people enjoy a high standard of living; stable and harmonious social structure is constructed; balanced and harmonious relationships are established between human and nature, ecology and resources. Agricultural modernization covers the following aspects: (1) scale expansion and technicalization of agriculture; (2) non-agriculturisation of farmers; (3) urbanization of the rural population; (4) marketisation of economic activities; (5) rationalisation of resource utilisation; (6) normalization of social relations; (7) integration of human and nature.

现代化的最终目标是:人民具有较高的思想道德素质、科学文化素质和健康素质;人们过上高质量的生活,社会结构实现持久的和谐与稳定;人与自然、生态和资源之间形成均衡与协调关系。这也是中国农村现代化的最终目标。其内涵是:(1)农业规模化和科技化;(2)农村职业非农化;(3)农村人口城镇化;(4)经济活动市场化;(5)资源利用合理化;(6)社会关系规范化;(7)人与自然一体化。

aground sweeping 拖底扫海

It means that two survey vessels (boats), which are kept a certain width apart in parallel, drag the cables along a predetermined course to detect whether there are navigation obstructions, such as reefs and sunken vessels, in the surveyed area while all bottom cables of a sweeper used for sweeping surveys are required to arrive at the bottom of the sea.

将扫海测量的扫海具底索全部着底,由两条测量船(艇)沿预定的航向保持一定宽度平行拖行,探测测区内是否存在礁石、沉船等航行障碍物。

air-sea interface 海-气界面

It refers to an interface where sea and air connects, interacting, restricting and adapting to each other mutually.

海洋与大气相接的面,出现海-气相互影响、相互制约、彼此适应的作用。

alerting services of marine disasters 海洋灾害预警服务

It refers to service activities of the alerting and forecasting for marine disaster.

对海洋灾害的预警、预报服务活动。

alga 藻类

It is a generic term for lower aquatic plants that contain assimilative pigment and can therefore live an independent nutritional life.

泛指具同化色素而能进行独立营养生活的水生低等植物的总称。

algal reef 藻礁

It is a reef formed after algal calcification.

藻类钙化后而形成的礁体。

alkali products manufacturing 氯碱产品制造

It is the manufacturing of caustic soda, chlorine and hydrochloric acid by using sea salt or sea salt brine.

以海盐或海盐卤水为原料生产烧碱、氯气、盐酸的活动。

alluvial deposit 冲积物

It refers to materials such as silt, sand and gravel carried and deposited by rivers flowing throughout the year when rivers are running slowly. Alluvial deposits are usually most extensive in the lower part of a river course, forming floodplains and deltas, but they may form at any point where a river overflows its banks.

常年流水(主要是河流)所挟带、搬运的碎屑物,当水流能量降低时而堆积下来的物质。

alluvial fan 冲积扇

It is a fan-shaped or cone-shaped deposit of sediment accumulated by streams. As a stream's gradient decreases from the mountain pass down to the flat areas, it drops coarse-grained materials. This reduces the capacity of the channel and forces it to change direction and gradually build up a slightly mounded or shallow conical fan shape.

山地河流从出山口进入平坦地区以后,因坡降骤减,水流搬运能力大为减弱,部分挟带的碎屑物堆积下来,形成从出口顶点向外辐射的扇形堆积体。

alluvial island 冲积岛

It refers to low-lying islands formed by the accumulated alluvial deposit of rivers.

河流冲积物堆积成的岛屿,地势低平。

alluvial plain 冲积平原

It is a flat landform created by the deposition of sediment over a long period of time by one or more rivers coming from highland regions due to slow flowing speed and reduced energy.

河流搬运的碎屑物,因流速减缓、能量降低而逐渐堆积下来所形成的平原。

alluvion 冲积层

It describes the increase in the area of land due to sediment (alluvium) deposited by a river. This changes the size of a piece of land (a process called accession) and thus its value over time.

冲积物组合在一起而形成的沉积层。

alternating tidal current 往复流

It is a periodic flow of tide, evolving from an elliptic trend to a linear trend.

潮流椭圆蜕化为直线的周期性流动。

altitudinal regional economic cooperation 垂直区域经济合作

It refers to the cooperation in which the production elements provided by the cooperative parties have different processing depth and technology due to the large gap between their economic and technological level.

合作双方经济技术水平差距较大,所提供生产要素的加工深度和技术层次不同的合作活动。

ambient noise of the sea 海洋环境噪声

It refers to noise received by a hydrophone in the sea except self-noise, which includes ocean noise, biological noise, seismic noise, rain noise, artificial noise (from navigation, industry, and drilling), etc.

在海洋中由水听器接收到的除自噪声以外的一切噪声。包括海洋噪声、生物噪声、地震噪声、雨噪声和人为噪声(航海、工业、钻探)等噪声。

amount of island freshwater 海岛淡水资源量

It refers to the total amount of island freshwater resources in a reporting period, including surface water and underground water. The measurement unit is cubic metre.

报告期内海岛淡水资源的总量,包括地表水和地下水。计量单位:立方米。

amphidromic point 无潮点

Also called *amphidromic region*, it is a point without tidal undulation.

又称"无潮区",指海面无潮位升降的海域。

amphidromic region 无潮区

Also called *amphidromic point*, it refers to a region without tidal undulation.

又称"无潮点",指海面无潮位升降的海域。

amplitudes of groundwater level fluctuation 地下水水位变幅

It refers to the difference value between the maximum value and minimum value of the groundwater level in a certain time period.

某一时间期内地下水水位最大值与最小值的差值。

An Inconvenient Truth《难以忽视的真相》

It is a documentary film on climate change adapted from a book written by former United States Vice President Al Gore. It reveals the data and makes predictions about climate change. Al Gore illustrated his belief about the cruel future that the relationship between human civilization and the planet's ecosystem will change significantly based on three aspects like population explosion, science and technology revolution, and the basic way to deal with climate crisis.

《难以忽视的真相》是根据美国前副总统阿尔·戈尔所著书籍改编的一部有关气候变迁的纪录片,其中揭露了气候变迁的资料并对此做出预测,通过人口爆炸、科技革命、应对气候危机的基本途径三个方面阐述他相信人类文明与地球生态系统之间关系将发生重大变化的残酷的未来。

anchorage 锚地

It refers to a location at sea for ship berthing(casting an anchor or mooring a buoy) and carrying on offshore operation (such as joint inspection, formation and dismissal of barges, and lightering).

供船舶停泊(抛锚或系浮筒)和进行各种水上作业(例如联检、编解队、过驳)的水域。

anniversary winds 季节风

They are local winds or large-scale system of winds circulating once annual basis, such as monsoon.

以年为周期循环的局地风或大尺度风系(如季风)。

annual producing capacity of sea salt 海盐年生产能力

It refers to the comprehensive balance capacity of all equipments owned by enterprises for producing sea salt. Sea salt producing which requires open work is influenced by climate. Therefore, the production capacity of mature beach land is calculated by the actual average yield per unit area of production within 10 years multiplied by the production area of mature beach land this year, while the production capacity of new beach lands is calculated by potential production based on the designed capacity and maturity of beach land. The unit of measurement is ten thousand tons.

企业生产海盐的全部设备的综合平衡能力。海盐生产露天作业,受天气影响,因而计算生产能力时,成熟滩田按10年实际平均单位生产面积产量乘以本年成熟滩田生产面积计算,新滩田按设计能力及滩田成熟程度可能达到的产量计算。计量单位:万吨。

anthropogenic climate change 人致气候变化

It refers to the climate changes caused by human activities (such as deforestation, aeroplane flights, vehicle emissions, and industrial and agricultural production).

由于人类活动(如森林砍伐,飞机飞行、汽车排放、工农业生产)而造成的气候变化。

antinode 波腹

A node is a point along a standing wave where the wave has the minimum amplitude. The opposite of a node is an antinode, a point where the amplitude of the standing wave is a maximum.

驻波在空间内特定量振幅为最大值处的点或轨迹。

appreciation 升值

It refers to an increase in the value of one country's currency measured by the amount of foreign currency that can be purchased.

按所能购买到的外国货币量衡量一国货币的价值增加。

apron 海幔

It refers to a gentle slope on the bottom of the sea with a relatively small surface undulation formed by surrounding islands, archipelagos, seamounts and seamount groups.

环绕岛、群岛、海山和海山群形成的海底平缓斜坡,表面起伏较小。

aquaculture 水产养殖

It is the farming of economic aquatic organisms by using a variety of waters or tidal flats.

利用各种水域或滩涂养殖经济水产动植物的生产活动。

aquaculture stress 养殖压力

It is the stress on the material cycle of a marine ecosystem by output materials from aquaculture production.

通过养殖生产输出物质造成对海洋生态系统物质循环的压力。

aquatic ecosystem 水生生态系统

It is a unitary system, in which organisms and organisms, organisms and non-organisms, interact with each other in water system.

水域系统中生物与生物、生物与非生物成分之间相互作用的统一体。

aquatic environment 水生环境

It is the external environment medium for aquatic organisms to survive. It includes lotic and lentic environments. Flowing water like rivers, streams, springs and ditches are called lotic; still water like ponds, lakes, swamps and reservoirs are lentic. Different aquatic environments have different physical and chemical properties.

水生生物生存的外部环境介质。有静水和流水环境,前者如池塘、湖泊、沼泽、水库,后者如江河、溪流、泉水、沟渠。不同的水生环境理化性质不同。

aquatic fauna 水生动物

They are heterotrophic organisms living in the water. They cannot make food themselves, and they acquire nutrition from plants, other animals and organic residues.

在水中生活的异养生物。它们自身不能制造食物,营养靠摄食植物、其他动物和有机残体获得。

aquatic life 水生生物

It refers to animals and plants living, as a whole or in part, in various waters, including freshwater organisms and marine organisms.

全部或部分生活在各种水域中的动物和植物。包括淡水生物和海洋生物。

aquatic organism 水生生物

It refers to animals and plants living as a whole or in part, in various waters, including freshwater organisms and marine organisms.

全部或部分生活在各种水域中的动物和植物。包括淡水生物和海洋生物。

aquatic plants 水生植物

They are plants required to live at least part of their life in aquatic environments. They are also referred to as hydrophytes or aquatic flora.

至少有一部分生命阶段是在水中度过的植物。

aquatic products 水产品

They are seawater or freshwater economic animals and plants, as well as their processed products.

海水或淡水经济动植物及其加工品。

aquatic species protection zones 水产种质资源保护区

They are established in the main growth and reproduction areas of aquatic species with a high economic value and genetic breeding so as to conserve them and their natural habitats. Without the approval of *the Department of Fishery Administration under the State Council*, any unit or individual is not permitted to be engaged in fishing activities in aquatic species protection zones.

国家保护水产种质资源及其生存环境,并在具有较高经济价值和遗传育种的水产种质资源的主要生长繁殖区域建立水产种质资源保护区。未经国务院渔业行政主管部门批准,任何单位或者个人不得在水产种质资源保护区从事捕捞活动。

Arabian Sea 阿拉伯海

It is a marginal sea located in the northwest of the Indian Ocean between the Arabian Peninsula and the Indian Peninsula.

位于印度洋西北部,亚洲阿拉伯半岛和印度半岛之间,印度洋的边缘海。

archipelagic baseline 群岛基线

It refers to a broken line connecting adjacent base points of islands (the outermost islands of archipelago and the outermost points of a drying reef) and the state baseline of a territorial sea defined by archipelago states. *United Nations Convention on the Law of the Sea* stipulates regulates that an archipelagic state may draw straight archipelagic baselines joining the outermost points of the outermost islands and drying reefs of the archipelago. But the main islands and area should be within such baselines, the length of such baselines shall not exceed 100 nautical miles, except that up to 3 percent of the total number of baselines enclosing any archipelago may exceed that length, up to a maximum length of 125 nautical miles.

一条连接群岛相邻各基点(群岛最外缘岛屿和干礁的最外缘各点)构成的折线。群岛国确定的国家领海基线。《联合国海洋法公约》规定:群岛国可划定连接群岛最外缘各岛和各干礁的最外缘各点的直线群岛基线,但这种基线应包括主要的岛屿和一个区域,这种基线的长度不应超过100海里。但围绕任何群岛的基线总数中至多3%可超过该长度,最长以125海里为限。

archipelagic principle 群岛原则

It refers to the principle of addressing related archipelago issues. According to the *United Nations Convention on the Law of the Sea*, the contents of archipelagic principle mainly include: islands of archipelagic state, waters and other natural features forming geographical, economic and political entity, or which historically have been regarded as such; the sovereignty of an archipelagic state extends to the air space over the archipelagic waters, as well as their bed and subsoil, and the resources contained therein; all other states enjoy the right of innocent passage and archipelagic sea lanes passage through archipelagic waters; an archipelagic state should respect existing agreements with other states and should recognise traditional fishing rights and other legitimate activities of the immediately adjacent neighbouring states in certain areas within archipelagic waters (Article 51 from the *United Nations Convention on the Law of the Sea*).

解决有关群岛问题的原则。按照《联合国海洋法公约》规定,群岛原则的内容主要包括:组成群岛国的各岛屿和其他自然地形应当在本质上构成一个地理、经济和政治的实体,或在历史上已被视为这种实体;群岛国主权及于群岛水域及其上空、海床和底土,以及其中所包含的资源;在群岛水域内,其他国家享有无害通过权、群岛海道通过权。群岛国应尊重与其他国家间的现有协定,并应承认直接相邻国家在群岛水域的某些区域内的传统捕鱼权利和其他合法活动(《联合国海洋法公约》第五十一条)。

archipelagic waters 群岛水域

It refers to the waters enclosed by archipelagic baselines, whose sovereignty extends to the air space over the archipelagic waters, as well as their bed and subsoil, and the resources contained therein.

群岛基线所包围的水域,该水域的海床、底土和上空,以及其中资源均属于群岛国的主权范围,受群岛国主权的支配。

archipelago 群岛

It refers to a group of islands, including parts of islands, interconnecting waters and other natural features, which are closely interrelated to form geographical, economic and political entity, or which historically have been regarded as such.

一群岛屿,包括若干岛屿的若干部分、相连的水域或其他自然地形,彼此密切相关,以致这种岛屿、水域和其他自然地形在本质上构成一个地理、经济和政治的实体,或在历史上已被视为这种实体。

archipelago state 群岛国

It refers to a state constituted wholly by one or more archipelagos or other islands.

全部由一个或者多个群岛构成的国家,并可包括其他岛屿。

Arctic Ocean 北冰洋

Located among Asia, Europe and North America and mostly in the Arctic north polar region, it is the smallest and shallowest of the world's five major oceanic divisions.

位于亚洲、欧洲和北美洲之间,地球最北端,且面积是最小、最浅的大洋。

area 区域

It refers to a large space which holds multiple kinds of resources and where a variety of productive and non-productive social and economic activities can be operated.

拥有多种类型的资源,可以进行多种生产性和非生产性社会经济活动的一片相对较大的空间范围。

area for special use 特殊利用区

It refers to a designated sea area to meet the needs of specific purposes such as scientific research, the dumping of dredged materials and wastes. It includes scientific research test areas, and dumping areas, etc.

为满足科研、倾倒疏浚物和废弃物等特定用途需要而划定的海域。包括科学研究试验区和倾倒区等。

area of marine nature reserve 海洋自然保护区面积

It refers to an area for marine nature reserves established with approval in accordance with the *Nature Reserves Regulations* and *Marine Nature Reserves Management Procedures*, which is classified statistically by national level and regional level. The unit of measurement is square kilometres (km^2).

依照《自然保护区条例》和《海洋自然保护区管理办法》,批准建立的海洋自然保护区的面积。按照国家级和地方级分类统计。计量单位:平方千米。

area of marine special reserves 海洋特别保护区面积

It refers to the area for marine special reserves, established with approval in accordance with *Marine Special Reserves Management Procedures*, which is classified statistically by national level and regional level. The unit of measurement is square kilometres (km^2).

依照《海洋特别保护区管理办法》,批准建立的海洋特别保护区的面积。按照国家级和地方级分类统计。计量单位:平方千米。

artificial beach 人工沙滩

It refers to a beach restored by artificial sand-filling or a new beach formed by dumping fill in coast erosion area, which is generally built in places where there are plenty of sand sources nearby or by hydraulic reclamation in combination with dredging engineering. Artificial beaches are often used to construct bathing beaches or seaside resorts.

在海岸受冲刷地段采用人工补沙方法恢复沙滩原貌或抛填成新的沙滩,称为人工沙滩。一般在附近有大量沙源时采用,或结合疏浚工程吹填。人工沙滩也用于建造海滨浴场或海滨休闲地。

artificial coast 人工海岸

It refers to a coast with a certain inclination, which is constructed artificially by using stones, concretes, bricks, etc.

用石块、混凝土和砖石人工修筑的海岸,具有一定的倾斜度。

artificial fish reef 人工鱼礁

It is a reef which is formed by throwing stones, concrete blocks, waste vehicles, etc. into the bottom of the sea to attract marine organisms to live and reproduce so as to increase fish catch.

有目的地向海底投放石块、混凝土块、废旧车船等物体而形成的暗礁。可吸引海洋生物来此繁衍生息,增加渔获量。

artificial island 人工岛

It refers to an island constructed by human beings or by expanding existing islets. According to the *United Nations Convention on the Law of the Sea*, "artificial islands, installations and structures do not possess the status of islands. They have no territorial sea of their own, and their presence does not affect the delimitation of the territorial sea, the exclusive economic zone or the continental shelf" (Item 8 of Article 60 from *United Nations Convention on the Law of the Sea*); "due notice must be given of the construction of such artificial islands, installations or structures, and permanent means for giving warning of their presence must be maintained. Any installations or structures which are abandoned or disused shall be removed to ensure safety of navigation" (Item 3 of Article 60 from *United Nations Convention on the Law of the Sea*).

人工建筑或拓固礁石形成的岛屿。《联合国海洋法公约》规定:"人工岛屿、设施和结构不具有岛屿地位。它们没有自己的领海,其存在也不影响领海、专属经济区或大陆架界限的划定"(《联合国海洋法公约》第六十条第8款);"这种人工岛屿、设施或结构的建造,必须妥为通知,并对其存在必须维持永久性的警告方法。已被放弃或不再使用的任何设施或结构,应予以撤除,以确保航行安全"(《联合国海洋法公约》第六十条第3款)。

artificial lake 人工湖

An artificial lake, namely a reservoir, is a hydraulic engineering building which is constructed for flood retention, water storage, water flow regulation, irrigation, electricity generation and fish-farming as well. In some places, artificial lakes are built as a landscape or architecture, ect.

人工湖即水库,用于拦洪蓄水和调节水流的水利工程建筑物,可以用来灌溉、发电和养鱼。在某些地方,人工湖是以景观或建筑等方式存在的。

artificial sea water 人造海水

It refers to an aqueous solution formulated by using chemical reagents to simulate the chemical composition of seawater.

用化学试剂模拟海水的化学成分配制的水溶液。

assessment services of marine development 海洋开发评估服务

It is about services of assessment for marine development.

对海洋开发评估的服务活动。

astronomical point 天文点

It is the ground point which determines the astronomical latitude and longitude.

测定天文经纬度的地面点。

astronomical tides 天文潮

They are the tidal phenomena caused

by tidal forces of celestial bodies.

由天体引潮力所引起的潮汐现象。

Atlantic Ocean 大西洋

It is the world's second largest ocean which is situated among Europe, Africa, Antarctic, South America and North America.

位于欧洲、非洲、南极洲和南、北美洲之间的世界第二大洋。

atmospheric tide 大气潮

It refers to the periodic atmospheric pressure occurring in low-latitude areas in a half day cycle and with amplitude being about 2 hPa.

发生于低纬度地区的周期为半天、振幅约为 2 百帕的周期性气压变化。

atmospheric water 大气水

It refers to the water existing in the atmosphere, mainly in the troposphere in gaseous, liquid or solid states.

以气态、液态或固态存在于大气圈并主要存在于对流层中的水。

atoll 环礁

It is a ringlike, elliptical or horseshoe-shaped coral reef that nearly or entirely encloses a lagoon in the ocean. Formerly known as *stone embark*.

大洋中呈环状、椭圆状或马蹄状生长的围绕潟湖发育的珊瑚礁。古称"石塘"。

auxiliary services of sea-related insurance 涉海保险辅助服务

They refer to the auxiliary activities of offering insurance agencies, assessment, supervision and consulting services for marine production and management.

为海洋生产和管理服务的保险代理、评估、监督、咨询的辅助活动。

avalanche 崩塌

It is a sudden fall of unstable rock (soil) mass down a steep slope or cliff due to gravity.

陡坡或悬崖上的不稳定岩(土)体在重力作用下突然下坠滚落的现象。

average groundwater level 地下水平均水位

It refers to the mean value of the groundwater level over within a certain observation period.

在某一观测时段内,地下水水位的平均值。

B

back-arc basin 弧后盆地

Back-arc basin, in the back of island arc, is a deep basin with oceanic crustal structure on the land side of trench-arc system.

又称"边缘海盆地",是指沟-弧体系陆侧具有洋壳结构的深水盆地,位于岛弧后方。

backshore 后滨

It is the area of shore covered by sea water between average high-tide mark of spring tides and storm surge.

从大潮平均高潮位到风暴潮期间为海水覆盖的陆上地带。

backward connection effect 后向关联效应

It refers to connection effect between departments, which occurs when products from leading industries serve as raw material, fuel or production facility for other enterprises or enter consumer sectors directly.

主导产业在进行生产之后,其产品成为许多产业的原料、燃料或生产设备,或直接进入消费部门而产生的部门关联效应。

balanced trade 贸易平衡

It refers to a state where the total imports and exports of a country are equalized in a particular year.

一国在特定年度内外贸进、出口总额基本上趋于平衡。

Balintang Channel 巴林塘海峡

Located between the Batan Islands and the Babuyan Islands of Philippines, it is an important channel, connecting the South China Sea and the Pacific Ocean.

位于菲律宾巴坦群岛与巴布延群岛之间,沟通南海与太平洋的重要通道。

Bali Roadmap 巴厘路线图

The Bali Roadmap is one of the agendas passed by the *United Nations Climate Change Conference* on December 15, 2007, which is concerned with the key themes for responding to climate change negotiations before 2009. These themes include: the action to adapt to the negative effect of climate change; the methods to reduce the emission of greenhouse gases; the methods to widely use climate-friendly technology and the action to subsidise the measures to adapt to and alleviate climate change.

联合国气候变化大会于2007年12月15日通过的关于2009年前应对气候变化谈判的关键议题所确立的议程。议题包括:适应气候变化消极后果的行动,减少温室气体排放的方法,广泛使用气候友好型技术的方法,以及对适应和减缓气候变化的措施进行资助。

Baltic Sea 波罗的海

It is an intracontinental sea of the Atlantic Ocean east of the Scandinavian Peninsula in northern Europe and the Jutland Peninsula. It is the largest brackish waters of the world with the lowest salinity.

位于欧洲北部斯堪的纳维亚半岛和日德兰半岛以东的大西洋的陆内海,是世界上最大的半咸水水域,也是世界上盐度最低的海。

bank protection work 护岸工程

It is the construction work to support bank slopes and slope bases with impact-resistant materials for preventing invasion by water flows, waves and others.

对河、湖、海堤的岸坡和坡脚用耐冲材料保护,防止水流、波浪等侵袭破坏的工程。

bank protection 护岸

It refers to artificial structure for protecting bank slopes.

防护岸坡的人工建筑物。

bank shoal 浅滩

It refers to the loose sediment deposit lying above an adjacent seabed and below the sea surface, which is generally flat on the top with a water depth of 20–200 metres.

高出邻近海底而未露出海面的松散沉积物堆积体,顶部较为平坦,一般水深20~200米。

bar 沙坝

It is a long and narrow coastal deposition landform, which comes into being with the development of coastal sand or pebbles along the coast in the shape of a dam.

滨海沙砾或卵石顺延海岸方向所构成的狭长形海岸堆积地貌,形如堤坝,故名沙坝。

Barabási-Albert network model (BA network model) BA 网络模型

It is a scale-free network model put forward by Barabási and Albert in 1999 to explain the producing mechanism of power law. Compared to the previous model, it focuses more on the growth and preferential attachment of natural and human-made network. Growth means that the number of nodes in the network increases over time. Preferential attachment means that the more connected a node is, the more likely it is to receive new links. Nodes with higher degree have stronger ability to grab links added to the network.

该模型是 Barabási 和 Albert 为了解释幂律的产生机制,于1999年提出的无标度网络模型(BA 模型),与以前模型相比强调真实网络的增长性和择优选择性。所谓增长性是指网络规模是在不断增大的,即在研究的网络当中,网络的节点是不断增加的,择优连接性是指网络中不断产生的新节点更倾向于和那些连接度较大的节点相连接。

barrier 沙坝

It is a long and narrow coastal deposition landform, which comes into being with the development of coastal sand or pebbles along the coast in the shape of a dam.

滨海沙砾或卵石顺延海岸方向所构成的狭长形海岸堆积地貌,形如堤坝,故名沙坝。

barrier island 堡岛

It is an accumulative landform made of gravels which is parallel to a mainland coast separated by lagoons. A barrier is normally located above high tide level, the outside of which is open sea and the inside is closed or semi-enclosed brackish lagoon, connected by one or several tidal passages.

与海岸平行,其间被潟湖隔开的长条状沙质或砂砾质堆积体。堡岛一般位于高潮位以上,外侧为开放海,内侧是封闭或半封闭的半咸水潟湖,两者之间常以一个或多个潮汐通道相连接。

barrier reef 堡礁

Also called *an offshore reef*, it refers to coral reefs separated from the adjacent coast by a channel or lagoon.

又称"离岸礁",指与海岸有一定距离的珊瑚礁群。

barrier reef offshore 离岸礁

Also called *barrier reef*, it refers to coral reefs separated from the adjacent coast by a channel or lagoon.

又称"堡礁",指与海岸有一定距离的珊瑚礁群。

baseline of territorial sea 领海基线

It is a starting line by which coastal states delimit their inner territorial seas, including normal baselines, straight baselines, mixed baselines and other baselines for optional use under all the sovereignty of all coastal states.

沿海国据以划定其领海内侧的起算线。包括正常基线、直线基线、混合基线和其他基线4种,由沿海国行使主权选用。

baseline point of the territorial sea 领海基点

It refers to a point which determines straight baselines. The *United Nations Convention on the Law of the Sea* does not specify the extreme distance between two adjacent baseline points, but describes that "the drawing of straight baselines must not depapt to any appreciable extent from the general direction of the coast, and the sea areas lying within the lines must be sufficiently closely linked to the a land domain to be subject to the regime of internal waters." (Item 3 of Article 7 in the *Convention*)

又称"基线转向点",指确定直线基线的点。《联合国海洋法公约》并未明确规定两个相邻基点间的极限距离,只要求"直线基线的划定不应在任何明显的程度上偏离海岸的一般方向,而且基线内的海域必须充分接近陆地领土,使其受内水制度的支配"(《公约》第七条第3款)。

Bashi Channel 巴士海峡

The Bashi Channel, which is located between the Taiwan island of China and the Batan Islands of Philippines, is an important channel, connecting the South China Sea and the Pacific Ocean.

位于中国台湾岛与菲律宾巴坦群岛之间,连接南海与太平洋的重要通道。

basic industry 基础产业

Belonging to secondary industry, it is an industry about the producing of medium and upstream products first level, including mining industry and raw material industry.

工业中上游产品的生产,包括采掘业和原材料工业,属第二产业。

basic surveying and mapping 基础测绘

It covers establishing nationally uniform surveying and mapping datums and systems prescribed by the State, carrying out basic aviation photography, acquiring basic geographical information and remote sensing data, mapping and updating national basic scale topographical map, shadowgraph and digitalized products, and establishing and updating basic geographical information system.

建立全国统一的测绘基准和测绘系统,进行基础航空摄影,获取基础地理信息和遥感资料,测制和更新国家基本比例尺地图、影像图和数字化产品,建立、更新基础地理信息系统。

basin 盆地

Also called *intermountain basins*, they refer to the low-lying and relatively closed area with a flat bottom between seamounts or sea knolls.

又称"山间盆地",指海山或海丘之间地形低洼、相对封闭的区域,底部地形平坦。

bathing beach services 海滨浴场服务

They refer to the services provided for bathing beaches.

为海上游泳场所进行的服务活动。

bathyal benthos 深海底栖生物

They refer to benthos living in the deep sea, where the water depth is between 200-2,000 metres.

生活于水深介于 200~2000 米间的深海底栖带的底栖生物。

bathyal zone 深海带

It refers to the deep sea area far away from a continent, whose depth is 2,000-6,000 metres.

远离大陆的深海地带,深度为 2 000 ~6 000 米。

bathypelagic zone 深层

It refers to the aquifer lying below the mesopelagic area with average temperature, salinity and density.

大洋中中层以下温度、盐度、密度较均匀的水层。

bay barrier 海坝

It refers to a sand bank in a bay area. It is an accumulated body formed by the decrease of sediment flow capacity as a result of energy decrease due to refraction occurring when waves enter the bay.

海湾区的沙坝。波浪进入海湾发生折射引起能量降低,也使泥沙流容量降低而形成的堆积体。

bay culture 港湾养殖

It refers to the production activity for breeding marine economic animals and plants in an encircled water area by taking advantage of ports and harbours, or embanking the tidal-flat area and the low-lying area near the sea or estuary.

利用港、湾或在海边、河口附近的滩涂、洼地拦闸筑堤,围成一定的水面,养殖海水经济动、植物的生产活动。

bay 海湾

[marine science and technology] It refers to a body of water connected to an ocean or sea by an inlet of land or between continents and islands.

[*resource technology*] It refers to a body of water surrounded by land with an area no smaller than a semicircle area with the diameter of a harbour entrance width.

[海洋科技]海或洋伸入大陆或大陆与岛屿之间的一部分水域。

[资源科技]被陆地环绕且面积不小于以口门宽度为直径的半圆面积的海域。

bay head beach 湾顶海滩

It refers to a beach distributed within a bay, which is opposite to a port.

分布于海湾内,与港口相对部位的海滩。

Bay of Bengal 孟加拉湾

It is an adjacent sea of the Indian Ocean which is located between Burma and India, opening toward the south.

位于缅甸与印度之间开口向南的印度洋的附属海。

bay side beach 湾侧海滩

It is a triangular beach group accumulated inside a bay.

海湾内侧堆积的三角滩群。

beach 岸滩

It refers to an accumulate landform along the shoreline of an ocean, a sea, a lake, or a river, which is covered by rock, sand, gravel, mud, shingle, cobblestones and biological remains.

被岩石、沙、砾石、泥、生物遗骸覆盖的河流、湖泊、海洋沿岸堆积地面。

beach berm 滩肩

It is a gentle terrace or platform on a backshore, which is formed through storm deposits and dips towards the land.

后滨上由风暴堆积而成的、向陆地倾斜的平缓阶地或台地。

beach cycle 海滩旋回

It is a phenomenon about periodic beach changes caused by periodic ocean-dynamic changes.

海洋动力周期变化引起海滩周期性变化的现象。

beach face 滩面

It is the surface of a beach, generally referring to a place with a relatively steep beach slope on the upper part of a foreshore.

海滩的表面,一般指前滨上部滩坡较陡的部位。

beach mineral resource 滨海矿产资源

They refer to minerals, stones and sediment distributed on the sea floor of an offshore beach area, which can be used by human beings. They mainly include of beach placer, oil, natural gas, sand, gravel, apatite, baritite nodule, calcareous shell, coal, iron, sulphur, halite, sylvite, barite, tin ore, ect.

分布在离岸较近的滨海地区海底,可以被人类利用的矿物、岩石和沉积物。主要有海滨砂矿、石油、天然气、砂、砾石、磷灰石、硫酸钡结核、钙质贝壳、煤、铁、硫、岩盐、钾盐、重晶石和锡矿等。

beach placer 滨海砂矿

It is an accumulation of valuable minerals formed in loose seabed sediments under the action of inland rivers, waves, tides, tidal currents and ocean currents in coastal zones. It includes two types, one is a metal placer, the other is a non-metal placer.

又称"海底砂矿"。滨海地带内河流、波浪、潮汐、潮流和海流等作用,使重矿物富集于海底松散沉积物中而形成的矿产,包括海滨金属砂矿和非金属砂矿两种类型。

beach profile 海滩剖面

It refers to a beach transverse section perpendicular to the coastline.

与海岸线垂直的海滩横断面。

beach rock 海滩岩

It is a rock formed when fragmentary substances such as sand and gravel are cemented by carbonate minerals.

海滩上砂、砾等碎屑物质经碳酸盐胶结作用而形成的岩石。

Beibu Gulf 北部湾

It is a semi-enclosed bay extending westward, located in the northwest of the South China Sea.

位于南海西北部,并向西凸出的半封闭海湾。

benefit of marine resource exploitation 海洋资源开发效益

It is a generic term for economic benefit obtained from exploitation of marine resources as well as the resultant eco-environmental benefit and social benefit.

开发海洋资源所获得的经济效益以及所产生的生态环境效益和社会效益的总称。

benthic division 海底区

It refers to seabed and subsoil covered by seawater.

海水覆盖区域的海床和底土。

benthic organism 底栖生物

They are the community of sessile or crawling organisms living on or under the bottom of bodies of water. Now, they generally refer only to benthonic animals.

生活在水域底上或底内、固着或爬行的生物。现在一般只用于表示底栖动物。

benthic-pelagic coupling 海底－水层耦合

It refers to the circulation and shift of nutrients between the benthos system and the plankton system.

底栖生物系统和浮游生物系统之间营养物质的循环转移。

benthology 底栖生物学

It refers to a discipline which studies the classification, distribution, quantitative variation, population dynamics, biological phenomena such as reproduction, growth and ingestion of organisms living on and under the bottom of water column, and the correlation between organisms and environmental factors as well.

研究水域底内和底上生物的分类、分布、数量变动、种群动态、群落结构和繁殖、发育、摄食等生命现象,及其生物与环境因子间相互关系的学科。

benthos 底栖生物

They are the community of sessile or crawling organisms living on or under the bottom of bodies of water. Now, they generally refer only to benthonic animals.

生活在水域底上或底内、固着或爬行的生物。现在一般只用于表示底栖动物。

Bering Sea 白令海

It is a marginal sea to the north of the Pacific Ocean, which is located between Alaska of northern America, the Aleutian Islands and the Asian continent. The Bering Strait connects it to the Arctic Ocean.

位于亚洲大陆与北美洲阿拉斯加、阿留申群岛之间,经白令海峡与北冰洋相通,太平洋北部的边缘海。

berm 栈道

It refers to a channel erected on a steep section in ancient times for pedestrians and supplies transportation.

特指古代架设于陡峻地段提供给行人、物资运输的通道。

berth 泊位

It is a designated location in a port or harbour used for mooring vessels usually for the purposes of loading and unloading.

港区内供船舶安全停泊并进行装卸作业所需要的水域和相应设施。

bight 海湾

[marine science and technology] It refers to a body of water connected to an ocean or sea by an inlet of land or between continents and islands.

[resource technology] It refers to a body of water surrounded by land with an area no smaller than a semicircle area with the diameter of a harbour entrance width.

[海洋科技]海或洋伸入大陆或大陆与岛屿之间的一部分水域。

[资源科技]被陆地环绕且面积不小于以口门宽度为直径的半圆面积的海域。

biochemical oxygen demand (BOD) 生化需氧量

It refers to the amount of dissolved oxygen consumed in the biological and chemical processes when organisms in water are decomposed by microorganism under aerobic conditions.

在有氧条件下,水中有机物在被微生物分解的生物化学过程中所消耗的溶解氧量。

bioclimatics 生物气候学

It is a discipline which studies the interrelationship between organisms and climatic conditions.

研究生命有机体与气候环境条件相互关系的学科。

biocoene 生物群落

It refers to the aggregation of various biological populations which gather in a certain area or environment in the same period of time.

在相同时间聚集在一定地域或生境中各种生物种群的集合。

biocoenosis 生物群落

It refers to the aggregation of various biological populations which gather in a certain area or environment in the same period of time.

在相同时间聚集在一定地域或生境中各种生物种群的集合。

biocycle 生物带

It refers to the horizontal and vertical zonal distribution of marine organisms with each zone having unique animal and plant communities.

海洋生物水平和垂直的带状分布,

各带具有独特的动物和植物群落。
biodiversity 生物多样性

[*marine science and technology*] It is the generic term for the diversity of three levels of genetic genes, species and biological systems.

[*biology*] It refers to the diversity of a hierarchical structure and functions of biological groups, including genetic diversity, species diversity, biological diversity and landscape diversity.

[海洋科技]遗传基因、物种和生态系统三个层次多样性的总称。

[生态学]生物类群层次结构和功能的多样性。包括遗传多样性、物种多样性、生态系统多样性和景观多样性。

bioerosion 生物侵蚀

It refers to the basal decomposition caused by various organisms, such as the decomposition of coral calcium carbonate.

各类生物对基底的分解,如对珊瑚礁碳酸钙的分解。

biogeochemical cycles 生物地球化学循环

It refers to a process in which inorganic matter is synthesized into organic matter through autotroph, and organic matter eventually goes into the environment through the food chain and is then recycled.

环境中的无机物通过自养生物合成有机物,后者经过食物链最终又进入环境,再被循环利用的过程。

biological genetic resources 生物遗传基因资源

They refer to the biological resources which can produce physiologically active substances.

能够产生生理活性物质的生物资源。

biological liquid fuels 生物液体燃料

They refer to the liquid fuels which are produced by exploiting biomass resources such as methanol, ethanol, biodiesel, etc.

利用生物质资源生产的甲醇、乙醇和生物柴油等液体燃料。

biological noise 生物噪声

It refers to the noise produced by some animals in the sea.

海洋中一些动物发出的噪声。

biological oceanography 生物海洋学

It is a discipline which studies the interrelationship among the production, development, motion and variation of marine organisms and the ocean water, basement structure and various dynamic processes.

研究海洋生物发生、发展、运动变化和海洋水体、基底结构及各种动态过程间相互关系的学科。

biological pump 生物泵

It refers to the whole process of transfer or sedimentation of carbon from the surface to the deep sea. Carbon which is produced by organisms and formed by a series of biological processes including consumption, transfer and decomposition.

由有机体所产生的,经过消费、传递和分解等一系列生物学过程构成的碳从海洋表层向深层转移或沉降的整个过程。

biological resources 生物资源

They refer to genetic resources, organisms, population, biological systems and any components included, which have a practical or potential value and use for human beings.

对人类具有实际的或潜在的价值与用途的遗传资源、生物体、种群或生态系统及其中的任何组分的总称。

biomass 生物量

It refers to the quantity of organisms per unit area or volume, which is generally calculated by wet weight or dry weight. Broadly speaking, the density of organisms, volume thickness and coverage area are all representations of biomass.

单位面积或体积内生物的量,一般以湿质量或干质量计。广义上,生物的密度、体积厚度、覆盖面积等也都是生物量的一种表示方法。

biomass energy 生物质能

It refers to the energy transformed from plants and faeces in nature, faeces and organic wastes in urban and rural areas.

利用自然界的植物、粪便以及城乡有机废物转化成的能源。

biome 生物群落

It refers to the aggregation of various biological populations which gathers in a certain area or environment in the same period.

在相同时间聚集在一定地域或生境中各种生物种群的集合。

biosphere 生物圈

It refers to the circle of organisms existing on the earth, including the lower stratum of atmosphere, upper stratum of the lithosphere and the whole hydrosphere and pedosphere.

地球上存在生物有机体的圈层。包括大气圈的下层、岩石圈的上层、整个水圈和土壤圈全部。

bioturbation 生物扰动

It refers to the phenomenon when layers and chemical components are displaced by movements and feeding activities of benthic animals.

软底沉积物层次和化学成分被底内动物运动和摄食活动所搅和的现象。

biozone 生物带

It refers to the horizontally and vertically zonal distribution of marine organisms with each zone having unique animal and plant communities.

海洋生物水平和垂直的带状分布,各带具有独特的动物和植物群落。

bi-seaside coordination 海陆统筹

In a broad sense, it means that sea and land should be taken into account as a whole based on the full awareness of the fact that our country is not only a big marine but land country as well. In accordance with domestic development needs and changes of international situation, we need to coordinate a sea-land relationship, balance the strategy of sea-land development, synthesise ocean and land (and the various attached benefits, value and civilizations) into developing a national and social economy, safeguarding the national interests, and bringing the overall advantage of possessing both land and ocean to full play.

In a narrow sense, it means strengthening the circulation of essential productive factors (labour forces, resources, funds, technologies, information, etc.) between a sea and land area in order to achieve an effective and reasonable allocation of resources and enhance the sustainable, fast and healthy development of a regional economic society on condition that a regional ecological environment system and social economic system can function normally, in accordance with regional natural conditions and social economic development situations. The bi-seaside coordination falls under the category of regional economy, and it has become the dominant idea in leading the economic development of coastal regions.

广义上指在充分意识到我国具有陆地大国和海洋大国双重属性的客观事实基础上,把海洋与陆地作为整体考虑,根据国内发展需求以及国际形势变化,协调海陆关系,平衡海陆发展战略,将海洋和陆地(以及附着其上的各种利益、价值和文明)统合进国家经济社会发展和维护民族利益的过程中,充分发挥我国海陆兼备的整体优势。

狭义上指以区域自然条件与社会经济发展状况为依据,以保障区域生态环境系统与社会经济系统正常运行为前提,在市场机制与政府宏观调控的共同作用下加强生产要素(劳动力、资源、资金、技术、信息等)在海洋与陆域之间的流通,实现资源有效、合理配置,从而促进区域经济社会持续、快速、健康发展。属于区域经济范畴,成为引领沿海区域经济发展的主导思想。

bi-seaside economic structure 海陆二元经济结构

It refers to an economic structure in which a coastal economy characterised by a marine industry coexists with the inland economy characterised by a non-marine industry.

以海洋产业为特色的沿海经济与以非涉海产业为特色的内陆经济并存的经济结构。

Black Sea 黑海

It is an inland sea between southeastern Europe and Asia Minor Peninsula. It is bounded by Europe, Anatolia and the Caucasus and is ultimately connected to the Atlantic Ocean via the Mediterranean and the Aegean Seas and various straits.

欧洲东南部与小亚细亚半岛之间的陆内海。

Black stream 黑潮

It is the strongest warm current among world oceans. It starts from the North Equatorial Current in the Pacific Ocean, flows from east to west and arrives at east coast of Philippines where it is blocked and formed when turning northward, with blue-black color from which the name is derived. Also called Kuroshio Current.

世界大洋中最强的暖流。源于太平洋北赤道流,自东向西流动,在菲律宾东海岸受阻后,向北转向而成。呈蓝黑色,故名黑潮。

blue economy 蓝色经济

It is an economic form or development concept featured by marine economy, which aims to build marine advantage industry, realise scientific development and comprehensive utilisation of marine resources, maintain sustainable development of harmonious relationships between human and ocean relationships by the integration of sea and land, scientific and technological innovation and ecological management.

以海洋经济为显著特征,通过海陆统筹、科技创新和生态化管理,打造海洋优势产业,实现海洋资源的科学开发与综合利用,人-海关系和谐可持续发展的经济形态或经济发展理念。

blue industry 蓝色产业

It refers to the various industries which are developed by utilizing the advantages and resources of the ocean and coastal location.

利用海洋和海岸区位优势和资源所发展的各种产业。

blue revolution 蓝色革命

It is a technological revolution of producing large amount of food protein for human beings by controlling and utilizing sea areas with high-tech methods.

用高科技手段控制和利用海洋水域,为人类大量生产食物蛋白的技术革命活动。

blue sea action 碧海行动

The *Bohai Sea Blue Sea Action Plan*, approved by the State Council, one of the key environmental protection programs implemented by China during the period of the Tenth Five-Year Plan, is the first national *Land-Source Pollution and Offshore Area Protection* Plan in China. This Plan will last for 15 years and falls into three phases: near term, medium term, and long term. It is composed of 427 subprograms with a total investment of 55.5 billion yuan.

"渤海碧海行动计划"是中国"十五"期间实施的环境保护重点项目,是国务院批准的"渤海绿色运动",是中国最早实施的国家级"陆源污染及近岸海域保护"计划。该计划分为近期、中期、远期3个阶段,共15年,总投资555亿元,由427个项目构成。

blue state territory 蓝色国土

It is a vivid general term for inland waters, territorial waters and sea areas under the national jurisdiction of a coastal country, including contiguous zones, exclusive economic zones, continental shelves, historic waters or traditional sea frontiers outside the territorial sea.

是一个沿海国家的内水、领海和管辖海域的形象统称。管辖海域包括领海以外的毗连区、专属经济区、大陆架、历史性海域或传统海疆等。

bog solonchak 沼泽盐土

Distributed sporadically at such topographic positions as lake depressions, joint depressions, or fan-margin groundwater overflow belts, in various saline soil areas, it is mainly evolved from salt deposition dried apin various marsh soils, salt swamps or salt marsh, where hygrophilous and salt-tolerant vegetation grow. Gleisation starts from the surface layer, forming black sapropel and coarse organic matters. White salt efflorescence or salt crust often lies on the land surface. Under the salic horizon are black slimy humus horizon and green red gley horizon.

各盐渍土区的湖泊洼地、交接洼地或扇缘地下水溢出带等地形部位都有零星分布。主要由各种沼泽土、盐泽或盐沼干涸积盐演变而成，生长喜湿耐盐植被。从表层起就有潜育现象，形成黑色腐泥及粗有机质，地表常有白色盐霜或盐结皮，积盐层以下为黑色糊状的腐殖质层和青赤色的潜育层。

Bohai Sea 渤海

Also called *Bohai Gulf* or *Bo Hai*, it is the innermost gulf of the Yellow Sea on the coast of Northeastern and North China. It is bounded by the Liaodong and Shandong Peninsulas.

中国大陆东部由辽东半岛与山东半岛所围绕的、近封闭的浅海，是中国的内海，也称渤海湾或渤海。

Bohai Strait 渤海海峡

Located between the Liaodong and Shandong Peninsulas, it is the only waterway connecting the Bohai Sea and the Yellow Sea.

位于中国辽东半岛与山东半岛之间，沟通渤海与黄海的唯一通道。

boiling point of sea water 海水沸点

It refers to the critical temperature at which sea water boils. It rises with the increase of the seawater salinity. With a 10‰ increase of seawater salinity, the boiling point of sea water rises by 0.16℃.

海水沸腾时的临界温度。随海水盐度的增加而升高，盐度每增加 10，沸点温度升高 0.16℃。

bottom water 底水

It refers to the water at the bottom of an oil-bearing layer.

位于含油层底部的水。

boundary mark 界址点

It refers to an inflection point which defines the scope and boundary of sea lot (unit of sea within the sovereignty of boundary line) and its internal elements.

用于界定宗海及其内部单元范围和界线的拐点。

boundary point 界址点

It refers to an inflection point which defines the scope and boundary of sea lot (unit of sea within the sovereignty of boundary line) and its internal elements.

用于界定宗海及其内部单元范围和界线的拐点。

bounded rationality 有限理性

It is the idea that in decision-making, the rationality of individuals is limited by the information they have, the cognitive limitations of their minds, and the finite amount of time they have to make a decision.

博弈者有一定的统计分析能力和对不同的策略下相关得益的事后判断能力，但缺乏事前预见、预测和判断能力。

breakwaters 防波堤

It refers to the offshore structure constructed in the harbour waters or outside some part of the harbour waters. Breakwaters protect a harbour from the direct intrusion of a wave, thereby creating relatively calm waters inside the harbour where vessels can moor and stevedore safely.

建在港口水域或其某部分外围,阻挡波浪直接侵入港内,使港内水面相对平静、船舶能安全靠泊和装卸的建筑物。

brine 卤水

It refers to underground saline water whose salinity is greater than 50 grams/litre.

矿化度大于50克/升的地下咸水。

brine water intrusion 卤水入侵

It refers to a phenomenon of which the underground water becomes salty due to the intrusions of brine water into freshwater resources. It is mainly caused by groundwater over-drilling or geological engineering activities, which leads to the decline of freshwater levels and the disruption of underground water system balance in coastal plain areas.

人为超采地下淡水或地质工程活动,导致地下淡水水位的下降,破坏了滨海平原区域地下水力系统平衡,卤水浸染淡水资源,使地下淡水变咸的现象。

brittleness property 脆弱性

It is a property that the structure and function of the system tend to be changed due to the sensitiveness of the system (subsystem, component parts) to the disturbance inside and outside the system and lack of responding ability.

由于系统(子系统、系统组分)对系统内外扰动的敏感性以及缺乏应对能力从而使系统的结构和功能容易发生改变的一种属性。

budget deficit 预算赤字

Also called *fiscal deficit*, it is the difference between fiscal spending and fiscal revenue when the former exceeds the latter.

又称"财政赤字",是财政支出大于财政收入而形成的差额。

budget surplus 预算盈余

Opposite of budget deficit, it is the difference between fiscal revenue and fiscal expenditure when the former exceeds the latter.

与预算赤字相对应,是财政收入超出支出的部分。

business cycle 经济周期

It refers to the short-term fluctuation of output and employment.

产出与就业的短期波动。

butterfly effect 蝴蝶效应

Butterfly effect is the sensitive dependency on initial conditions in which a small change at one place in a deterministic nonlinear system can result in large differences in a later state.

初始值的极微小的扰动而会造成系统巨大变化的现象。

C

cable laying of submarine cable and fiber 海底电缆、光缆的铺设

It refers to the construction activities of laying cable and fiber on the seabed.

将电缆、光缆铺设在海底的施工活动。

capital flight 资本外逃

It refers to a large-scale exodus of financial assets and capital from a nation due to events such as political or economic instability, currency devaluation or the imposition of capital controls. Capital flight occurs when domestic and overseas investors in a country or economy massively undersell domestic financial assets and transfer funds abroad because of their concern about a possible recession or other economic or political uncertainty in the country.

一国或经济体的境内及境外投资者由于担心该国将发生经济衰退或其他经济或政治的不确定性而大规模抛出该国国内金融资产,将资金转移到境外的情况。

carbon acquisition 碳获取

It is the process by which plants acquire carbon by fixing carbon dioxide (CO_2) via photosynthesis.

植物通过光合作用固定二氧化碳而获得碳的过程。

carbon assimilation 碳同化

It is the incorporation of carbon from atmospheric carbon dioxide (CO_2) into organic molecules, which occurs during photosynthesis.

大气中的碳被生物系统吸收并被转化成它们自身的过程。

carbon barrier 碳壁垒

It refers to the regulations and standards influencing product trade, including carbon tax, border carbon tax adjustment, carbon mark and carbon standard, which are designed and implemented for the carbon generated from production, transport, consumption and disposal of products.

针对产品在生产、运输、消费和处置环节中产生的碳而设计和实施的碳税、边境碳税调整、碳标志和碳标准等影响产品贸易的规章和标准。

Carbon Capture and Storage (CCS) 碳捕集和埋存

It is a carbon removing technology of capturing carbon dioxide (CO_2) from the burning of fossil fuels, such as coal, oil, natural gas, etc. and burying it in deep strata. The aim is to prevent the release of CO_2 into the atmosphere and mitigate the contribution of its emissions to global warming.

捕集来自煤、石油、天然气等化石燃料燃烧产生的二氧化碳,并埋存在地层深部,防止二氧化碳排放到大气中和全球变暖的一种去碳技术。

carbon capture 碳捕捉

It means to capture the carbon dioxide

released into the atmosphere and then compress it back into depleted oil and gas fields, or other safe underground places.

捕捉释放到大气中的二氧化碳,压缩之后,压回到枯竭的油田和天然气田或者其他安全的地下场所。

carbon circulation 碳循环

Also called carbon cycle. It is one of the major cycles of chemical elements in the environment. Carbon (as carbon dioxide (CO_2)) is taken up from the atmosphere and incorporated into the tissues of plants during photosynthesis. It may then pass into the bodies of animals as the plants are eaten. During the respiration of plants, animals, and organisms that bring about decomposition, carbon dioxide is returned to the atmosphere.

绿色植物(生产者)在光合作用时从大气中取得碳,合成糖类,然后经过消费者和分解者,通过呼吸作用和残体腐烂分解,使碳又返回大气的过程。

carbon emission 碳排放

It is a generic or a short term for greenhouse gas emissions. It falls into renewable carbon emissions and non-renewable carbon emissions. The former is the normal carbon cycle of all plants and animals on the earth's surface, including carbon emissions from various renewable energy sources. The latter refers to carbon emissions from the development and consumption of fossil energy.

关于温室气体排放的一个总称或简称,分为可再生碳排放和不可再生碳排放。可再生碳排放是地球表面的各种动植物正常的碳循环,包括使用各种可再生能源的碳排放;不可再生碳排放指开发和消耗化石能源产生的碳排放。

carbon finance 碳金融

It refers to low carbon economy investment and financing activities initiated by the *Kyoto Protocol*. More specifically, it is financial activities, including direct investment and financing, carbon index trading, and bank loans, which serve the technologies and projects for curbing greenhouse gas emission, etc. The rise of carbon finance originated from changes in international climate policies and two significant international conventions—*the United Nations Framework Convention on Climate Change*(UNFCCC) and *Kyoto Protocol*.

又称"碳融资",是指由《京都议定书》而兴起的低碳经济投融资活动,即服务于限制温室气体排放等技术和项目的直接投融资、碳指标交易和银行贷款等金融活动。"碳金融"的兴起源于国际气候政策的变化以及两个具有重大意义的国际公约——《联合国气候变化框架公约》和《京都议定书》。

carbon footprint 碳足迹

It is used to mark the "carbon consumption" of an individual or group, i.e. the total set of greenhouse gas emissions caused by an organization, event, product or individual through transportation, food production and consumption, as well as various production processes, etc. Carbon footprint includes carbon dioxide (CO_2) emissions caused both directly by the use of fossil energy (the primary carbon footprint) and indirectly by the use of various products (the secondary carbon footprint).

被用来标示一个人或者团体的"碳消耗量",即指企业机构、活动、产品或个人通过交通运输、食品生产和消费以及各类生产过程等引起的温室气体排放的集合。碳足迹既包括因使用化石能源而直接排放的二氧化碳(第一碳足迹),也包括因使用各种产品而间接排放的二氧化碳(第二碳足迹)。

carbon labeling 碳标签

It is applied to mitigate climate change, reduce greenhouse gas emissions and promote low-carbon technology. Specifically, the greenhouse gas emission quantity during the production of a commodity is marked in a quantitative index on the product label; in other words, the carbon information is notified to the consumer in the form of a label.

为了缓解气候变化,减少温室气体排放,推广低碳排放技术,把商品在生产过程中所排放的温室气体排放量在产品标签上用量化的指数标示出来,以标签的形式告知消费者产品的碳信息。

carbon loss 碳损失

It occurs when plants consume carbon through dark respiration, photorespiration, etc.

植物通过暗呼吸、光呼吸等消耗碳的过程。

carbon neutral 碳中和

It refers to achieving net zero carbon emissions by balancing a measured amount of carbon released with an equivalent amount sequestered or offset, or buying enough carbon credits to make up the difference. It is used in the context of carbon dioxide (CO_2) releasing processes associated with transportation, energy production, and industrial processes. The best practice for organizations and individuals seeking carbon neutral status entails reducing and/or avoiding carbon emissions first so that only unavoidable emissions are offset.

Carbon neutral status is commonly achieved in two ways:

Balancing carbon dioxide (CO_2) released into the atmosphere from burning fossil fuels, with renewable energy that creates a similar amount of useful energy, so that the carbon emissions are compensated, or alternatively using only renewable energies that don't produce any CO_2 (also called *a post-carbon economy*).

Carbon offsetting by paying others to remove or sequester 100% of the CO_2 emitted from the atmosphere, for example by planting trees, or by funding 'carbon projects' that should lead to the prevention of future greenhouse gas emissions, or by buying carbon credits to remove (or 'retire') them through carbon trading.

又称"碳中立",人们算出自己日常活动直接或间接产生的二氧化碳排放量,并算出为抵消这些二氧化碳所需的经济成本或所需的碳"汇"数量,然后个人付款给专门企业或机构,由他们通过植树或其他环保项目来抵消大气中相应的二氧化碳量。

carbon politics 气候政治

It refers to the international politics

formed among various countries regarding the issue of greenhouse gas emission; internationally, greenhouse gas emission is calculated in terms of carbon dioxide, so it is called carbon politics.

又称"碳政治",是指各国围绕温室气体排放问题所形成的国际政治,而国际上关于温室气体排放又按照二氧化碳来计算,故称之为"碳政治"。

carbon pool 碳池

It refers to a bunker (e.g. an ocean) where carbon is conserved. It plays an important role in the biogeochemical cycle.

又称"碳库",指保存碳的贮藏库(如海洋),在生物地球化学循环中起重要作用。

carbon sink 碳汇

[*atmospheric science*] It refers to a carbon storage bank which receives carbon from other carbon storage banks and so the storage increases over time.

[*ecology*] It is a system or area where the amount of organic carbon absorption is greater than the amount of release, such as atmosphere, ocean, etc.

[大气科学]一个碳贮库,它接收来自其他碳贮库的碳,因此贮量随时间增加。

[生态学]有机碳吸收超出释放的系统或区域。如大气、海洋等。

carbon source 碳源

[*atmospheric science*] It is a carbon storage bank which provides carbon for other carbon storage banks and so the storage decreases over time.

[*ecology*] It is a system or area where the amount of organic carbon release exceeds that of absorption; for example, tropical deforestation, the burning of fossil fuels, etc.

[大气科学]一个碳贮库,它向其他碳贮库提供碳,因此贮量随时间减少。

[生态学]有机碳释放超出吸收的系统或区域。如热带毁林、化石燃料燃烧等。

carbon storage 碳封存

It is a process of injecting captured carbon dioxide (CO_2) into an underground geological structure or deep sea, or solidifying the captured CO_2 into inorganic carbonate through an industrial flow.

将捕捉的二氧化碳注入地下地质构造中、深海里,或者通过工业流程将其凝固在无机碳酸盐中的过程。

carbon tariffs 碳关税

They are special carbon dioxide (CO_2) emission tariffs levied on imported high energy consumption products. This concept was first put forward by former French President Chirac with an intention that the EU countries levy duties on imported goods from countries not complying with the *Kyoto Protocol*; otherwise, after the EU carbon emission trade mechanism started, the commodities produced by EU countries will face unfair competition, especially the domestic steel industry and high energy consuming industries. On July 4, 2009, the Chinese government clearly expressed its opposition to carbon tariffs.

对高耗能的产品进口征收特别的二氧化碳排放关税。这个概念最早由法国前总统希拉克提出,用意是希望欧盟国家应针对未遵守《京都议定书》的国家课征商品进口税,否则在欧盟碳排放交易机制运行后,欧盟国家所生产的商品将遭受不公平之竞争,特别是境内的钢铁业及高耗能产业。2009年7月4日,中国政府明确表示反对碳关税。

carbon tax 碳税

It refers to a tax levied on the emission of carbon dioxide. For the purpose of environmental protection, a tax will be collected on the carbon content based on the proportion of the fossil fuel products like coal, oil, petrol, aviation fuel and natural gas, etc. to reduce the consumption of fossil fuels and carbon emission to mitigate global warming.

针对二氧化碳排放所征收的税。它以环境保护为目的,通过对燃煤和石油下游的汽油、航空燃油、天然气等化石燃料产品碳含量的比例征税来减少化石燃料消耗和二氧化碳排放,以减缓全球变暖。

carbon trading 碳交易

It refers to a market mechanism adopted to enhance the emission reduction of the greenhouse gases and curb the global emission of carbon dioxide. According to *Kyoto Protocol*, the new mechanism is regarded as a new pathway to solve the problem of greenhouse gas emission which is represented by carbon dioxide. In other words, CO_2 emission right, simply called "carbon trading", can be traded as a commodity. The basic principles of carbon trading lies in that one contract party pays the other party for greenhouse gas emission reduction quota, so that the buying party can use it to mitigate the greenhouse effect and achieve its emission reduction target.

为了促进全球温室气体减排,减少全球二氧化碳排放所采用的市场机制。《京都议定书》把市场机制作为解决二氧化碳为代表的温室气体减排问题的新路径,即把二氧化碳排放权作为一种商品,从而形成了二氧化碳排放权的交易,简称碳交易。碳交易基本原理是:合同的一方通过支付另一方获得温室气体减排额,买方可以将购得的减排额用于减缓温室效应从而实现其减排的目标。

carcinology 甲壳动物学

It is a discipline which studies crustacean classification, shape, reproduction, development, ecology, physiology, biochemistry, geological distribution and its relationship with human beings.

研究甲壳动物的分类、形态、繁殖、发育、生态、生理、生化、地理分布及其与人类关系的学科。

cargo throughput 货物吞吐量

It is the quantity of cargos passing into and out of the seaports by water through loading and unloading, including mails check-in packages and parcels, and supplies of fuel, material and fresh water for transportation ships. The unit of measurement is ten thousand tons.

经由水路进、出沿海港区范围并经过装卸的货物数量。包括邮件、办理托运手续的行李、包裹以及补给运输船舶的燃料、物料和淡水。计量单位:万吨。

Caribbean Sea 加勒比海

It is a sea of the Atlantic Ocean, an intercontinental sea, bounded by South America, Antilles and Central American Isthmus.

位于南美大陆、安的列斯群岛、中美地峡之间的陆间海,大西洋的附属海。

carrying capacity of land resources 土地资源承载力

It is the maximum amount of biological survival allowed by unit area of land, based on the premise that the ecological and environmental qualities will be maintained not to deteriorate.

在保持生态与环境质量不致退化的前提下,单位面积土地所容许的最大限度的生物生存量。

carrying capacity of population 人口承载能力

It refers to complex synthesis composed of factors such as population, resources, environment, social and economic development, etc., which can control and affect the population carrying capacity of a certain area in the future. Resources, environment, social and economic development are the basis and foundation of the population carrying capacity, which are influenced by human consumption activities through the reaction of a sub system. Per capita consumption or possession is one of the significant factors which directly determines the population carrying capacity.

在一定地域范围内由未来各时期制约和影响可承载人口数量的人口、资源、环境、社会及经济发展等因素构成的复杂综合体。资源、环境、社会及经济发展等因素是人口承载能力的基础和依托,人类消费活动则通过对其他子系统的反作用,影响人口承载能力。人均消费或占有量是直接决定可承载人口数量的重要因素之一。

carrying capacity of regional resources 区域资源承载力

It refers to the supporting capacity of regional resources to population increase, economic development and ecological balance under certain technological and economic conditions in a specific period.

在一定时期、一定的技术经济条件下,某地区资源对人口增长和经济发展以及生态平衡的支持能力。

catastrophe theory 突变论

It is a mathematical theory developed by a French mathematician René Thom in 1972 in the book, *Structural Stability and Morphogenesis*, which is concerned with modelling non-linear interactions within systems that can produce sudden and dramatic effects from apparently small changes in one variable. It has been used to explain how, through the interaction of various factors, a small change in one of the factors affecting a system can lead to a catastrophic change in the system.

1972年法国数学家雷内·托姆(René Thom)的《结构稳定性和形态发生学》一书提出突变论。突变论的研究非线性系统从一种稳定组态以突变的形式转化到另一种稳定组态的现象和规律。

catchment area 汇水面积

It is the area surrounded by a drainage divide. In other words, waters falling on the area converge to a watercourse along the slope at the exit of the basin where waters join another water body. Size and shape of drainage area directly affect the formation of river runoffs.

又称"受水面积"、"流域面积",指流域分水线所包围的面积,即降落在流域面积上的降水都沿着地面斜坡汇入河道,经流域出口断面流出。流域面积的大小和形状,直接影响河流径流形成的过程。

catch-up effect 追赶效应

It refers to the effect in economies that poorer economies tend to grow faster than richer economies at the initial stages under the same conditions.

其他条件相同情况下,开始时贫穷的国家倾向于比开始时富裕的国家增长更快的特征。

cartographic generalization of relief 地貌综合

It refers to the compilation operation of selecting and generalizing a geomorphologic shape when compiling a cartograph. Its purpose is to maintain the essential features and the corresponding accuracy of the real morphology on smaller areas when being scaled down. Currently, isoheight is more often used to represent a landform, therefore, the generalization of the isoheight is the main content of the cartographic generalization of relief.

编绘地图时选取和概括地貌形态的编绘作业。目的是当比例缩小后,在较小的面积上仍能保持实地形态的基本特征和相应的精度。现代普通地图上多以等高线表示地貌,因此,对等高线图形的概括是地貌综合的主要内容。

cellular automata(CA) 元胞自动机

It is an effective model used to simulate various complex phenomena including self-organizing structure. With the methods of time discretization, space discretization and state discretization, various units (cell and grid) run continuously in parallel in time and space through simple connection and simple computing rules to simulate complex but rich phenomena including self-organizing phenomenon, dissipative structure, and synergistic effect, etc.

是模拟包括自组织结构在内的复杂现象的一种强有力的方法。在研究上采用时间离散化、空间离散化和状态离散化的方法,通过大量个体(元胞或网格点)的简单连接和简单运算规则,在时空中并行地持续运行,以模拟出包括自组织现象、耗散结构、协同效应等在内的复杂而丰富的现象。

central bank 中央银行

It is a country's highest monetary and financial management institution entrusted by its government to formulate and implement monetary policy, exercise macro-control over the national economy, supervise and administer financial institutions and even the financial sector.

是国家赋予其制定和执行货币政策,对国民经济进行宏观调控,对金融机构乃至金融业进行监督管理的一国最高的货币金融管理机构。

Changjiang River 长江

It is the longest river in China and the third longest river in the world, just second to the Nile in Africa and the Amazon in South America. Rising in Geladandong Peak the Tanggula Mountains on the Qinghai-Tibet Plateau, it flows southeast through 11 provinces and municipalities including Qinghai, Tibet, Sichuan, Yunnan, Chongqing, Hubei, Hunan, Jiangxi, Anhui, Jiangsu before emptying into the East China Sea at Shanghai. The total length is 6,397 kilometres and it is regarded as the mother river of China as is the Yellow River.

中国第一大河,世界第三长河。仅次于非洲的尼罗河与南美洲的亚马孙河。它发源于青藏高原唐古拉山主峰各拉丹冬雪山,流经青海、西藏、四川、云南、重庆、湖北、湖南、江西、安徽、江苏、上海等11个省(自治区、直辖市),最后在上海注入东海。长江全长6 397千米和黄河并称为中华民族的"母亲河"。

channel 海峡

It is a relatively narrow water channel connecting two seas or oceans between continents.

两块陆地之间连通两个海或洋的宽度较狭窄的水道。

chaos 混沌

It is a seemingly chaotic and random phenomenon that is generated by a system described by deterministic equations. To be more precise, chaos is pseudo-random in a deterministic system, which is not simply chaotic but apparently periodical and symmetrical, with an ordered structure of rich internal hierarchy. It is only generated in a non-linear system, which is the inherent characteristic and new existence form of a non-linear system. When the system has a change namely in self-organized critically state, it will exhibit unique external feature, chaos.

由确定性方程描述的系统产生的一种貌似无规则、类似随机的现象。更确切地讲,混沌是决定性系统的伪随机性,混沌不是简单的无序而是没有明显的周期和对称,但却有丰富的内部层次的有序结构。混沌只能在非线性系统中产生,它是非线性系统的固有特性,也是非线性系统中的一种新的存在形式。当系统发生改变即发生自组织临界状态时,系统就表现出独特的外在特征,即混沌。

chaotic phenomena 混沌现象

They are the phenomena in which seemingly random and irregular movements within a deterministic system described by a deterministic theory, exhibit uncertain, non-repeatable and unpredictable behaviors.

发生在确定性系统中的貌似随机的不规则运动,一个确定性理论描述的系统,其行为却表现为不确定性、不可重复、不可预测的现象。

chart 海图

Also called *a nautical chart*, it is a geographic representation of a maritime area and adjacent land, mainly describing the ocean. It has three types: nautical chart, general chart and thematic chart based on the content.

又称"海洋地图",以海洋及其毗邻的陆地为描绘对象的地图,描绘对象的主体是海洋。按表示内容分为航海图、普通海图和专题海图。

chart accuracy 海图精度

As an important index for measuring chart quality, it is embodied by a magnitude of plane and vertical errors of chart elements. The errors include chart projection error, measuring and compilation error, chart printing error, deformation error caused by various materials, etc.

海图的精确程度,是衡量海图质量的重要指标,以海图要素的平面和高程误差大小来体现,包括海图投影误差、测量和编绘作业中产生的误差,海图印刷过程中产生的误差以及各种材料引起的变形误差等。

chart compilation 海图编制

It is about designing and compiling the original chart for publishing.

海图出版原图的设计和制作。

chart composition and drawing 海图编绘

It refers to the whole process of drawing and publishing the original chart, among which drawing is the central part of the charting process.

制作出版原图的全过程。其中,编图作业为整个海图生产的中心环节。

chart contents 海图内容

It is a generic term for various chart information, including mathematical basis, geographic information and auxiliary information.

海图各种信息的统称。包括数学基础、地理信息和辅助信息三大类。

chart of bottom quality 海底底质图

It refers to the thematic chart which indicates the properties of bare bed rock at the bottom of the sea floor and the surface layer sediments.

表示海底底部裸露的基岩、表层沉积物特性的专题海图。

chart of marine resources 海洋资源图

It refers to the thematic chart, which shows distribution of marine resources, including marine biological chart, marine minerals chart, etc.

反映海洋资源分布情况的专题海图,包括海洋生物图和海洋矿产图等。

China Green Carbon Foundation 中国绿色碳汇基金

It was established on August 30, 2010, where predecessor was the *Green Carbon Foundation* established on July 20, 2007 under the China Green Foundation. The *Green Carbon Foundation*, with "reducing carbon emission to tackle climate change" as the main target, is the first national public fund in China, which is committed to promoting afforestation, forest management, deforestation reduction, and other related carbon sink increase and emission abatement activities in response to climate change, popularise relevant knowledge, raise public awareness and the ability to address climate change, and support and perfect ecological benefit compensation mechanism in China. The *Foundation* uses a brand-new running mode, namely, enterprises and individuals make donations to this *Foundation* for carrying out activities such as carbon sink afforestation and forest management, and publicly posting online the enterprise and individual carbon sink accounts which include the actual amount of carbon dioxide absorbed by the planted trees; farmers obtain employment opportunities, increase their revenue, and improve the life quality through participation in activities such as afforestation and forest management, thus exerting the effect of "industry nurturing agriculture and city nurturing country".

2010年8月30日成立了中国绿色碳汇基金,其前身是2007年7月20日在中国绿化基金下设立的绿色碳基金。绿色碳汇基金是中国第一家以"碳汇减排、应对气候变化"为主要目标的全国性公募基金,其宗旨是致力于推进以应对气候变化为目的的植树造林、森林经营、减少毁林和其他相关的增汇减排活动,普及有关知识,提高公众应对气候变化意识和能力,支持和完善中国生态效益补偿机制。该基金运用一种全新的运行模式,即企业和个人捐资到该基金会开展碳汇造林、森林经营等活动,并将实际所植树木吸收的二氧化碳量计入企业和个人的碳汇账户,在网上予以公示;农民则通过参加造林与森林经营等活动获得就业机会并增加收入,提高生活质量,由此起到"工业反哺农业、城市反哺农村"的作用。

chlorinity 氯度

It is a measure of the amount of chlorine in seawater. It is 0.3285234 times as the ratio of the mass of sterling silver (atomic weight) to the mass of seawater when precipitating chloride in seawater samples.

沉淀海水样品中含有的卤化物所需纯标准银(原子量)的质量与海水质量之比值的0.3285234倍,以符号"Cl^-"表示。

circular economy 循环经济

[*resources technology*] It is an economic production mode in which the resources required for production are retrieved and regenerated to regain their value in use, thus achieving recycling and reducing waste emissions.

[*ecology*] It is the short term for imitating the overall, cooperative, cyclic and self-adaptive functions of nature to plan, organise and manage the production, consumption, circulation, restoration, adjustment and control activities of human society. It is a type of network and evolution based compound eco-economy integrating authigenic, symbiotic and competitive economy with a high-efficient resource metabolism process and a complete system coupling structure.

[资源科技]将生产所需的资源通过回收、再生等方法再次获得使用价值,实现循环利用,减少废弃物排放的经济生产模式。

[生态学]模仿大自然的整体、协同、循环和自适应功能去规划、组织和管理人类社会的生产、消费、流通、还原和调控活动的简称,是一类融自生、共生和竞争经济为一体、具有高效的资源代谢过程、完整的系统耦合结构的网络型、进

化型复合型生态经济。

circulation 环流

It refers a phenomenon about large-scale movement of sea water or atmosphere above the sea or in the atmosphere, which forms a closed loop through head-tail connecting.

海面上(或大气内)海水(或大气)大范围流动的现象,通常首尾相接,成一闭合环路。

circum-Bohai Sea economic zone 环渤海经济区

It refers to economic zone consisting of coastal areas surrounding Bohai Sea (and partly Huanghai Sea), including sea and land areas in Liaoning Province, Hebei Province, Tianjin, and Shandong Province.

环绕着渤海(包括部分黄海)的沿岸地区所组成的经济区域。主要包括辽宁省、河北省、山东省、天津市,即三省一市的海域与陆域。

city district planning 城市分区规划

It refers to further arrangements made on the aspects of regional land use, population distribution, public facilities, urban infrastructure configuration on the basis of the overall urban planning.

在城市总体规划的基础上,对局部地区的土地利用、人口分布、公共设施、城市基础设施的配置等方面所作的进一步安排。

city (or county) marine functional zoning 市、县级海洋功能区划

It is approved by the maritime administration department of the people's government at a city or county level and the relevant departments of the people's government at the corresponding level. Based on the superior marine functional zoning, it divides waters and islands under the jurisdiction of the people's government at the corresponding level.

市、县级人民政府海洋行政主管部门会同本级人民政府有关部门,依据上级海洋功能区划开展的,以本级人民政府所辖海域及海岛为划分对象,以海洋功能区为划分单元的海洋功能区划。

city's ternary structure 城市的"三元结构"

Namely, (1) the urban social structure to satisfy and organise community life; (2) the urban economic structure to satisfy and promote community production and circulation; (3) the urban space structure to satisfy and maintain community ecological balance.

即:(1)满足和组织社区生活的城市社会结构;(2)满足和促进社区生产和流通的城市经济结构;(3)满足和维持社区生态平衡的城市空间结构。

city's "three functions" 城市的"三生"功能

Namely, (1) to ensure the survival and development of urban residents; (2) to ensure the operations of urban production and circulation; (3) to transform from ecological unbalance to ecological balance.

即:(1)确保城市居民的生存和发展;(2)确保城市生产和流通的运行;(3)努力从生态失衡走向生态平衡。

classification of sea area use 海域使用分类

It refers to the classification of sea area use and definition of the way the sea is used based on certain principle.

按照一定的原则,划分海域使用类型并界定其用海方式。

clast 碎屑

[ecology] It refers to the broken granular organic matter from the decomposition of plant and animal residues.

[geology] It refers to fragmentary material mainly from the parent rock in a land-source area and formed through physical weathering (mechanical disruption), and it is also called *terrigenous detritus*. Detritus is a component of sedimentary rock or sediment; it can be monomineral or rocky, and the former is called mineral detritus while the latter is called rock detritus.

[生态学]植物和动物残体被分解成的破碎的颗粒状有机物质。

[地质学]是指主要由来自于陆源区的母岩经过物理风化作用(机械破碎)所形成的碎屑物质,又称陆源碎屑。碎屑是沉积岩或沉积物的一种组分,可以是单矿物的,也可以是岩石质的,前者称为矿物碎屑,后者称为岩屑。

Clean Development Mechanism (CDM) 清洁发展机制

It allows for the project cooperation between developed countries and developing countries by means of investment and technology transfer. The greenhouse gas reduction can be achieved through the project, which can help the contracting party of developed countries to accomplish the emission reduction regulated by the *Kyoto Protocol*.

发达国家通过提供资金和技术的方式,与发展中国家开展项目合作,通过项目所实现的温室气体减排量,可以由发达国家缔约方用于完成《京都议定书》中的减排承诺。

clearer energy 清洁能源

It refers to the energy which does not produce harmful substances in the process of production and use, including renewable energy, energy that can be restored after consumption, non-renewable energy (such as wind power, hydropower, natural gas etc.) and energy processed by clean technology (such as clean kerosene).

在生产和使用过程不产生有害物质排放的能源。可再生的、消耗后可得到恢复,或非再生的(如风能、水能、天然气等)及经洁净技术处理过的能源(如洁净煤油等)。

clearer production 清洁生产

It refers to the practical production methods and measures which are employed to satisfy human needs, make rational use of natural resources and energy, and protect the environment, including four aspects: clearer energy, clearer raw material resources, clearer production and clearer products.

既可以满足人们的需要又可以合理使用自然资源和能源并保护环境的实用生产方法和措施。包括清洁的能源、清洁的原料资源、清洁的生产过程、清洁的产品等四个方面的内容。

clearer production technology 清洁生产技术

It is a technology which can reduce the impact of the product on the environment over its life cycle, including saving raw materials, removing poisonous raw materials and reducing the amount and toxicity of any emissions and wastes.

减少整个产品生命周期对环境影响的技术。包括节省原材料、消除有毒原材料和削减一切排放和废物数量与毒性。

climate change 气候变化

It refers to the change of climate caused by direct or indirect human activities on the global atmosphere except natural climate change over a certain period of observation.

经过相当一段时间的观察,在自然气候变化之外由人类活动直接或间接地改变全球大气组成所导致的气候改变。

climate damage 气候灾害

It is a climate phenomenon which brings disaster to human activities and production.

对人类生活和生产造成灾害的气候现象。

climate variability 气候变率

It is a value which reflects changes of climate elements. The average variance or average absolute deviation of the elements can be used as index.

反映气候要素变化大小的量,可用该要素的平均方差或平均绝对偏差等作为指标。

climatic deterioration 气候恶化

It refers to climatic environment changes against human survival due to natural environmental changes or human activities.

因自然环境变化或人类活动而造成的气候环境向不利于人类生存发展的变化。

climatic fluctuation 气候振动

Besides tendency and discontinuity or irregular climatic variation, it covers at least two maximum values (or minimum values) and one minimum value (or maximum value).

除去趋势与不连续以外的规则或不规则气候变化,至少包括两个极大值(或极小值)及一个极小值(或极大值)。

climatic oscillation 气候振荡

It refers to a high-frequency climate change recurring with the regularity of several years, such as quasi-biennial oscillation.

时间尺度为几年的高频气候变化,如准两年振荡。

climatic resources 气候资源

They are generic terms for materials, energies, conditions and phenomena of climate elements, which are the basis for the survival of human beings, all living things and social development.

人类和一切生物生存所依赖的和社会发展可能开发利用的气候要素中的物质、能量、条件及其现象的总体。

climatic evolution 气候演变

It refers to climatic changes caused by geotectonic movements (such as continental drift, orogeny and large-scale changes of land and sea distribution) and solar variations (of over 10^6 years) over a long time scale.

由于地壳构造的活动(如大陆漂移、造山运动、陆海分布的大尺度变化等)和太阳变化引起的较长的时间尺度(超过10^6年)的气候变化。

closed economy 封闭经济

In contrast to an open economy, it refers to a self-sufficient economy which has no economic contacts with outside economies. In an economic sense, it means that a country has no economic interaction with foreign countries in its economic activities, such as the exchange of international trade, international finance and labour forces.

与开放经济相对,是指没有和外部发生经济联系的经济。在经济学意义上是指一国在经济活动中没有与国外的经济往来,如没有国际贸易或国际金融、劳动力的交流等。

closed fishing areas 禁渔区

It is an area which is defined to protect fishery resources and ecological environments by forbidding any fishing or using of some fishing gears.

为保护渔业资源和生态环境所划定的,禁止一切捕捞生产活动或某些渔具作业的水域。

closed fishing season 休渔期

It is the moratorium period prescribed during the breeding and growing season of main fishing targets in order to protect fishery resources.

为了保护渔业资源,在主要捕捞对象繁殖、生长季节规定禁捕的时期。

Club of Rome 罗马俱乐部

Founded in April 1968 with headquarters in Rome of Italy, it is an international civil academic group which studies futurology, and a think tank which explores global issues. It aims to investigate the impact by future scientific and technological revolution on human development and illustrate main obstacles facing humanity to raise the attention of policy-makers and public opinions. Currently, Club of Rome mainly deals with publicity, prediction and investigation of global issues.

罗马俱乐部成立于1968年4月,总部设在意大利罗马,是关于未来学研究的国际性民间学术团体,也是一个研讨全球问题的智囊组织。以研究未来科学技术革命对人类发展的影响为宗旨,阐明人类面临的主要困难以引起政策制定者和舆论的注意,目前主要从事有关全球性问题的宣传、预测和研究活动。

co-production 合作生产

It refers to production activity by enterprises in various areas to work together for products, that is, cooperative enterprises take on the production of some products or procedures among the whole project and jointly complete the entire project in the end.

不同地区的企业共同完成某项产品的生产活动,通常表现为合作企业各自承担总项目中部分产品或部分工序的生产,最后共同完成全部项目。

coast landslide 海岸滑坡

It refers to the collapse of an upper

part of the coast caused by water erosion and artificial exploitation of the coastal base.

由于水力淘蚀或人工开采海岸基部造成的海岸上部滑塌的现象。

coast of emergence 上升海岸

It refers to the coast formed when land rises or sea levels fall, or the rising of land exceeds that of sea level. Coastal terraces and sea caves above the present sea level often come into being along an elevated coast; a silt coast or sandy coast develops when a shallow seabed emerges above the sea level.

因陆地上升、海面下降或因陆地上升超过海面上升量而形成的海岸。上升的沿岸常形成海岸阶地和高出现代海面的海蚀洞等;斜缓的浅海底,上升后露出水面形成淤泥质或沙质海岸。

coast of submergence 下沉海岸

It is a coast whose formation is due to the subsidence of land, or rise of sea levels, or the fact that the amount of land uplift is less than that of a sea level rise. It is often featured by a relatively tortuous coastline, lots of capes, peninsulas and islands, as well as deep waterways and good harbours.

因陆地下沉、海面上升或因陆地上升小于海面上升量而形成的海岸。一般具有较曲折的海岸线,多岬角、半岛和岛屿,并有深水道和良港。

coastal area 沿岸海域

It is a sea area close to a continental coast within nearshore waters, whose hydrological elements are affected greatly by terrestrial meteorological conditions and runoff. Note: It generally refers to the waters within 10 kilometres off a continental coast.

近岸海域之内靠近大陆海岸,水文要素受陆地气象条件和径流影响大的海域。注:一般指距大陆海岸10千米以内的海域。

coastal city 沿海城市

It is a city with a coastline, including municipalities, prefecture-level cities, as well as waters and islands under their jurisdiction.

有海岸线的沿海城市,包括直辖市和地级市以及所辖海域、海岛等。

coastal climate 滨海气候

Also called *marine climate*, it refers to a mild, humid and cloudy climate in coastal areas and islands where the climate is much more influenced by the ocean than by continent.

又称"海岸带气候"、"海洋(性)气候",指陆地沿海和海岛气候受大陆影响小,受海洋影响大,气候温和、湿润、多云。

coastal construction projects 海岸工程建设项目

They refer to new construction, reconstruction and expansion engineering projects which will exert an effect on the marine environment. They are located on or connected to the coast, and their main structures are located on the landward side of the coastline.

位于海岸或者与海岸连接,工程主体位于海岸线向陆一侧,对海洋环境产生影响的新建、改建、扩建工程项目。

coastal current 沿岸流

It is a current flowing along a coast with no direct relation with the waves.

沿海岸流动的与海浪无直接关系的海流。

coastal dune 海岸沙丘

It refers to the undulating sand deposits formed on the seacoast under the action of the wind.

风力吹扬作用下,海岸砂积成的呈波状起伏的砂质堆积体。

coastal effect 海岸效应

It refers to the effect of the coast and islands on the electromagnetic measurements of the sea and land.

海岸和岛屿对海洋大地电磁测量产生的影响。

coastal engineering 海岸工程

It refers to the measures taken and the corresponding structures built based on various coastal environments for the exploitation and utilisation of coastal resources.

为海岸资源开发利用,针对各种海岸环境所采取的措施及构建的相应的建筑物。

coastal erosion 海岸侵蚀

It refers to a recession of the coastline, the downcutting, the narrowing down or the steepening of an intertidal zone and the downcutting of an offshore slope caused by natural factors, human factors or the combination of the two factors.

由自然因素、人为因素或者两种因素叠加而引起的海岸线位置的后退、潮间带下蚀、变窄、变陡以及水下岸坡下蚀。

coastal ferry 滨海轮渡

It refers to a ferry service for transporting cargoes and vehicles across a body of water such as a bay or channel by ferryboat.

用渡船将滨海客货、车辆渡过港湾或海峡的轮渡服务。

coastal freight transport 沿海货物运输

It includes the activities specialised in coastal cargo transport, and freight-oriented coastal transport activities.

包括专门从事沿海货物运输的活动,以货物为主的沿海运输活动。

coastal geohazard 海岸带地质灾害

It is a disastrous geological incident caused by natural geological process or human activities in a coastal zone (including islands) and its adjacent zone. Coastal geohazards mainly refer to coastal erosion (siltation), seawater intrusion, collapse, landslide (including submarine landslide), debris flow, land subsidence, surface collapse, ground fracture, earthquake, sand liquefaction, soil salinization, shallow gas, submarine sand dune (waves), submarine tidal sand ridge, fluctuation of sea level, etc.

在海岸带(包括海岛)及其近邻区域由自然地质过程或人为作用造成的灾害性地质事件。本标准主要指海岸侵蚀(淤积)、海水入侵、崩塌、滑坡(含海底滑坡)、泥石流、地面沉降、地面塌陷、地裂缝、地震、砂土液化、土地盐渍化、浅层气、海底活动沙丘(波)、海底潮流砂脊和海平面变化等。

coastal hotel 滨海旅馆

It refers to hotels in coastal areas which do not have evaluation eligibility for tourist hotels or restaurants at the same level, including all types of hotels, hostels and inns.

不具备评定旅游饭店和同等水平饭店的滨海旅馆,包括各种旅馆、旅社、客栈等。

coastal land area 沿海陆域

It is an area connected to a coast or discharging pollutants and their related stuff to the sea directly or indirectly through pipes, ditches, and other facilities.

与海岸相连,或者通过管道、沟渠、设施,直接或者间接向海洋排放污染物及其相关物质的区域。

coastal line 海岸线

A coastal line refers to the sea-land demarcation line. In China, it refers to the sea-land demarcation line of the mean high water mark of springs tide over many years.

海陆分界线,在我国系指多年大潮平均高潮位时的海陆分界线。

coastal morphology 海岸地貌

It refers to topographic relief caused by the effect of wind, tide and coastal current on lands of the coastal zone. Coastal morphology can be categorized into sea erosion landform caused by marine erosion, and marine accumulation geomorphy caused by marine deposition.

由波浪、潮汐和沿岸流作用于海岸带陆地而形成的地形起伏。分为由海蚀作用形成的海蚀地貌和由海积作用形成的海积地貌。

coastal night fog 海岸夜雾

It refers to the fog that appears at night off the coast.

海岸附近夜间发生的雾。

coastal ocean dynamics 近海海洋动力学

It is a discipline which explores offshore ocean dynamics and thermodynamics, including the study of various ocean motions at different scales of space-time, distributions and changes of ocean hydrological state parameters, such as seawater temperature-salinity and density and their interactions.

研究发生在近海中的海水动力学和热力学过程,其中包括不同类型和不同时空尺度的海水运动规律、海水的温盐度和密度等海洋水文状态参数的分布和变化,以及它们之间相互作用机制等的学科。

coastal passenger transportation 沿海旅客运输

It includes coastal passenger vessel transportation activities and passenger-oriented coastal transportation activities.

包括沿海客轮的运输活动和以客运为主的沿海运输活动。

coastal plain 沿海平原

It refers to a plain which is exposed when the coastal crust rises and sea surface drops. It is characterized by a flat earth surface, dipping gently toward the sea and clear stratification of the deposition.

又称"海岸平原",指沿海地壳上升

或海面下降出露的平原,具有地表平坦、向海微倾、沉积物层理清晰的特点。

coastal protection 海岸防护

It refers to the various engineering measures which can be taken to protect coastal cities and towns, farmlands, wetlands and coastal beaches, etc. from the attack and destruction of storm surges. Under the action of waves, tides, tidal currents, etc., the seacoast can be easily washed and denuded which can cause a coastline recession and jeopardise the stablization of the shoreline and safety of the coastal structures. Generally, common coastal protection structures are submerged breakwaters, breakwaters groynes, longitudinal dikes, sea walls and dykes. Besides, methods like biological deposition promotion and artificial beach nourishment can be taken for the purpose of coastal protection.

采取的各种工程措施来保护沿岸城镇、农田、湿地及滨海浴场等不受风暴潮的侵袭和破坏,称为海岸防护。海岸在波浪、潮汐、潮流等动力因素作用下,易受冲刷、剥蚀,引起岸线后退,危及岸滩的稳定和海岸建筑物的安全。一般较常见的护岸建筑物有潜堤、防波堤、丁坝、顺坝、挡浪墙、堤坝,也可采用生物促淤和人工补沙方法进行海岸防护。

coastal public transportation 滨海公共交通运输

It refers to an urban public transportation service provided for the coastal tourists.

为滨海游客提供的城市公共交通运输服务。

coastal reef 岸礁

Also called *fringing reef*, it is a type of coral reef which grows and develops along the edge of a continent or an island. The surface of the reef has a generally equivalent height to the low tide of sea surface, tilting towards the sea from the coast.

又称"裙礁"、"裾礁",珊瑚礁的一种,沿大陆或岛屿的边缘生长发育的珊瑚礁。礁石表面大致与低潮水位相当,从海岸逐渐向海倾斜。

coastal saline soil 滨海盐土

It refers to the soil formed by saline silt in coastal areas with characteristics as follows: the saline water mainly comes from sea water, and the salt content of the surface layer is $0.6\% - 1.0\%$, while the salt content of the lower layer is between $0.4\% - 0.8\%$; the ground water embedded depth is $1-2$ metres, and the mineralization degree is generally $10-30$ g/L, with the highest reaching $30-50$ g/L. The saline water is mainly composed of sodium chloride, with an amount of chloride ions accounting for above $80\% - 90\%$ of the anions.

沿海地区由盐渍淤泥形成的土壤。特点是盐分主要来自海水,表层含盐量大多在 $0.6\% \sim 1.0\%$,下层在 $0.4\% \sim 0.8\%$,地下水埋深 $1\sim2$ 米,矿化度一般为 $10\sim30$ 克/升,高者 $30\sim50$ 克/升以上。盐分组成以氯化钠为主,氯离子约占阴离子总量的 $80\% \sim 90\%$ 以上。

coastal scenic spot for religious tourism 滨海宗教旅游景区

It refers to coastal temples, mosques,

churches and other religious service areas, such as the Matsu temple.

滨海地区寺庙、清真寺、教堂等宗教服务区域,如妈祖庙。

coastal states 沿海国

They refer to countries with part or all of a land territory adjacent to the ocean.

又称"沿岸国",指陆地领土的一部分或全部邻接海洋的国家。

coastal tourism industry 滨海旅游业

It refers to the activities of marine sightseeing tourism, leisure and entertainment, vacations and accommodations as well as sports activities in coastal areas. It is also a tourism industry including urban construction and commercial activities in coastal areas and tourism-related catering, housing, transportation, tourism, shopping, entertainment, etc.

沿海地区开展的海洋观光游览、休闲娱乐、度假住宿和体育运动等活动所产生的现象和关系的总和,是包括沿海地区的城市建设、商务活动等与旅游活动相关的食、住、行、游、购、娱等诸要素所形成的旅游产业。

coastal tourism management 滨海旅游业管理

It refers to management activities that government departments at all levels have implemented to marine tourism-related affairs.

各级政府部门对海洋旅游相关事务的管理活动。

coastal tourism resort 滨海旅游度假村

It refers to a building complex used for leisure and entertainment in coastal areas covering a natural landscape, buildings and artificial scenery.

滨海地区一个涵盖自然地貌、建筑物及人工景物,用作休闲娱乐的建筑群。

coastal tourism restaurant 滨海旅游饭店

It refers to a tourism restaurant evaluated in accordance with relevant national provisions or with the same quality and standard, including restaurants, guesthouses, hotels, spas and tourism resorts.

按国家有关规定评定的或具有同等质量、水平的滨海旅游饭店,包括饭店、宾馆、酒店、疗养所、度假村等。

coastal zone 海岸带

[*oceanography*] It refers to an area where land meets the sea or ocean, or a line that forms the boundary between the land and the ocean. The modern coastal zone covers the farthest line that the seawater movement can reach, its adjacent land and range that the seawater can exert influence on the sediment changes at the bank slope of a subtidal zone. There is no single standard for the width limit, which varies with different topographic features and research fields.

[*economic*] A coastal zone is featured by coastlines, including coastal provinces, autonomous regions, municipalities, as well as waters and islands under their jurisdiction, etc.

[海洋学]海洋与陆地相互作用的地带。现代海岸带包括现代海水运动对

于海岸作用的最上限及其邻近的陆地,以及海水对于潮下带岸坡剖面冲淤变化所影响的范围,其宽度的界限无统一标准,随海岸地貌形态和研究领域不同而异。

[经济学]又称"沿海地带",是指有海岸线的沿海省、自治区和直辖市以及所辖海域、海岛等。

coastal zone resources 海岸带资源

They refer to the materials, energies and spaces which are distributed in areas where oceans and lands interact and can be utilized by human beings.

分布在海陆相互作用地区的、可以被人类利用的物质、能量和空间。

coastline length occupied by development zone 开发区占用岸线长度

It refers to the length of a coastline which is practically occupied by development zones of various kinds and levels in coastal cities. The unit of measurement is kilometres.

沿海城市各级各类开发区实际占用的海岸线长度。计量单位:千米。

coastline resources 岸线资源

They refer to territorial resources which cover a certain range of water and land space. They are special resources with a combination of water and land and are usually divided into coastline resources and inland water-front resources.

占用一定范围水域和陆域空间的国土资源,是水土结合的特殊资源,分为海岸线资源和内河岸线资源。

cold water mass 冷水团

It refers to a water mass whose temperature is lower than that of surrounding ones.

较周围水体温度低的水团。

cold water tongue 冷水舌

It refers to cold water whose isothermal line is distributed from low to high in the shape of a tongue on an ocean temperature distribution map.

在海洋水温分布图上,等温线从低到高呈舌状分布的冷水。

collection amount of marine area use payment 海域使用金征收额

It refers to the actual amount of marine area use payment collected within the reporting period, which is equal to the amount already collected to the national treasury and the total sum of collection amount of new project and collection amount of original project. The unit of measurement is ten thousand yuan.

报告期内实际征收的海域使用金金额,等于已缴国库金额,是"新增项目征收金额"和"原有项目征收金额"的合计值。计量单位:万元。

commodity money 商品货币

It refers to money which has value in itself and whose commodity value equals monetary value. It is a physical commodity which can be used as medium of exchange.

本身具有价值,商品价值与货币价值相等的货币,并能作为交换媒介的实物商品。

common heritage of mankind 人类共同继承财产

It refers to the high seas, outer space, the Antarctic continent and their natural resources which can not be occupied or exercised sovereignty by any state, legal person or natural person confirmed by international law. The so-called confirmation of international law indicates the regulation of *The United Nations Convention on the Law of the Sea*, *the Outer Space Treaty* and *the Antarctic Treaty*.

国际法确认的任何国家、法人或自然人不得据为己有或行使主权的公海、外层空间、南极大陆及其自然资源。所谓国际法确认的是指由《联合国海洋法公约》、《外层空间条约》和《南极条约》所规定的。

common resources 公共资源

They refer to the natural resources which exist on the earth and can be used by everyone, whose ownership is impossible to be defined or ownership has not been defined.

地球上存在的，不可能划定所有权或尚未划定所有权从而任何人都可以利用的自然资源。

commonly shared resources 共享资源

They refer to the biological resources which inhabit more than two exclusive economic zones of the coastal states and high seas.

栖息于两个以上沿海国专属经济区及公海中的生物资源。

community ecology 群落生态学

It is a discipline which studies the composition characteristics of all species aggregation living in one area, interactions between species and species and environment, formation of community structure, changing mechanisms, etc.

研究栖息于同一地域中所有种群集合体的组成特点、彼此之间及其与环境之间的相互关系、群落结构的形成及变化机制等问题的学科。

community evenness 群落均匀度

It refers to the degree of uniformity in the distribution of species among biocoenosis.

生物群落中各物种间数量分布的均匀程度。

community factor evaluation of tourist resources 旅游资源共有因子评价

It refers to a value and degree evaluation of individual tourism resources by co-owned factors according to types of tourism resources.

按照旅游资源基本类型所共同拥有的因子对旅游资源单体进行的价值和程度评价。

compensation current 补偿流

It refers to the current formed by adjacent water flowing to compensate for the loss of sea water in a particular water, including a horizontal direction and vertical direction compensation current.

由于某一海区的海水流失，邻近的海水随即流去补充而形成的海流，包括水平方向的补偿流和垂直方向的补偿流。

compensation trade 补偿贸易

It is a form of barter between two parties in a transaction. The principle exports such as machines, equipment, technology, raw materials or labour forces are paid for with goods, products or labour forces from the importing market on installment within certain periods of time based on credit.

一方在信贷的基础上,从国外另一方买进机器、设备、技术、原材料或劳务,约定在一定期限内,用其生产的产品、其他商品或劳务,分期清偿贷款的一种贸易方式。

complex adaptive system (CAS) 复杂适应系统理论

The Santa Fe Institute (SFI) is dedicated to the study of Complex Adaptive Systems (CAS) and it focuses on the interdisciplinary and inter-realm study of complexity theory. The fundamental principles of CAS can be generalized as follows: the components of the system are collectively called agents and the agents are dynamic and adaptive "living" entities. The adaptability of the so-called agents are adaptive, which means that they can interact with the environment and other agents constantly, and then "learn" and "accumulate" experiences from them. In addition, with the knowledge and experience they have learned, they can change their own structures and the way they act to adjust to changes in the environment and coordinate with other agents, and stimulates the development and evolution of the whole system as well. According to the theory of CAS, the complexity of the system originates from its adaptability.

美国圣塔菲研究所的工作,其宗旨是开展跨学科、跨领域的复杂性研究。复杂适应系统的基本思想可概括为:将系统的组成元素统称为主体,并强调主体是具有主动性、适应性的"活"的实体。所谓主体具有适应性,就是指它能够与环境以及其他主体进行持续不断的交互作用,从中不断地"学习"或"积累经验",并且能够利用所学到的知识经验改变自身的结构和行为方式,以适应环境的变化以及与其他主体协调一致,并能促进整个系统的发展、演化或进化。该理论认为系统的复杂性来源于适应性。

complex fertilizer manufacturing 复混肥料制造

It is the manufacturing of complex fertilizers with marine chemical products as the raw material, such as potassium and magnesium.

利用钾、镁等海洋化工产品,制造复混肥料的生产活动。

complex network 复杂网络

It refers to a network which is composed of substantial generalized nodes and links. Nodes indicate the system elements, and the links between the nodes indicate the intricate interaction between the elements.

由数量巨大的广义结点和连线共同构成的网络结构,其中结点表示系统元素,结点间的连线表示元素间错综复杂的相互作用。

complex system 复杂系统

It is a system which shows structures with variables from different scaled hierarchies or a dynamical system which is composed of substantial mutually different units. Generally, a complex system can show its complexity, and it may also show its simplicity.

具有变量来自不同标度层次的结构,或由大量相互之间有差别的单元构成的动态系统。通常表现出复杂性,但也可能出现简单性。

complex system theory 复杂系统论

It is a science which studies the complex behaviours, properties and laws manifested during the interaction process of different components in a complex system.

研究复杂系统中各组成部分之间相互作用所涌现出复杂行为、特性与规律的科学。

compounding 复利

It refers to the annual addition of earned interest to a capital sum of borrowed or loaned money at the existing market interest rate, including the interest earned by the accumulated interest; the calculated future value of an existing capital sum.

由本金和前一个利息期内应记利息共同产生的利息。

comprehensive utilization of seawater chemical resource 海水化学资源的综合利用

It refers to the extraction of chemical elements and chemicals from seawater and deep processing, etc.

从海水中提取化学元素、化学品及深加工等。

comprehensive utilization of natural resources 自然资源综合利用

It refers to the multi-level and multi-purpose development and utilisation of each element of natural resources with advanced science, technology and method.

以先进的科学技术与方法,对自然资源各组成要素进行的多层次、多用途的开发利用。

condition of ground water recharge 地下水补给条件

It refers to the recharge resource, recharge mode, recharge area and boundary, and the recharge amount of the groundwater.

地下水的补给源、补给方式、补给区面积及边界、补给量等。

confluence 汇流

[*atmospheric science*] It is the fluid movement during which adjacent fluids converge towards the direction of primary fluid.

[*geography*] It is the process of movement during which slope flows formed by rainfall converge with channel flows.

[大气科学]相邻流体向主流体运动方向辐合(有向中心流动的分量)的流体运动。

[地理学]降水产生的坡面与河槽径流汇集流动的过程。

conservation ecology 保护生态学

It is a discipline which studies biodiversity conservation from the perspective protecting of biological species and their living environments.

研究生物多样性保护的科学,即研究从保护生物物种及其生存环境着手来保护生物多样性的学科。

constant elements resources 常量元素资源

They refer to various constant elements resources in sea water. It has been found so far that sea water, a mixed liquid with a complex chemistry composition, includes more than 80 chemical elements. An element, whose content in sea water is above 100 mg/L, is called a constant element. There are mainly 11 constant elements including oxygen, sodium, magnesium, sulphur, calcium, potassium, bromine, carbon, strontium, boron and fluorine, approximately accounting for 99.8% - 99.9% of the total content of chemical elements.

海水中所含的各种常量元素资源。海水是一种化学成分复杂的混合溶液，迄今已发现的化学元素超过 80 种。每升海水超过 100 毫克的元素，称为常量元素。最主要的常量元素有氧、钠、镁、硫、钙、钾、溴、碳、锶、硼、氟 11 钟，约占化学元素总含量的 99.8%~99.9%。

constrained generating procedures, CGP 受限生成过程

It is an effective description of emergence. The model generated by emergence is dynamic, and it is reflected in a state of "process". The constant dynamic change of the model results from the "constraint" or "limitation" and new behaviours of the model are induced or generated by interaction between different components.

受限生成过程是对涌现现象一种有效的描述方法。涌现生成的模型是动态的，体现为一种"过程"状态，支撑这个生成模型不断动态变化的是各个组元之间机制的"约束"或"限制"，通过相互作用诱发或生成了模型的新行为。

consumer price index, CPI 消费物价指数

It is a measure of changes in the purchasing power of a currency and the rate of inflation. The CPI reflects the current prices of a basket of goods and services in terms of the prices during the same period in a previous year, to show the effect of inflation on purchasing power.

普通消费者所购买的物品与劳务的总费用的衡量指标。

consumption 消费

It is a process of using social products to satisfy various human needs. It includes productive consumption and personal consumption. The former refers to the use and consumption of means of production and living labour in the process of material production; while the latter refers to the behaviour and process of using manufactured material goods and intellectual products to meet the needs of personal life needs. Generally speaking, consumption we talk about refers to the latter, i.e. personal consumption.

利用社会产品来满足人们各种需要的过程，分为生产消费和个人消费。前者指物质资料生产过程中的生产资料和生活劳动的使用和消耗；后者是指人们把生产出来的物质资料和精神产品用于满足个人生活需要的行为和过程。通常讲的消费是指个人消费。

consumption of comprehensive energy 综合能源消费量

It refers to the total net value of all kinds of energy actually consumed by industrial enterprises in industrial production activities within a reporting period. Various energy resources used need to be converted into standard fuel first before the calculating of comprehensive energy consumption. The unit of measurement is tons of standard coal.

报告期内工业企业在工业生产活动中实际消费的各种能源的总和净值。计算综合能源消费量时,需要先将使用的各种能源折算成标准燃料后再进行计算。单位:吨标准煤。

contiguous zone 毗连区

It refers to a particular area extending from the outer edge of a territorial sea, within which a state can exert its right to its customs, fiscal, immigration or sanitary laws and regulations within its territory or territorial sea. According to the Article 33 of the *United Nations Convention on the Law of the Sea*, the contiguous zone may not extend beyond 24 nautical miles from the baselines from which the breadth of the territorial sea is measured.

毗连沿海国领海,并在领海以外的一定宽度、供沿海国行使关于海关、财政、卫生和移民等方面管制权的一个特定区域。《联合国海洋法公约》第三十三条规定,毗连区的宽度从领海基线量起不超过 24 海里,按照确定宽度形成的水域外缘为毗连区的外部界限。

continent 洲

It is a generic term for the vast area of land and all countries on its nearby islands.

面积广阔的陆地及其附近岛屿上所有国家的总称。

continental block 大陆块

It refers to the block-shaped surface configuration in the ocean.

海洋中,呈块状的地表形态。

continental drift 大陆漂移

It refers to a large-scale horizontal movement between continents and between continents and oceanic basins.

大陆与大陆之间、大陆与大洋盆地之间的大规模水平运动。

continental margin 大陆边缘

It refers to the wide transitional zone existing between the continental surface and the ocean bottom which comprises the continental shelf, continental slope, and continental rise, excluding the ocean floor and ocean ridge of the deep ocean, and subsoil.

大陆表面和大洋底面之间存在的一个广阔过渡带,由陆架、陆坡和陆隆的海床和底土构成,不包括深洋洋底及其洋脊,也不包括其底土。

continental plain 陆架平原

As the main part of a continental shelf, it is a flat, spacious and large-scale geographic entity, whose average slope is generally less than $0°10'$.

大陆架上地形平坦、广阔的大型地理实体,为大陆架的主体,平均坡度一般小于 $0°10'$。

continental shelf 大陆架

[*oceanography*] It refers to a submerged border of a continent that slopes gradually and extends to a point of steeper descent into the ocean. The average slope is approximately 0.5 degrees.

[*United Nations Convention on the Law of the Sea*] The continental shelf of coastal countries includes all the natural extensions from their continents and beyond their territorial waters, reaching the seabed and subsoil of the seabed areas of the continental margin, to 200 or less than 200 nautical miles or less from the base line which is used to measure the width of the territorial waters to the edge of the continental margin.

[海洋学]又称"陆架"、"陆棚"、"大陆棚",指大陆向海的自然延伸的浅水海底,自海底海岸低潮线起地形缓缓倾斜至海底坡度突然增加处(坡折线),平均坡度一般为0.5°左右。

[《联合国海洋法公约》]沿海国的大陆架包括其领海以外依其陆地领海的全部自然延伸,扩展到大陆边缘海底区域的海床和底土,或者从测算领海宽度的基线量起到大陆边缘外缘的距离不到200海里,则扩展到200海里的距离。

continental solonchak 内陆盐土

It is a generic term for salinized soil of all types in arid and semi-arid areas with the exception of seashore solonchak. Due to an arid climate, low rainfall, large evaporation and strong saltification, salt incrustation, salt crust and loose poly salt layers come into being at the surface of a continental solonchak in an arid area, whose surface is 1-5 centimetres thick, salinity is generally between 5%-20%, groundwater level depth is 1-3 metres and degree of mineralisation is 3-20 grams/litre.

除滨海盐土以外的干旱、半干旱地区各种盐渍土的统称。干旱地区的内陆盐土,由于气候干旱,雨量稀少,蒸发量大,积盐强烈,地表常形成盐结皮、盐结壳和疏松的聚盐层,表层1~5厘米,含盐量通常在5%~20%,地下水位埋深一般1~3米,矿化度3~20克/升。

contribution rate of scientific and technological progress 科技进步贡献率

Also called *total factor productivity*, it is an index of analysing the economic benefit of scientific and technological progress, referring to the ratio of effective or useful fruits number to resource consumption and occupancy, namely the ratio between output and input, or that between what is produced and what is consumed. Also called Total Factor Productivity contribution (TFP contribution) in academia, it includes the contribution share of total factors except capital and labour input and represents the share of total factors' increase in economic growth as an evaluation index instead of statistical index, which is of great reference or guiding significance not only in analysing the long-term developing trend and interrelationship between economic growth, scientific and technological progress, labour and capital, but also in developing national development strategies and implementing macroeconomic management.

又称"综合要素生产率"。分析科技进步经济效益的一个指标。指有效或有用成果数量与资源消耗及占用量之比,即产出量与投入量之比,或所得量与所费量之比。它包含除资本与劳动投入之外的其他全部综合要素投入的贡献份额,学术界又称为综合要素贡献率(TFP贡献率),表达了在经济增长中,综合要素的增长所占的份额。它是一个评价指标,不是统计指标。这一指标对分析经济增长与科技进步、劳动和资本的长期发展趋势与相互关系,对制定国家发展战略和宏观经济管理具有重要的参考指导意义。

control 控制

It is a process of changing conditions based on definite objectives to narrow down possibility space of matters so that they can develop in a certain direction. Therefore, all control processes involve three fundamental steps: (1) to get to know the possibility space confronted; (2) to choose some states from possibility space as targets; (3) and to control conditions to move toward stated objective.

人们根据自己的目的,改变条件,使事物的可能性空间缩小,沿着某种确定的方向发展,从而形成控制。因此,一切控制过程,实际都是由3个基本环节构成的:(1)了解事物面临的可能性空间是什么;(2)在可能性空间中选择某一些状态为目标;(3)控制条件,使事物向既定的目标转化。

controlled ecosystem experiment (CEPEX) 控制生态系统实验

It is an experimental method of exploring structures, functions and changing rules of some natural marine ecosystems by utilizing experimental ecosystem equipment under artificial control.

又称"围隔式生态系统实验",运用建立的实验生态系统装置,在人为控制条件下,研究某一自然海洋生态系统的结构、功能及其变化规律的一种实验方法。

conversion efficiency 转换效率

It is the ratio between the output of a trophic level and that of a lower trophic level.

一个营养级的生产量与较低一个营养级的生产量之比。

convective mixing 对流混合

It refers to the mixing phenomenon of seawater from different water layers moving to each other vertically.

海水在垂直方向上做相向运动造成不同水层的海水混合现象。

Convention on Protection of the Mediterranean from Pollution 保护地中海免受污染公约

It is the regional international convention concluded for the protection of the oceanic environment of the Mediterranean. It was passed in Barcelona on February 16, 1976 and went into effect on February 12, 1978, with 17 countries and the European Economic Community signing the treaty. The convention stipulates as follows: each contracting party should take all measures to prevent and eliminate the pollution within its own territory caused by waste dumping of ships and aeroplanes, ship discharge, the exploration and development of seabeds and subsoils, the effusions of rivers and coastal facilities, or other land sources. All the contracting parties should cooperate and take measures to deal with pollution caused by any reason, establish regional pollution detection plan, and carry out scientific research related to oceanic pollution; and they should jointly formulate the procedure of sentencing for accountabilities and indemnification due to the damage caused by the violation of the convention and its protocols.

为保护地中海海洋环境而订立的区域性国际公约。1976 年 2 月 16 日在巴塞罗那通过,1978 年 2 月 12 日生效,有 17 个国家及欧洲经济共同体参加缔约。公约规定:缔约国应采取一切措施,以防止和消除该国领土内因船舶和飞机倾废、船舶排出物、勘探和开发海床及底土、来自河流及沿岸设施的流出物,或其他陆地来源造成的污染,应合作采取措施应付任何原因引起的污染,建立区域污染检测方案,进行有关海洋污染的科技研究;联合制订关于因违反公约和议定书发生损害的责任和赔偿的判决程序。

conventional energy resources 常规能源

It refers to the energy which has been massively produced and widely utilized at present.

在现阶段已经大规模生产和广泛使用的能源。

coral island 珊瑚岛

A coral island is a type of oceanic island, referring to a rocky island made of coral reefs or a sandy island formed by coral fragments on coral reefs.

海洋岛的一种。由珊瑚礁构成的岩石岛或珊瑚礁上的珊瑚碎屑形成的沙岛。

coral reef 珊瑚礁

They are underwater structures made from calcium carbonate secreted by corals. Coral reefs are colonies of tiny living animals found in marine waters that contain few nutrients. Most coral reefs are built from stony corals, which in turn consist of polyps that cluster in groups.

造礁珊瑚及其他造礁生物对生成礁的钙物质长期积累沉积,形成珊瑚礁。

coral reef coast 珊瑚礁海岸

It refers to a coast accumulated by coral reefs. Dead corals condense into solid coral reefs in some sections of the coast.

珊瑚礁堆积组成的海岸,有的岸段由珊瑚遗体凝结成坚固的珊瑚礁体。

corange line 等潮差线

Corange line is a line on a tidal (or tidal constituent) chart connecting points of equal tide range.

在潮汐(或分潮)分布图上,潮差(或振幅)相等点的连线。

core area 核心区

It is the heartland of a nation with dominant status, fast economic growth and high-quality development in national or regional economic development. It is the main area where a nation or region can realize economic development and its GDP per capita is much higher than the national average level.

在全国经济或者区域经济发展中居于主导地位、经济增长快、发展质量高的地区,是一个国家或地区实现经济发展的主要地区,通常其人均 GDP 要大大高于全国人均水平。

core inflation 核心通货膨胀

It represents the long-run trend in the price level. In measuring long-run inflation, commodity and service prices, which tend to be volatile and frequent and to cause transitory price shocks, are usually excluded from core inflation calculations. The measure of core inflation uses the consumer price index (CPI).

剔除一些价格变动巨大且频繁或短期性质价格波动的商品或服务后的通货膨胀,以核心 CPI 来表示。

corrosion 溶蚀

It refers to the chemical dissolution and erosion of soluble rocks, which is mainly composed of carbonate rocks with the combination of groundwater and surface water.

地下水和地表水相结合,对以碳酸盐岩为主的可溶性岩石的化学溶解和侵蚀作用。

costal museum 滨海博物馆

It refers to comprehensive museums, exhibitions and so on in coastal areas, such as marine science and technology museums.

沿海地区的综合类博物馆、展览馆等,如海洋科技馆等。

cost of marine resource exploitation 海洋资源开发成本

It refers to the sum of costs or prices on material capital investment, human capital investment and natural capital consumption during marine resources exploitation.

开发海洋资源过程中的物质资本投放、人力资本投入和自然资本耗用的费用或代价的总和。

cotidal line 等潮时线

It is a line on a tidal (or tidal constituent) chart where the points of the same tidal levels are connected.

在潮汐(或分潮)分布图上,具有相同潮汐位相点的连线。

counter urbanization 逆城市化

Due to the deterioration of city problems such as traffic congestion, a growing number of crimes and heavy pollution, urban population begins to move to suburbs and country areas, causing the hollowness of a city centre, which is also called counter urbanization.

又称"反城市化"。由于交通拥挤、犯罪增长、污染严重等城市问题日益增加,城市人口开始向郊区乃至农村流动,市区出现"空心化"的现象。

coupling 耦合

It refers to a relationship through which all elements consisting of a system are mutually conditioned and interdependent in nature and existence.

构成系统的各个部分在性质和存在上互为条件、互为因果的关系。

crowd out 挤出

It is an economic theory stipulating that rises in public sector spending drive down or even eliminate private sector spending. Though it is a general term, it is often used in reference to the stifling of private spending in areas where government purchasing is high.

政府支出增加所引起的私人消费或

投资降低的效果。

crowding-out effect 挤出效应

It is an economic theory stipulating that rises in public sector spending drive down or even eliminate private sector spending. Though it is a general term, it is often used in reference to the stifling of private spending in areas where government purchasing is high.

政府支出增加所引起的私人消费或投资降低的效果。

crude oil production 原油产量

It is the amount of crude oil that can be directly used in sales and production for private use, calculated according to the net volume of crude oil. The unit of measurement is ten thousand tons.

按净原油量来计算的,能直接用于销售和生产自用的原油量。计量单位:万吨。

cumulative causation theory 循环积累因果原理

Changes in a certain social economic factor will lead to changes in another social economic factor. The latter changes in turn strengthen the previous changes, and cause the social economic process to develop along the direction of the initial changes, thus forming a cumulative circular development tendency.

某一社会经济因素的变化,会引起另一社会经济因素的变化,这后一因素的变化,反过来又加强了前一个因素的那个变化,并导致社会经济过程沿着最初那个因素变化的方向发展,从而形成累积性的循环发展趋势。

currency 通货

Currency, namely the currency in circulation, serves as a medium of exchange in the process of commodity circulation, including legal tender, such as banknotes, coins, etc, and credit currency.

即流通货币,在商品流通过程中充当一般等价交换物,包括纸币、铸币等有形实体货币和信用货币。

current observation 海流观测

It refers to an observation process of the flow conditions of ocean water.

观测海水流动状况的过程。

current year value added tax payable 本年应交增值税

In accordance with the regulations of tax law, the tax paid by companies engaged in selling and processing products, labor services of repairing and replacement which add value to the products is the value added tax payable of the company within a reporting period. The unit of measurement is ten thousand yuan. Calculation formulae: current year value added tax payable = (output tax - input tax - transfer-out of input tax) - export deduct domestic sales goods tax - allowance + export tax rebate.

企业按税法规定,从事货物销售或提供加工、修理修配劳务等增加货物价值的活动本期应交纳的税金,指企业在报告期应交增值税额。计量单位:万元。计算公式为:本年应交增值税=销项税额-进项税额-进项税额转出)-出口抵减内销产品应纳税额-减免税款+出口退税。

cuspate bar 尖头沙坝

It is cuspate-shaped sand bar formed when coastal spit is folded by waves and bends towards the bar connecting the land and coast, or one formed when two spits extend towards the sea and intersect there.

又称"尖角坝",海岸沙嘴被浪流所折,弯向陆与岸相连的坝,或由两个沙嘴向海延伸交汇而形成的沙坝。呈尖角形。

cuspate delta 尖(形)三角洲

It is a delta formed when a river drops sediment onto a straight shoreline with strong waves. Waves force the sediment to spread outwards in both directions from the river's mouth, making a pointed tooth shape with curved sides by regular opposing, gentle water movement.

尖嘴向海明显凸出的三角洲或由河道及两侧河口沙嘴组成的三角洲。

cyclical unemployment 周期性失业

It is unemployment caused by the falling demand in recession or depression of an economic cycle.

经济周期中的衰退或萧条时因需求下降而造成的失业。

D

damage by tide 潮灾

It refers to the damage done on land when it is inundated by seawater under the joint action of astronomical tides and storm surges.

天文潮和风暴潮共同作用下,海水涌上陆地所造成的灾害。

debris flow 泥石流

It is an intermittent flood carrying a large amount of mud and stone blocks.

携带大量泥沙、石块的间歇性洪流。

Decentralized Experiment Evolution Model, DEEM 去中心化试验演化模型

The evolution of a whole system network, which includes complicated structures, plenty of subsystems and a main body of institutional innovation with various statuses and innovative abilities, involves all systems and limited rational agents with a large scale, high-level complexity, higher risks and higher costs. Therefore, all system subjects should be allowed to make innovations independently within the system network or optionally in some systems to promote the effective and successful innovative results of the entire system, which can reduce the uncertainty and risks of evolution effectively.

对于结构比较复杂,拥有较多的子系统,制度创新主体具有不同的地位(连接度)和创新能力的制度网络系统,整个制度系统的演化涉及所有的制度、所有的有限理性主体,规模大、复杂程度高、风险大、成本高,因此,应该让不同的制度主体在网络系统内各自进行有效的创新,或有选择地在一些制度主体中进行制度创新的试验,然后再逐步平面式地把有效的、成功的制度创新结果推广到整个制度系统,这样可以有效地减小演化中的不确定性,降低制度系统演化的风险。

deep clay 深海黏土

It refers to the clay in brown red or marron color in the deep sea and the sediment distributed in the deep sea, whose particle size is no more than 0.0039 millimetres.

又称"远洋黏土"、"褐黏土"、"红黏土",深海呈褐红色、棕红色分布的黏土,分布于深海,粒径不大于0.0039毫米的沉积物。

deep fosse 海渊

Deep is an abyssal depression with distinct outline, whose maximum depth is over six thousand metres. Deep, mostly the deepest part of the ocean trench, is usually named after the ship that discovers it.

深度超过6 000米、轮廓清楚的深海凹地。多数位于海沟中,海沟中已测得的最深陷部分,通常以发现它的船只命名。

deep layer 深层

It refers to the aquifer lying below the mesopelagic area with average temperature, salinity and density.

大洋中中层以下温度、盐度、密度较均匀的水层。

deep ocean floor 深海洋底

It refers to a surface or ocean ridge of deep ocean floors outside the continental margin, which excludes the surfaces or ocean ridges of deep ocean floors or subsoils.

大陆边缘以外的深洋底的表面及洋脊,大陆边缘既不包括深洋底的表面及洋脊,也不包括其底土。

deep sea 深海

It refers to the continental slope and marine band outside the continental shelf. In practice, the depth of deep sea is measured by different standards in various industries. For example, the part of the ocean which is below 2,000 metres is considered deep sea by marine professionals, while marine aquaculture regards part of the ocean below 200 metres as deep sea aquaculture.

大陆架以外的大陆斜坡和海洋带。实践中,深海的深度常因不同的行业有不同的标准。海洋专业一般把2000米以下作为深海,但养殖业将200米以下即作为深海养殖。

deep-sea ecology 深海生态学

It is a discipline which explores the habitation of organisms living in deep waters and seabeds under the conditions of high pressure, no light and low temperature and their interrelationship with environmental factors as well.

研究在大陆架以外深层水域及海底生活的生物在高压、无光、低温条件下的栖息活动及其与环境因子间相互关系的学科。

deep-sea ecosystem 深海生态系统

It refers to a unified whole formed by the material exchange and energy transfer between all marine biotic communities existing in the benthic division and pelagic division in deep waters outside the continental shelf and a dark environment with low temperature, high pressure and no plant distribution.

大陆架以外深水水域的海底区和水层区所有海洋生物群落与其周围无光、低温、压力大而无植物分布的环境进行物质交换和能量传递所形成的统一整体。

deep-sea organisms 深海生物

They refer to organisms living below 200 metres in the ocean or in the benthic division and pelagic division deeper than the continental shelf.

生活于海洋中200米以深的生物或比大陆架更深的海底区和水层区生物。

deep-water aquiculture 深水养殖

It refers to a cultivation of aquatic economic plants and animals in a sea area which is below 83 metres.

在水深83米以下海域进行水产经济动植物养殖的生产活动。

deficiency of natural resources 自然资源稀缺性

Natural resources that can be used by people within a certain time and space are limited, but people's desire for material needs is infinite. The contradiction between the two constitutes the scarcity of resources.

在一定的时空范围内能够被人们利用的自然物(资源)是有限的,而人们对物质需求的欲望是无限的,两者之间的矛盾构成资源的稀缺性。

degree of perception coefficient 感应度系数

It is the degree of the demand perception of a certain department when one unit is added and applied in departments of national economy. In other words, it refers to the output this given department is supposed to supply for the production of other departments.

国民经济各部门每增加一个单位最终使用时,某一部门由此而受到的需求感应程度,也就是需要该部门为其他部门生产而提供的产出量。

delta 三角洲

It is a landform that forms from deposition of silts carried by rivers in the estuary when rivers enter the receptions basin and interact with the receving water, leading to flow diffusion and flow velocity decrease.

河流流入受水盆地(海洋或湖泊)与受水体相互作用,因水流扩散,流速降低,河流携带的泥沙在河口附近发生堆积,形成的突出的堆积体。

demand 需求

It refers to a specific quantity of a good or service for which consumers are willing and able to pay a price over a specific period.

一种商品的需求是指消费者在一定时期内在各种可能的价格水平意愿而且能够购买的该商品的数量。

demand function 需求函数

It indicates the correlation between the quantity of a particular commodity demanded and various factors affecting the quantity demanded.

表示一种商品的需求数量和影响该需求数量的各种因素之间的相互联系。

demand schedule 需求表

It is a digital sequence table that shows the relationship between various prices of a certain commodity and the quantity of the commodity demanded at those various prices.

表示某种商品的各种价格水平和与各种价格水平相对应的该商品的需求数量之间关系的数字序列表。

demolition of marine structures 海上建筑物拆除

It refers to the demolition activities of man-made marine structures abandoned at sea.

人工建造的海上滞留物的拆除活动。

density in situ 现场密度

It refers to the density at a marine specific point.

海水特定点的密度。

deposition 淤积

It is a process in which sediment particles from a current sink down to a river bed, causing bottom elevation uplift or river bank siltation.

水流挟带的泥沙颗粒沉落到河床上致使河底高程抬升或河岸淤涨的过程。

deposition island 堆积岛

It refers to an island formed due to the accumulation or deposition of rivers, tides and waves in an estuary area or coastal zones.

在河口区及海岸带中由于河流、潮流和波浪的堆积作用而形成的岛屿。

deposit mining of coastal ironstone 海滨黑色金属矿采选

It refers to the deposit mining activities of coastal ironstone.

对滨海地区铁矿石的采选活动。

deposit mining of coastal nonferrous metal 海滨有色金属矿采选

It refers to the deposit mining activities of coastal copper ore, lead zinc ore, nickel cobalt mine, tin ore, magnesium ore and other nonferrous metal ores.

对海滨地区的铜矿、铅锌矿、镍钴矿、锡矿、镁矿及其他有色金属矿的采选活动。

deposit mining of coastal precious metal 海滨贵金属矿采选

It refers to the deposit mining activities of a coastal gold mine, silver mine and other precious metal mines.

对海滨地区金矿、银矿及其他贵金属矿的采选活动。

deposit mining of coastal rare earth metal 海滨稀有稀土金属矿采选

It refers to the deposit mining activities of coastal lanthanide series metal and metal ore similar to the property of lanthanide series metal.

对海滨地区镧系金属及与镧系金属性质相近的金属矿的采选活动。

deposit mining of coastal sand and gravel for construction 海滨建筑用砂、砾石开采

It refers to the exploration activities of coastal limestone, architectural ornament stone and refractory clay and stone.

对海滨地区石灰石、建筑装饰用石、耐火土石的开采活动。

depreciation 贬值

It refers to the value decrease of a country's currency according to the foreign currency amount it can purchase.

又称"折旧",按所能购买到的外国通货量衡量的一国通货的价值减少。

depreciation of fixed assets 固定资产折旧

It refers to an allowance made for a decrease in value of fixed assets which were shifted to products due to the wear and tear, usually calculated on the basis of the original value of the fixed assets (or the net book value of the fixed assets for enterprises which calculate the depreciation according to double declining balance method) and the decided rate of depreciation. Accumulated depreciation refers to the cumulative depreciation of fixed assets of a company over the years that has been claimed at the end of the reporting period. Depreciation of the current year refers to the total depreciation of fixed assets of a company claimed during the reporting period. The unit of measurement is 10,000 yuan.

对固定资产由于磨损和损耗而转移到产品中去的那一部分价值的补偿。一般根据固定资产原价(选用双倍余额递减法计提折旧的企业,为固定资产账面净值)和确定的折旧率计算。"累计折旧":指企业在报告期末提取的历年固定资产折旧累计数。"本年折旧":指企业在报告期内提取的固定资产折旧合计数。计量单位:万元。

depth of groundwater table 地下水埋藏深度

It refers to the vertical depth from the ground surface to the phreatic water level or confined water level.

从地表面至地下水潜水面或承压水面的垂直深度。

desertification 荒漠化

It refers to the process in which desert environment extends to or invades arid or semiarid regions, caused by climatic changes, human influence, or both.

由气候变化、人类活动或两者共同作用所引起的荒漠环境向干旱或半干旱地区延伸或侵入的过程。

desertization 荒漠化

It is the process in which desert environment extends to or invades arid or semiarid regions, caused by climatic changes, human activities, or combined action of above two.

由气候变化、人类活动或两者共同作用所引起的荒漠环境向干旱或半干旱地区延伸或侵入的过程。

detailed planning of urban construction 城市修建性详细规划

It refers to planning and design formulated to guide the construction of buildings and engineering facilities in a city based on comprehensive urban planning and regulatory planning.

以城市的总体规划或控制性详细规划为依据,制定用以指导城市各项建筑和工程设施及其施工的规划设计。

detailed planning of urban control 城市控制性详细规划

It refers to the planning requirements made to determine the control index of the nature and intensity of land use in urban construction areas, control position of the roads and utilities pipeline and space environmental control on the basis of comprehensive urban planning.

以城市的总体规划为依据,确定城市建设地区的土地使用性质和使用强度的控制指标、道路和工程管线控制性位置以及空间环境控制的规划要求。

detritus 碎屑

[ecology] It refers to the broken granular organic matter from the decomposition of plant and animal residues.

[geology] It refers to fragmentary material mainly from the parent rock in a land-source area and formed through physical weathering (mechanical disruption), and it is also called *terrigenous detritus*. Detritus is a component of sedimentary rock or sediment; it can be monomineral or rocky, and the former is called mineral detritus while the latter is called rock detritus.

[生态学]植物和动物残体被分解成的破碎的颗粒状有机物质。

[地质学]是指主要由来自于陆源区的母岩经过物理风化作用(机械破

碎)所形成的碎屑物质,又称陆源碎屑。碎屑是沉积岩或沉积物的一种组分,可以是单矿物的,也可以是岩石质的,前者称为矿物碎屑,后者称为岩屑。

develop marine resources by relying on science and technology 科技兴海

Based on scientific and technological progress, this concept was put forward to enhance the development of ocean resources and marine industry. As multi-level social system engineering, it relates to many fields of scientific research, development, popularisation, production, environmental protection and management. It aims to promote rapid and sustainable development of marine economy, increase the proportion of marine industry in GDP and facilitate the sustainable development of the ocean.

我国依靠科技进步推动海洋资源开发和海洋产业发展而提出的一项涉及科研、开发、推广、生产、环保、管理等领域的多层次、多环节的社会化系统工程,其目标是推动海洋经济快速、持续发展,提高海洋产业产值在国内生产总值中的比重,促进海洋的可持续利用。

development crisis 发展危机

Also called *economic growth without development*, it generally refers to an unhealthy phenomenon or mode of economic growth with only increase of economic yield but lack of economic structure changes in the economic construction of developing countries.

又称"无发展的经济增长"。一般指发展中国家经济建设中的一种仅有产量的增加而无结构上变化的不健康的经济增长现象或增长方式。

Diaoyu Islands 钓鱼岛

Diaoyu Dao and its affiliated islands, consisting of Diaoyu Dao, Huangwei Yu, Chiwei Yu, Nanxiao Dao, Beixiao Dao, and other islands and reefs, are located 92 nautical miles to the northeast of Keelung, Taiwan, in the East China Sea waters. The total landmass of these islands is approximately 6.3 square kilometres. The Diaoyu Islands boast important strategic location, and abundant resources like submarine oil, minerals, and fishery resources. Since ancient times, China enjoys indisputable sovereignty over Diaoyu Dao and its affiliated islands. After the Sino-Japanese War in 1895, Japan robbed the Diaoyu Islands away from the Qing government. After Japan was defeated in 1945, the Diaoyu Islands were supervised by America and then America gave them over to Japan in 1962. Hereafter, Diaoyu Islands have caused more heated disturbances which has also became the focus of attention for China, America, Japan and Taiwan of China.

钓鱼岛列岛位于台湾基隆市东北约92海里的东海海域,主要由钓鱼岛、黄尾屿、赤尾屿、南小岛和北小岛及一些礁石组成,总面积约6.3平方千米,战略位置重要,海底石油、矿产、渔业资源丰富。自古以来,中国对钓鱼诸岛及其附近海域拥有无可争辩的主权。中日甲午战争之后,钓鱼岛被日本从清政府手中抢去,1945年日本战败后,钓鱼岛由美国监管,1972年美国又将钓鱼岛移交给日本。此后,钓鱼岛成为中、美、日及台湾所关注的焦点。

diffusion effect 扩散效应

It refers to a phenomenon indicating that all surrounding areas of an economic expansion centre will have access to capital and human resources from central area with the improvement of infrastructure to be stimulated to develop and catch up with the central area.

所有位于经济扩张中心的周围地区,都会随着与扩张中心地区的基础设施的改善等情况,从中心地区获得资本、人才等,并被刺激促进本地区的发展,逐步赶上中心地区。

digital chart 数字海图

It is a nautical chart stored in a digital form on tape, disk, CD-ROM, or other media.

以数字形式存储在磁带、磁盘、光盘等介质上的航海图。

digital city 数字城市

In urban planning, construction and operation management, as well as urban life and production, digital information processing technology and network communication technology are fully applied to integrate and utilise all kinds of urban information resources. Actually, digital technology, information technology and network technology have infiltrated every nook and cranny of urban activities, thus making digital life become the mainstream of the city. Digitisation is essentially informatization by its nature.

在城市规划、建设与运营管理以及城市生产和生活中,充分利用数字化信息处理技术和网络通信技术,把城市的各种信息资源加以整合并充分利用。究其实质,就是使数字技术、信息技术、网络技术渗透到城市活动的每一个方面、每一个角落,使数字成为城市活动的神经系统。数字化本质上是信息化。

dike 堤

It refers to the water retaining structure built along the banks of rivers, lakes and seas. The structures built on both sides of the rivers can be called "river embankments" or "river levees"; the structures built on the coast of the sea can be called "sea dikes" or "seawalls".

沿江、河、湖、海的边岸修建的挡水建筑物,建在江、河两岸的,称"江堤"或"河堤";建在海边的,称"海堤"或"海塘"。

diluvial 洪积

It refers to a term for deposits formed when weathering products by bedrock are affected by ephemeral torrent and carried to the valley outlet, which is found mostly in semi-arid climatic areas, shaped like a fan. It is primarily composed of gravel, sand, silt and clay, with selective and unclear stratification.

基岩的风化产物由山区暂时洪流的作用,携带到山谷出口处形成的沉积物,多半见于半干燥气候区,其形状如扇。主要由砾石、砂、粉砂和黏土物质组成,略具分选性和不清楚的层理。

diluvial solonchaks 洪积盐土

They are special solonchaks formed

during modern salt accumulation with the absence of underground water, which are widely distributed on piedmont proluvium fan and area attached to it in the desert. The reason leading to its formation is the re-accumulation of salts towards ground surface under the influence of strong evaporation of desert climate due to the sedimentation of salt-bearing strata exposed from flood dissolving zone with salt and silt in mineralized fissure spring in piedmont zone.

现代积盐成土过程中没有地下水参与而形成的一种特殊的盐土,在我国主要分布于漠境地区的山前洪积扇和附地上。由于地面洪水溶带山区出露的含盐地层和矿化裂隙泉水中的盐分与泥沙一起沉积在山前地带,在漠境气候强烈蒸发影响下盐分重新向地表聚积而形成。

diminishing returns 收益递减

Also called law of diminishing returns, it is the decrease in the marginal (incremental) output of a production process as the amount of a single factor of production is incrementally increased, while the amounts of all other factors of production stay constant.

随着投入量的增加,每一单位额外投入得到的收益减少的特性。

direct economic losses 直接经济损失

It refers to economic losses caused directly by a decrease or loss in the value of an affected body after it suffered damage from various marine disasters such as storm surges and sea ice, excluding any indirect economic losses and losses from intermediate links. The unit of measurement is ten thousand yuan.

受灾体在遭受风暴潮、海冰等各种海洋灾害的损毁后,其自身价值降低或者丧失所直接造成的经济损失,不含任何中间环节和间接的经济损失。计量单位:万元

direct investment 直接投资

It refers to an economic behaviour where an investor directly invests funds in a specific project of an investing area in the form of capital or physical objects, and always participates in or exercise controls over the entire production and operation process of the invested project, and finally obtains economic benefits.

投资方将资金以资本或事物的形式直接投放于受资区域的特定项目,并始终参与或控制投资项目的生产、经营权过程,最终获取经济利益的经济行为。

direction of seashore 海岸走向

It refers to the direction of coastal extension. The main factors which affect the direction of seashore include the natural coastal landform, the scouring of the sea water, man-made projects, etc.

海岸延伸的方向,影响海岸走向的主要因素包括自然海岸地形、海水的冲刷、人造工程等。

direct resources of sea water 直接海水资源

They are resources of seawater which can be directly used without any treatment for human life and production.

对海水不加任何处理,直接取海水为人类的生产生活所利用的海水资源。

discharge area 排泄区

It refers to the scope in which underground water in an aquifer discharges to the outside.

含水层的地下水向外部排泄的范围。

discharge resources 流量资源

Opposite to stock resources, they refer to resources which have no storage capability and will disappear with the lapse of time if they are not exploited by human beings.

"存量资源"的对称。指不具备储存功能,在人类不进行开采利用的情况下,随着时间的流逝而消失的资源。

discount 贴现

It is a method of financing. The note sellers (debtors) receive finance by selling notes at a price lower than its face value to a note buyer (creditor). The debtors then redeem the notes at par at maturity. The discount represents the difference (interest) paid to the creditors between the price paid for a note and the note's par value.

用票据进行短期融资的主要方式是出售票据一方融入的资金低于票据面值,票据到期时按面值还款,差额部分就是支付给票据买房(贷款人)的利息。这种融资的方式叫做贴现。

disordered chaos 混沌无序

It refers to a complex situation under which the state or change of a matter is relatively obscure and irregular.

事物的状态或变化相对模糊,无规律可循的复杂状态。

disposal of marine hazardous wastes 海洋危险废物治理

It refers to activities such as the treatment and disposal of hazardous wastes discharged into the sea by industrial activities.

对排放入海的工业活动产生的危险废物进行处理、处置等活动。

dissipative structure 耗散结构

It refers to a new organized system far from equilibrium state, created by exchange with the outside and non-linear relationship between components.

系统在远离平衡态条件下,通过与外界进行交换及组分间非线性关系所形成的一种新型有序的组织结构。

dissipative structure theory 耗散结构论

It was established by famous Belgian scientist Ilya Prigogine in 1969, based on the study of physics and chemistry. Several typical experiments on which the theory was established are: Benard fluid experiment, laser and chemical oscillating reactions, etc. The experimental phenomena and organisms have some common characteristic, that is formation and maintenance of organized structure require dissipative energy and substance. Ilya Prigogine called the structure dissipative structure while the theory is the study of formation, stabilization and evolution of the structure as well as its property. According to the theory, the spontaneous appearance and maintenance of a macro-organized state requires at least three conditions: (1) the system must be open because an open system is the premise for appearance of an ordered structure; (2) the system must be far away from equilibrium because non equilibrium is the source of order; (3) inside the system, there must be a proper non-linear feedback to achieve order by fluctuation.

比利时著名科学家普利高津(Ilya. Prigogine)于1969年创立的,来源于物理、化学研究。它赖以建立的几个典型实验是:贝纳德流体实验、激光和化学震荡反应等。这些实验现象和生物体有相同的特征,即有序结构的形成和维持需要耗散能量与物质,因此,普利高津把这类结构称为耗散结构,而耗散结构论就是研究耗散结构的形成、稳定、演化及其性质。按照耗散结构理论,一个宏观有序状态的自发产生和维持,至少需要三个条件:(1)系统必须是开放的,开放系统是产生有序结构的前提;(2)系统必须处于远离平衡的条件下,非平衡是有序之源;(3)系统内部必须存在适当的非线性反馈,通过涨落达到有序。

dissolved oxygen(DO) 溶解氧

It refers to the free oxygen dissolved in water.

溶解于水中的游离氧。

distant fisheries 远洋渔业

They refer to fishery production activities in waters far from a country's coast or fishery base, by using high seas or resources of other countries.

在远离本国海岸或渔业基地的海域,利用公海或他国资源的渔业生产活动。

distant fishing 远洋捕捞

It refers to fishing activities of all kinds of natural aquatic animals and plants in the sea area beyond the jurisdiction of a country (foreign exclusive economic zones, continental shelfs, or high seas). It includes ocean fisheries and trans ocean fisheries.

在非本国管辖海域(外国专属经济区、大陆架或公海)从事的对各种天然水生动植物的捕捞活动。包括大洋渔业和跨洋渔业。

distillation method 蒸馏法

It is a technical process of obtaining fresh water from sea water through water surface vaporisation.

利用水的表面汽化现象,从海洋水体中获取淡水的工艺过程。

disturbance 干扰

It refers to natural incidents which occur accidentally and unpredictably in different space or time. They exert direct influence on the evolution process of the ecological system and are destructive as well.

在不同空间和时间尺度上偶然发生的、不可预知的自然事件,它直接影响着生态系统的演变过程并具有破坏性。

diversification 多元化

Pluralism, or multiple co-existence, means that different components in the system can co-exist with other components positively and dynamically, relying on their own competitiveness so as to enhance the establishment of some contact or participation mechanism.

多元化即多元共存,指系统中不同组分依靠其自身竞争力,与其他组分积极和动态的共存,促进某种接触和参与机制的建立。

division of marine functional zonation 海洋功能区划

It is about the fundamental work of dividing sea areas and islands into various marine functional zones based on a standard of marine functional zones, to provide scientific basis for marine exploitation, protection and management.

按照海洋功能区的标准,将海域及海岛划分为不同类型的海洋功能区,为海洋开发、保护与管理提供科学依据的基础性工作。

dock 船坞

It refers to a large structure extending alongshore or out from the shore into a body of water, to which boats are built and repaired.

修、造船舶用的大型水工建筑物。

domain 区域

It refers to a large space which holds multiple kinds of resources and where a variety of productive and non-productive social and economic activities can be operated.

拥有多种类型的资源、可以进行多种生产性和非生产性社会经济活动的一片相对较大的空间范围。

dominant species 优势种

They are species which are more numerous than their competitors in an ecological community, or make up more of the biomass.

具有控制群落和反映群落特征、数量上所占比例较多的种群。

Dongsha Islands 东沙群岛

They are located in the middle of Guangdong China, Hainan island, Taiwan island and Luzon of the Philippines. These islands are the northernmost islands of the islands in the South Sea of China, which consist of Dongsha Dao, Dongsha Jiao (atoll), Nanwei Tan (submerged reef) and Beiwei Tan (submerged reef) and there are many hidden dunes or submerged reefs. Dongsha islands are located in the tropics with high temperatures all year round.

东沙群岛位居中国广东、海南岛、台湾岛及菲律宾吕宋岛的中间位置,是中国南海诸岛中位置最北的一组群岛,主要由东沙岛、东沙礁(环礁)、南卫滩(暗礁)和北卫滩(暗礁)所组成,附近海区还有不少暗沙和暗礁。属热带地区,终年高温。

double diffusion 双扩散

It refers to a phenomenon of microscale seawater motions caused by the difference between the molecular diffusion coefficient of heat and that of salinity.

由热量和盐度的分子扩散系数差异而引起的微尺度海水运动现象。

drag sweeping 拖底扫海

It means that two survey vessels (boats), which are kept a certain width apart in parallel, drag the cables along a predetermined course to detect whether there are navigation obstructions, such as reefs and sunken vessels, in the surveyed area while all bottom cables of a sweeper used for sweeping surveys are required to arrive at the bottom of the sea.

将扫海测量的扫海具底索全部着底,由两条测量船(艇)沿预定的航向保持一定宽度平行拖行,探测测区内是否存在礁石、沉船等航行障碍物。

drainage area 流域面积

It is the area surrounded by a drainage divide. In other words, waters falling on the area converge to a watercourse along the slope at the exit of the basin where waters join another water body. Size and shape of drainage area directly affect the formation of river runoffs.

又称"汇水面积"、"受水面积",指流域分水线所包围的面积,即降落在流域面积上的降水都沿着地面斜坡汇入河道,经流域出口断面流出。流域面积的大小和形状,直接影响河流径流形成的过程。

drinking water manufacturing 饮用水制造

It is a production activity of turning sea water into drinking water through a variety of desalination technologies.

利用各种淡化技术将海水处理为饮用水的生产活动。

dry salting processing of ocean surimi products and aquatic products 海洋鱼糜制品及水产品干腌制加工

It is the manufacturing of ocean surimi products and the dry and preservation processing of marine aquatic products.

海洋鱼糜制品制造,以及海洋水产品的干制、腌制等加工活动。

dry shoal 干出滩

It refers to the tidal waterways between the coastline and the dry line (0 metre depth contour). It is submerged at high tide and exposed at low tide.

海岸线与干出线(0米等深线)之间的潮浸地带。高潮时淹没,低潮时露出。

dual economic structure 二元经济结构

It refers to an economic system constituted due to the simultaneous existence of a traditional economic sector and a modern economic sector; the economically underdeveloped portion and the economically developed portion inside a society.

社会内部传统经济部门和现代经济部门、经济不发达部分和经济发达部分同时并存,共同构成的经济整体。

dumping amount allowed 批准倾倒量

It refers to an ocean dumping amount approved in a reporting period. The unit of measurement is ten thousand cubic metres, ton.

报告期内批准的海洋倾倒数量。计量单位:万立方米、吨。

dumping zone areas 倾倒区面积

It refers to an area of marine dumping zones for dumping waste into the ocean within a reporting period. The unit of measurement is square kilometres.

报告期内海域倾倒废弃物所使用的海洋倾倒区面积情况。计量单位:平方千米。

dune coast 沙丘海岸

It refers to a coast with the distribution of dunes in Beidaihe, Qinhuangdao, the Luanhe estuary, Penglai and Zhanjiang.

沙丘分布的海岸,分布于中国北戴河、秦皇岛、滦河口、蓬莱和湛江一带。

Durban conference 德班会议

It refers to the 17th session of the *Conference of the Parties* (*COP* 17) *to the United Nations Framework Convention on Climate Change* (*UNFCCC*) *and the 7th session of the Conference of the Parties serving as the meeting of the Parties* (*CMP* 7) *to the Kyoto Protocol*, which was convened in Durban, South Africa, from November 28th to December 9th 2011. The conference finally passed a resolution to build the platform of actions, create an ad hoc working group, implement the second phase of the *Kyoto protocol*, and start the green climate fund.

2011年11月28日至12月9日,在南非德班召开的《联合国气候变化框架公约》第17次缔约方会议暨《京都议定书》第7次缔约方会议。大会最终通过决议,建立德班增强行动平台特设工作组,实施《京都议定书》第二承诺期,并启动绿色气候基金。

dynamic mechanism of urbanization 城市化动力机制

It refers to an interrelation among or between various forces to promote an urbanization process in the process of urbanization.

在城市化过程中,能够推进城市化进程的各种力量及其之间的相互关系。

dynamical oceanography 动力海洋学

It refers to a discipline which studies the dynamic process and changes rule of seawater movement.

研究海水运动的动力过程及其变化规律的学科。

E

earth mantle 地幔

It is the middle layer in the interior of the earth. It is situated between the Mohorovicic discontinuity and the Gutenberg discontinuity, covering 83% of the total volume and 68.1% of the total mass of the earth.

地球内部的中间圈层。体积占地球总体积的 83%,质量占地球总质量的 68.1%。位于莫霍面与古登堡面之间。

Earth Observing System, EOS 地球观测系统

It is a program launched mainly by NASA, comprising a series of low earth orbit satellites designed for continuous and comprehensive observations of the land surface. It aims to deepen the understanding of the mutual interaction of natural processes and human activity, determine the degree, cause and effect of the global change and strengthen human beings' capability of predicting future global changes.

主要由美国宇航局启动的用一系列低轨卫星对地球进行连续、综合观测的计划,目的在于加深对自然过程和人类活动相互影响的理解,确定全球变化的程度、原因和影响后果,增强人类预报未来全球变化的能力。

earthquake subsidence 震陷

It is a phenomenon of foundation or ground subsidence caused by the earthquake-induced softening of highly compressible soil.

由于地震引起高压缩性土软化而产生地基或地面沉陷的现象。

East China Sea 东海

Also called *the Marginal Seas of the Western Pacific*, it is located between the Chinese mainland, Kyushu of Japan, the Ryukyu Islands and the Taiwan island of China. The East China Sea has an area of 770,000 square kilometres, with a mean depth of 0.38 kilometres, and the deepest point is 2.719 kilometres, south of Okinawa.

又称"东中国海"、"西太平洋边缘海"。位于中国大陆与日本九州岛、琉球群岛和中国台湾岛之间。面积约 77 万平方千米,最大深度(冲绳海槽南部)2.719 千米,平均深度 0.38 千米。

ebb tide 落潮

It refers to a process of sea level's decline from a high water mark to a low water mark.

海面从高潮位下降至低潮位的过程。

eco-city 生态城市

It refers to an intensive human habitation where society, economy and nature develop harmoniously; material, energy and information are employed effectively; culture and landscape integrates completely; potentials of human and nature are brought into full play; citizens maintain physical and mental health; and ecology develops continuously and harmoniously.

社会、经济、自然协调发展,物质、能量、信息高效利用,技术、文化与景观充分融合,人与自然的潜力得到充分发挥,居民身心健康,生态持续和谐的集约型人类聚居地。

eco-industrial park 工业生态园

It refers to a new type of industrial organisation form designed on the basis of the theory of circular economy and industrial ecology principles and it is a gathering space of ecological industries. Eco-industrial park follows the principle of reduction, reuse, and recycling of a circular economy and simulates a natural eco-system to set up a "producer-consumer-decomposer" circulation pathway and food chain in an industrial system. Then, it adopts the means of waste exchange, cleaner production, etc. and uses the by-products or wastes produced from one enterprise as the input or raw materials of another plant, to achieve a closed-loop flow of materials and multi-level development and utilization of energy. Therefore, a mutually dependent industrial ecological system which is similar to the process of the natural ecological food chain will be formed, in order to seek the balance between social economy, environment and the demands of human beings.

依据循环经济理论和工业生态学原理而设计成的一种新型工业组织形态,是生态工业的聚集场所。遵从循环经济减量化、再使用、再循环原则,模拟自然生态系统建立工业系统"生产者—消费者—分解者"的循环途径和食物链网,采用废物交换、清洁生产等手段,使一个企业产生的副产品或废物可用作另一个工厂的投入或原材料,实现物质闭环流动和能量多级开发利用,从而形成一个相互依存,类似自然生态系统食物链过程的工业生态系统,以寻求一种社会经济、环境和人类的需求三者之间的平衡。

ecological capacity 生态承载力

It refers to the maximum ecological service ability offered to human activities and coexistence of organisms continuously by ecological systems under certain conditions, especially the maximum capacity of resources and environment.

一定条件下生态系统为人类活动和生物生存所能持续提供的最大生态服务能力,特别是资源与环境的最大供容能力。

ecological civilization 生态文明

It is the concrete manifestation of material civilisation and spiritual civilisation in a natural and social ecological relationship, including the system rationality, scientific decisions, conservation of resources, environmental friendliness, simplicity of life, behaviour consciousness, public participation and system harmony in the aspects of the recognition of the relationship between man and nature, code of conduct, social and economic systems, production and consumption behaviour, material products and mental products related to the relationship between man and nature, spiritual outlook, etc.

物质文明与精神文明在自然与社会

生态关系上的具体体现,包括对天人关系的认知、人类行为的规范、社会经济体制、生产消费行为、有关天人关系的物态和心态产品、社会精神面貌等方面的体制合理性、决策科学性、资源节约性、环境友好性、生活俭朴性、行为自觉性、公众参与性和系统和谐性。

ecological compensation system 生态补偿制度

It is an economic system that encourages the investment in ecological protection and realizes ecological capital appreciation by a reasonable return to ecology investors. In a broad sense, ecological compensation includes compensation for a polluted environment and ecological functions. In a narrow sense, it mainly refers to the compensation for ecological functions, which is an important way for securing an effective supply of ecological products. The compensation objects include: (1) contributors to ecological protection; (2) victims in the process of ecological destruction, involving victims in the processes of ecological destruction and management; (3) contributors to reducing ecological damage. The compensation effects are classified as "blood transfusion type" and "hematopoiesis type". The former type indicates that the government or compensators transfer the collected compensation funds to the compensated ones regularly, whose advantage is that the compensated own great flexibility and whose disadvantage is that compensation funds might be transformed into non-productive expenditure and cannot help the compensated realize "the affluence deriving from the protection of ecological resources" in mechanism. The latter type generally adopts a mechanism combined with poverty alleviation and local development, whose advantage is to foster the sustainable development of the compensated, whose disadvantage is that the compensated lack flexible affordability and suitable subjects of investment.

通过对生态投资者的合理回报,激励人们从事生态保护投资,并使生态资本增值的一种经济制度。广义的生态补偿包括污染环境的补偿和生态功能的补偿。狭义的生态补偿仅指生态功能的补偿。是解决生态产品有效供给的重要途径,补偿对象可分为:(1)对生态保护做出贡献者;(2)生态破坏中的受损者,又分为生态破坏过程中的受害者和生态治理过程中的受害者;(3)减少生态破坏者。补偿的效果可分为"输血型"和"造血型"补偿。前者指政府或补偿者将筹集起来的补偿资金定期转移给被补偿方。优点是被补偿方拥有极大的灵活性,缺点是补偿资金可能转化为消费性支出,不能从机制上帮助受补偿方真正做到"因保护生态资源而富"。后者通常是与扶贫和地方发展相结合的机制,优点是可以扶植被补偿方的可持续发展,缺点是被补偿方缺少了灵活支付能力,而且项目投资还得有合适的主体。

ecological construction 生态建设

It refers to a process of artificially changing and cutting the impaired leading factor or process within an ecological system by adopting certain biological, ecological and engineering technologies and methods based on the ecological principle for adjusting, allocating and optimising the flowing process of material, energy and information inside and outside, and spatial and temporal order for recovering the structure and function of the ecological system and ecological potential to an original or higher level.

根据生态学原理,对受害或受损的生态系统,通过一定的生物、生态以及工程技术和方法,人为地改变和切断生态系统中受害或受损的主导因子或过程,调整、配置和优化系统内部及其与外界物质、能量和信息流动过程和时空秩序,使生态系统的结构、功能和生态学潜力,尽快地成功恢复到一定的或原有的乃至更高水平的过程。

ecological economics 生态经济学

It refers to both a transdisciplinary and interdisciplinary field of academic research that aims to address the interdependence and coevolution of human economies and natural ecosystems over time and space. It studies the composite structure, function and law of motion of ecological system and economic system from the perspective of economics.

从经济学角度研究生态系统和经济系统复合而成的结构、功能及其运动规律的学科。

ecological energetics 生态能量学

It is a discipline which studies energy conversion between various trophic levels in an ecological system.

研究生态系统不同营养级之间能量转换的学科。

ecological environment 生态环境

It refers to the total external conditions that influence the existence and development of humans and organisms, including biological factors (such as plants, animals, etc.) and non-biological factors (such as light, water, atmosphere, soil, etc.).

影响人类与生物生存和发展的一切外界条件的总和,包括生物因子(如植物、动物等)和非生物因子(如光、水分、大气、土壤等)。

ecological factor 生态因子

It refers to the surrounding environmental factors in a biological system or ecosystem.

生物或生态系统的周围环境因素。

ecological fishery 生态渔业

It refers to an aquaculture model of realizing an ecological balance and improving culture benefits by the exploitation of water and land material circulation systems, appropriate technologies and management measures based on the symbiosis theory between fish and other species.

根据鱼类与其他生物间的共生互补原理,利用水陆物质循环系统,通过采取相应的技术和管理措施,实现保持生态平衡,提高养殖效益的一种养殖模式。

ecological foot-print 生态足迹

[ecology] It is a measure of human demand on the Earth's ecosystem, the amount of natural capital used each year. It estimates the amount of biologically productive land and sea area necessary to sup-

ply the resources a human population consumes, and to assimilate the waste that population produces. At a global scale, this has been used by some ecological analysts to estimate how rapidly we are depleting limited resources, vs. using renewable resources.

[*resources of science and technology*] It refers to the biologically productive space occupied by the resources consumed and the waste assimilated by a production area or resource consumption unit.

[生态学]维持一个人、地区、国家或者全球的生存所需要的以及能够吸纳人类所排放的废物、具有生态生产力的地域面积,是对一定区域内人类活动的自然生态影响的一种测度。

[资源科技]生产区域或资源消费单元所消费的资源和接纳其产生的废弃物所占用的生物生产性空间。

ecological function regionalization 生态功能区划

It refers to the research process of dividing a regional space into various ecological function areas in terms of different natural geographic environments, varieties of ecological systems and current situations of the unbalanced economic and social development in a certain area. In combination with the strategies of natural resources protection and sustainable development, ecological function regionalization integrates and differentiates the ecological sensitivity to ecological system service functions caused by human activities.

生态功能区划指针对一定区域内自然地理环境分异性、生态系统多样性以及经济与社会发展不均衡性的现状,结合自然资源保护和可持续开发利用的思想,整合与分异生态系统服务功能对区域人类活动影响的生态敏感性,将区域空间划分为不同生态功能区的研究过程。

ecological impact assessment 生态影响评价

It determines the ecological load or environmental capacity of an area by quantitatively revealing and predicting the impact of human activities on ecology and analysis of its influence on human health and economic development.

通过定量揭示和预测人类活动对生态影响及其对人类健康和经济发展作用的分析确定一个地区的生态负荷或环境容量。

ecological monitoring of marine nature reserves 海洋自然保护区的生态监测

It refers to continuous observation and ecological quality assessment conducted on biotic population, community and abiotic environment within marine nature reserves, with comparable technology and method following predesigned time and space.

按照预先设计的时间和空间,采用可以比较的技术和方法,对海洋自然保护区内的生物种群、群落及其非生物环境进行连续观测和生态质量评价的过程。

ecological restoration from marine disasters 海洋灾害生态恢复

It refers to restoration and regulation

of marine eco-environmental damages from marine disasters, such as storm tide, red tide and tsunami.

对海洋灾害(如风暴潮、赤潮、海啸等)造成海洋生态环境破坏的恢复整治活动。

ecological restoration of marine pollutions 海洋污染生态恢复

It refers to the treatment of a polluted sea area to restore its ecological balance.

对污染海域的治理与生态恢复整治活动。

ecological security 生态安全

It refers to the state that environmental rights of human beings and their realization are protected, and the natural environment, human health and life activities are not threatened by ecological dangers, which is a significant component of national security.

人的环境权利及其实现受到保护,自然环境和人的健康及生命活动处于无生态危险或不受生态危险威胁的状态,是国家安全的重要组成部分。

ecological sensitive area 生态敏感区

It refers to an area which plays a dominant role and is likely to make changes and suffer from damages in maintaining regional ecological balance or functions.

在维持区域生态平衡或生态功能上具有重要作用并极易发生变动和遭受损害的区域。

ecological stress 生态压力

It refers to the natural interference from land, ocean and atmosphere and threats from human activities to ocean ecological systems as well.

来自陆地、海洋、大气的自然干扰和人类活动对海洋生态系统产生的胁迫。

ecological succession 生态演替

It refers to the continuous, unidirectional and orderly natural evolution process of a community in species composition, structures and functions in a certain area.

一定地区内,群落的物种组成、结构及功能随着时间进程而发生的连续的、单向的、有序的自然演变过程。

ecological sustainable theory 生态可持续论

This theory believes that we must consider the needs of an ecological system, emphasize the sustainable capacity of ecology and flexibility of the ecological system to maintain an ecological balance and realise the sustainable development of human society in the practice process of formulating and implementing the transformation of nature.

人们在制定改造自然的实践活动和实施改造自然的实践过程中,必须考虑到生态系统自身的需要,注重生态的可持续承载力与生态系统的弹性力,维护生态平衡,实现生态与人类社会的可持续发展。

ecological system 生态系统

It is an interacted and interdependent dynamic complex formed by the energy flow and material cycle between plants, animals, fungi, microbial community and their abiotic environment in a certain space.

在一定空间范围内,植物、动物、真菌、微生物群落与其非生命环境,通过能量流动和物质循环而形成的相互作用、相互依存的动态复合体。

ecological technology 生态技术

It refers to a technology and method that observes the ecological principle and the law of ecological economy to realise environment protection, maintain ecological balance, save energy and resources, promote harmony between human and nature, and achieve sustainable development of economy and society.

遵循生态学原理和生态经济规律,以保护环境,维持生态平衡,节约能源、资源,促进人与自然和谐、实现经济社会可持续发展的技术手段和方法。

ecology 生态学

It is a discipline which studies the interrelationship between a life system and its environment.

研究生命系统与其环境之间相互关系的学科。

ecology and resources recovery areas 生态与资源恢复区

It refers to an area where a fragile habitat, destroyed ecology and other marine resources can be restored or recovered by taking effective measures.

生境比较脆弱、生态与其他海洋资源遭受破坏需要通过有效措施得以恢复、修复的区域。

economic evaluation for marine resources 海洋资源经济评价

It refers to the evaluation and judgment, on economic value of marine resources, ecological and economic values of exploitation, with application of certain theories and methods, money being used as the calculating unit.

应用一定的理论和方法,对海洋资源的经济价值和开发利用的生态-经济效益进行以货币为计算单位的估价和评判。

economic feature of resources 资源经济特征

It includes the basic elements of capital, technology, economic structure, etc. and the phenomena caused by them, such as the balance of industry development, dependence on natural resources, vulnerability of local natural environment, and so on.

包括资本、技术、经济结构等基本要素及其引起的各种现象。如各产业发展平衡状况、对自然资源的依赖性及所在地自然环境的脆弱性等等。

economic globalization 经济全球化

It is a phenomenon indicating that with the free flowing and proper allocation of production factors, barriers between countries will be removed eventually and all countries in the whole world are united into an economic entity based on their mutual influence, interdependence and joint development.

生产要素在世界范围内自由流动和合理配置,逐渐以至最终完全消除国家之间的壁垒,使各国相互渗透、相互影响、相互依存、共同发展,从而在经济上把世界变成一个整体。

economic management of resources 资源经济管理

It refers to the management to the supply, demand, utilization and protection of natural resources by use of economic means such as prices, taxes and etc.

利用价格、税收等经济手段对自然资源的供给、需求、利用和保护等方面的管理。

economic regionalization 经济区划

According to the division of economic regions and national economic development objectives in a specific period, the whole country is subdivided into small regions to clarify economic development conditions, characteristics and difficulties and points out the status and directions for its future development in the national economic system, which provides scientific basis for central government's macro-control of regional economy, development planning of local government and regional analysis of enterprises.

在认识客观存在的经济区的基础上，根据特定时期国民经济发展的目标和任务，对全国区域进行分区划片，阐明各经济区经济发展的条件、特点和问题，指出它在国民经济体系中的地位和发展方向，最终为中央政府对区域经济进行宏观调控、地方政府制定区域发展规划、企业进行区域分析活动提供科学依据。

economic region 经济区

It is an economic area with multiple levels and distinctive characteristics, which is formed objectively on the basis of labour divisions.

以劳动地域分工为基础客观形成的不同层次、各具特色的经济地域。

economies of scale 规模经济

It refers to an economy in which with the increase in size of an operating unit and professional level of production, the cost of output per unit decreases, thus gradually reducing long-run average and marginal costs.

It applies to a variety of organizational and business situations and at various levels, such as a business or manufacturing unit, plant or an entire enterprise. For example, a large manufacturing facility would be expected to have a lower cost per unit of output than a smaller facility, all other factors being equal, while a company with many facilities should have a cost advantage over a competitor with fewer.

由于生产专业化水平的提高等原因，使企业的单位成本下降，从而形成企业的长期平均成本随着产量的增加而递减的经济。

ecosystem carrying capacity 生态承载力

It refers to the maximum ecological service ability offered to human activities and coexistence of organisms continuously by ecological systems under certain conditions, especially the maximum capacity of resources and environment.

一定条件下生态系统为人类活动和生物生存所能持续提供的最大生态服务能力，特别是资源与环境的最大供容能力。

ecosystems of offshore islands and their surrounding waters 海岛及其周边海域生态系统

They refer to the organic compound which is composed of the biocenosis like islands, coastlines, sand beaches, vegetation, freshwater, their surrounding waters, etc. and abiotic environments.

由维持海岛存在的岛体、海岸线、沙滩、植被、淡水和周边海域等生物群落和非生物环境组成的有机复合体。

ecotype 生态型

It is put forward by American botanist, G. W. Turesson in 1992, which refers to the cline of species phenotype in a specific habitat. It is the smallest unit of population in the same species which is below population.

由美国植物学家图雷森(G. W. Turesson)于1992年提出。指物种表型在特定的生境中产生的变异群,是同种中最小单位的种群,位于种群之下。

ecotope 生态区

Ecotope refers to a group of geographic symbiotic communities featured by having a series of unique top species, which reflects typical ecological conditions including climate, vegetation, soil, etc., such as ecological regions of a tropic forest, desert, etc. Species composition in one ecotope varies from region to region.

一组地理上共生的群落,反映了特定的气候、植被和土壤等生态条件,以具有一批独特的顶级物种为特征,如热带雨林、沙漠等生态域。同一生态区在不同地区物种成分有所不同。

edge water 边水

In reservoir structures, it is the subsurface water that surrounds gas or oil.

油气藏的含油气边界以外的水。

effect of agglomeration 聚集效应

It describes the reduction of production costs achieved by some industries and enterprises because they tend to congregate at a particular area, which is mainly realised through the division of labour and cooperation and increase of production scale, namely, combination and collaboration. In view of the conditions above, the benefits brought by agglomeration are much greater than extra freight and labour costs caused by the long distance from the places with lowest freight and labour costs.

由于某些产业部门、某些企业向某个特定地域集中所产生的使生产成本降低的效果,主要是通过企业间的分工协作、扩大生产规模等方法来实现,表现为联合化与协作化,在这些情况下,聚集所带来的效益要大于由于偏离运费最低点和劳动费最低点所增加的运费和劳动费。

effective coastline 有效岸线

It is a sea-land demarcation line that can be identified as a coastline formed by land reclamation activities.

通过填海造地活动形成的能被认定为海岸线的海陆分界线。

effective precipitation 有效降水量

It is precipitation that can produce infiltration recharge of groundwater within a certain time.

在某一时间范围内,能够产生入渗

补给地下水的降水量。

efficiency wages 效率工资

They are the wages higher than the equilibrium level paid by the employer to increase employees' productivity.

企业为了提高工人生产率而支付的高于均衡水平的工资。

elasticity coefficient of energy production 能源生产弹性系数

It refers to the ratio of annual growth rate of energy increase to that of GDP.

能源生产的年增长率与 GDP 的年增长率的比值。

electricity production 电生产量

It refers to the total amount of power generation at one time in a given period. The unit of measurement is ten thousand kilowatts per hour.

一定时期内一次电力生产量的总和。计量单位:万千瓦·小时。

elevated beach 上升海滩

It refers to a beach formed when the earth's crust rises and sea level falls.

地壳上升、海面相对下降形成的海滩。

elevated coast 上升海岸

It refers to the coast formed when land rises or sea level falls, or the rising of land exceeds that of sea level. Coastal terraces and sea caves above the present sea level often come into being along an elevated coast; a silt coast or sandy coast develops when a shallow seabed emerges above the sea level.

因陆地上升或海面下降或因陆地上升超过海面上升量而形成的海岸。上升的沿岸常形成海岸阶地和高出现在海面的海蚀洞等;斜缓的浅海底,上升后露出水面形成淤泥质或沙质海岸。

El Niño 厄尔尼诺

It refers to the unusual warming of cold surface waters of large parts of the eastern equatorial of the Pacific Ocean. When the unusual heat produced by this periodic marine incident enters the atmosphere, it will affect the global climate.

赤道东太平洋冷水域中海温异常升高现象。这种周期性的海洋事件产生的异常热量进入大气后将会影响全球气候。

embankment 堤

It refers to the water retaining structure built along the banks of rivers, lakes and seas. The structures built on both sides of the rivers can be called "river embankments" or "river levees"; the structures built on the coast of the sea can be called "sea dikes" or "seawalls".

沿江、河、湖、海的边岸修建的挡水建筑物,建在江、河两岸的,称"江堤"或"河堤";建在海边的,称"海堤"或"海塘"。

embayed coast 港湾海岸

A type of submerged coast of a primary coast, it refers to a rugged coast where the cape and bay are distributed crosswise.

又称"多湾海岸",原生海岸下沉海岸类型之一,指岬角与海湾交错分布的海岸,岸线曲折。

embeddedness 嵌入性

It refers to the degree to which the subjects within a connected group form some habitual and stable relationship in a long term association through which the decision making of subjects and behaviour intentions are influenced when taking relative actions.

具有一定连接关系的群体中,主体之间在长期的联系中形成了某些惯例和稳定的关系,并通过这种关系结构影响群体中主体的策略选择,或采取相关行动时的行为倾向。

emergency 涌现

It is a steady structure of a system formed by the interaction between various parts of the whole system under a certain condition of system environment.

构成系统整体的各个部分之间在一定的系统环境条件下通过相互作用所形成的系统的稳态结构。

emission trade(ET) 排放贸易机制

It is an approach of establishing a "cost-benefit" mode on the greenhouse gas emission principle through cooperation of developed countries, namely, by transforming a reduced greenhouse gas amount into the amount of a commodity (equivalent to the amount of CO_2), all organizations can trade to meet their emission reduction targets with the lowest cost.

发达国家间通过合作,使温室气体排放规则成为"成本-效益"形式,即通过将减排的温室气体量转化为一种商品量(相当于 CO_2 的量)使各组织之间可以进行交易,以最低成本满足其减排的指标义务。

employees remuneration 从业人员劳动报酬

It refers to the total remuneration the enterprises pay to the employees within a reporting period, including salaries, benefits, bonuses, subsidies and miscellaneous allowances. The unit of measurement is Yuan.

企业在报告期内支付给本单位从业人员的全部劳动报酬,包括工资、福利费、奖金、津贴及各种补助。计量单位:元。

employee's wages and welfare expenses 职工工资和福利费

They are various forms of remunerations and other relevant expenses paid by an enterprise to obtain services provided by employees, including employees' total wages and employees' welfare expenses. The former refers to the labour remunerations paid to all its employees by an enterprise within a reporting period, including wages, bonuses, allowances and subsidies, and it reflects the enterprise's accumulative total wages payable within this reporting period. The latter refers to various welfare expenses paid by an enterprise within a reporting period according to relevant national provisions, including basic endowment insurance funds, basic pension insurance premiums, unemployment insurance premiums, work-related injury insurance premiums, maternity insurance premiums, housing funds, supplementary pension insurance premiums, and supplementary medical insurance premiums held in escrow by the enterprise for its employees, as well as other cash or non-monetary collective welfare.

包括职工工资总额和职工福利费两部分,是企业为获得职工提供服务而给予的各种形式的报酬以及其他相关支出。其中:工资总额是指企业在报告期内支付给本单位全部职工的劳动报酬,包括工资、奖金、津贴和补贴,它反映企业报告期内累计应付的工资总额。工资总额根据企业会计核算中"应付工资"科目的本期贷方累计发生额填列。职工福利费:指企业在报告期内根据国家有关规定开支的各项福利支出,包括企业为职工提存的基本养老保险基金、基本医疗保险费、失业保险费、工伤保险费、生育保险费、住房公积金、补充养老保险费和补充医疗保险费,以及从成本费用。

enclosed seas 闭海

It refers to a bay, basin or sea surrounded by two or more countries with a narrow passage connecting to another sea or ocean, or constituted wholly or mainly by the adjacent seas of two or more coastal countries and exclusive economic zones.

两个或两个以上国家所环绕并由一个狭窄的出口连接到另一个海或洋,或全部或主要由两个或两个以上沿海国的邻海和专属经济区构成的海湾、海盆或海域。

end point of the sea boundaries 海域界线终点

In area where territorial sea base point has been publicized, the end point of administrative boundary between counties lies in the outer limit of territorial sea. In area where territorial sea base point has not been publicized, the end point of administrative boundary between counties extends 12 sea miles ocean ward from the coast along the direction roughly perpendicular to the coast.

在国家已公布领海基点的海域,县际间海域行政区域界线的终点止于领海外部界限。在国家尚未公布领海基点的海域,县际间海域行政区域界线的终点为自海岸线沿大体垂直海岸线方向向海延伸12海里。

energy 能值

It is the amount of available energy of one form that is used up in transformations directly and indirectly to make a product or service. Energy accounts for different forms of energy and resources, such as sunlight, water, fossil fuels, minerals, etc. Each form is generated by transformation processes in nature and each has a different ability to support work in natural and human dominated systems. It was put forward by American ecologist H. T. Odum in 1986. The energy of energies in different forms can be measured according to the total sum of solar energy utilized directly or indirectly in the process of production.

由美国生态学家奥德姆(H. T. Odum)1986年创立,他认为:一种流动或储存的能量中所包含的另一种类别能量的数量,称为该能量的能值。如属不同类的能,一般可以按照其产生或作用过程中直接或间接使用的太阳能的总量来衡量,以其实际能含量乘以太阳能转化率来比较。

energy conservation technology 节能技术

It is a technology which can improve

the efficiency and benefit of energy development and exploitation, decrease the influence on the environment and curb the waste of energy and resources, mainly including energy and resources optimisation technology, integration of single energy conservation and reconstruction technology with energy conservation technology, energy conservation production facilities and technologies, development of energy conservation materials and energy conservation management skills, etc.

提高能源开发利用效率和效益,减少对环境的影响,遏制能源资源浪费的技术。主要包括:能源资源优化开发技术,单项节能改造技术与节能技术的集成,节能生产设备与工艺,节能材料的开发利用,节能管理技术等。

energy crop 能源作物

They refer to herbal and ligneous plants which are specifically planted to produce raw materials of energy.

经专门种植,用以提供能源原料的草本和木本植物。

energy consumption for per unity unit products 单位产品能耗

It refers to the ratio between the total amount of various energy consumed in the production of a certain product and the output of the product within a reporting period.

报告期内生产某种产品所消耗的各种能源总量与该产品产量之比。

energy resources 能源资源

They refer to material resources in nature which can provide various energies such as heat, light, power, electricity etc.

自然界中能够提供热、光、动力和电能等各种形式的能量的物质资源。

energy sources 能源

They refer to material sources in nature which can provide human beings some kind of energy.

自然界中能为人类提供某种形式能量的物质资源。

Engel coefficient 恩格尔系数

It is the ration of expenditure on food to that on consumption. As people's income rises constantly, the proportion of income spent on food falls, even if actual expenditure on food rises. Consequently, the Engel coefficient is inversely proportional to people's income level.

食物支出与消费品支出的比值。随着人们收入水平的不断提高,食物支出在总消费支出中的比重不断下降,故恩格尔系数与人们的收入水平成反比。

engineering construction for marine protection 海洋防护性工程建筑

It refers to the construction of protective building for seacoast protection, such as seawall project construction.

为保护海岸修筑防护性建筑物的施工活动,如海上堤坝工程施工。

engineering construction of marine fishery facilities 海洋渔业设施工程建筑

It refers to construction activities for projects of seawater farms and large-scale artificial fish-reefs.

海水养殖场、大型人工鱼礁工程的建筑施工活动。

English Channel 英吉利海峡

Located between England and France is an important international shipping waterway connecting the Atlantic Ocean with the North Sea.

位于英国和法国之间,沟通大西洋与北海的重要国际航运水道。

Enteromorpha 浒苔

A generic term for algae under the genus Enteroporpha, it belongs to Chlorophyta, Chlorophyceae, Ulvales and Ulvaceae. Under the name of dried moss by *Compendium of Materia Medica* in china, it has other local names such as sea weed, green moss and shifa. In China, besides fresh eating, enteromorpha can be dried as edible Enteromorpha slices, or serve as feed and fertilizer for domestic animals.

浒苔属海藻的总称,属绿藻门、绿藻纲、石莼目、石莼科。中国《本草纲目》上称干苔,地方名有海苔、苔菜、石发等,在中国除鲜食外,也晒干制成苔条食用,亦可用作家畜饲料和肥料。

entropy 熵

It is a measure of the disorder or invalid state of a system, which usually represents the uncertainty of objects in a system.

系统中无序或无效能状态的度量。熵在系统中作为事物不确定性的表征。

environment 环境

It refers to all of the external conditions that affect and influence the growth, development, and survival of organisms.

影响生物机体生命、发展与生存的所有外部条件的总体。

environmental carrying capacity 环境承载能力

It is the maximum population size and social economic scale that a good ecological environment can sustain given that certain eco-environmental protection principles and criteria, certain social welfare level, economic level and technological level are maintained by a country, within a foreseeable period.

在可以预见期间,一个国家或地区在满足一定生态环境保护准则和标准、一定社会福利水平、一定经济、技术水平等条件下维系良好生态环境所能够支撑的最大人口数量及社会经济规模。

environmental chemistry of marine organic matter 海洋有机物环境化学

It is a discipline that study of the chemical phenomenon of organic matter in a marine environment and its influence on the environment and ecosystem.

研究有机物质在海洋环境中所发生的化学现象以及对环境和生态系统产生影响的学科。

environmental element 环境要素

Constitutes human environment, it is the basic part of substances that are individually independent, characteristically different but obedient to the whole evolution law.

构成人类环境整体的各个独立的、性质不同的而又服从整体演化规律的基本物质部分。

environmental factor 环境因素

It is the element of an organization's

activity, product and service, which can interact with environment.

一个组织的活动、产品或服务中能与环境发生相互作用的要素。

environmental impact 环境影响

It refers to the harmful or beneficial impact imposed on the environment, entirely or partially caused by organized activities, products or services.

全部或部分由有组织的活动、产品或服务给环境造成的任何有害或有益的变化。

environmental impact assessment 环境影响评价

It is the prediction and assessment conducted for potential environmental impact before implementing large-scale projects construction or regional development plan.

在大型建设项目或区域开发计划实施前对其可能造成的环境影响进行预测和评估。

environmental impact on marine geomorphology, erosion and accumulation 海洋地形地貌与冲淤环境影响

It means changes and environmental impacts that construction projects (including new projects, expansion projects and reconstruction projects) have brought on natural geographical conditions such as coasts, intertidal zones, ocean floors and bottom soils.

建设项目(包括新建、扩建、改建工程)对海岸、滩涂、海底和底土等自然地理条件的改变及其产生的环境影响。

environmental impact on marine hydrodynamics 海洋水文动力环境影响

It refers to the impact that construction projects (including new projects, expansion projects and reconstruction projects) have on a marine hydrodynamic (including waves, tides, ocean currents, etc.) environment.

建设项目(包括新建、扩建、改建工程)对海洋水文动力(包括波浪、潮汐、海流等)环境产生的影响。

environmental management system (EMS) 环境管理体系

It is a component of management system, which includes the organizational structure, planning, responsibility, convention, procedure, process and resources needed for formulating, implementing, realizing, evaluating and maintaining environmental policy.

整个管理体系的一个组成部分,包括为制定、实施、实现、评审和保持环境方针所需的组织结构、计划活动、职责、惯例、程序、过程和资源。

environmental parameter 环境参数

It refers to the basic parameter that depicts environmental conditions, including environmental quality index system parameter and environmental capacity index system parameter.

刻画环境状态的基本参变量,包括环境质量指标体系参数和环境容量指标体系参数。

equilibrium 均衡

It refers to a condition in which all competing influences in a system are balanced or relatively stationary in a specified space and time.

系统的各种力量在特定的时空上所达到的某种势均力敌的稳定或相对静止的状态。

equilibrium area 均衡区

It refers to a geohydrologic unit or section whose upper part above the base level has explicit borders in the process of water equilibrium calculation and equilibrium observation.

在水均衡计算中和均衡观测工作中,所选择的某一基准面以上具有明显边界的水文地质单元或地段。

equilibrium period 均衡期

It refers to a time interval when water is calculated in balance.

水均衡计算的时段。

equilibrium price 均衡价格

It refers to a state where economic forces such as supply and demand are balanced and in the absence of external influences the (equilibrium) values of economic variables will not change. 一种商品的市场需求量和市场供给量相等时的价格。

equilibrium tide 平衡潮

It refers to a hypothetical tide due to the tide-producing forces of celestial bodies, particularly the sun and moon. Under the influence of the earth gravity and rotation alone, the shape that the surface of the oceans would take is called geoid. The resultant force of the tidal force, additional tidal force and the gravity vector is called instantaneous gravity. The geopotential geoid which is formed by the sum of gravity of the earth, the tidal force and additional force is called instantaneous geoid. The undulation of the geoid to instantaneous geoid is called equilibrium tide.

从静力学平衡的角度出发,假设地球表面都被海水覆盖,海面在任何时刻都能够保持与重力和引潮力的合力处处垂直的理想化的海洋潮汐。

equitable principle 公平原则

It refers to the fundamental principle which coastal countries follow when delimiting continental shelf boundaries.

沿海国家在相互间划定大陆架界限时应遵守的基本原则。

estuarine delta 河口三角洲

It refers to a triangular or fan-shaped deposits pile, which is formed from the deposition of silt carried by the river near the estuary.

由河水所挟带的泥沙在河口一带沉积、淤积而形成的、多呈三角形或扇形的沉积物堆积体。

estuarine ecology 河口生态学

It refers to a discipline that studies of structure, function and development of biotic community as well as its interaction with environment (physicochemistry and biology) in estuarine area.

研究河口水域中生物群落结构、功能关系、发展规律及其与环境(理化、生物)间相互作用机制的学科。

estuary 河口湾

It is a funnel body of water, where the river enters the sea and the tides impact significantly, consequently causing water erosion to overwhelm accumulation.

河流的河口段因潮汐作用显著,使那里的侵蚀冲刷作用强于堆积作用而形成的漏斗状湾口。

EU Emission Trading Scheme(EU-ETS) 欧盟碳排放交易体系

It is the largest carbon emissions trading market in the world, which plays an exemplary role in the global carbon emissions trading market. Under the "cap and trade" principle, the total amount of greenhouse gas emissions of all participating installations is regulated, namely, allowances for emissions of all countries should not exceed the amount committed by Protocol. The allocation of emission quota will take account of factors of historical emissions, predictive emissions, emission standards of member states, etc.

欧洲碳排放交易体系是世界上最大的碳排放交易市场,在世界碳交易市场中具有示范作用。EU-ETS 属于限量和交易计划,该计划对成员国设置排放限额,即各国排放限额之和不超过《议定书》承诺的排量,排放配额的分配综合考虑成员国的历史排放、预测排放和排放标准等因素。

evaluation index system for sustainable development 可持续发展评价指标体系

It is a set of typical index systems which are established on some fundamental principles to realise the sustainable development and reflects characteristics of all factors (economy, science and technology, society, military, diplomacy, ecology, environment, etc.) and main elements.

为可持续发展的目标,依据一定基本原则进行设置的一组具有典型代表意义同时能全面反映可持续发展各要素(经济、科技、社会、军事、外交、生态、环境等)及子要素状况特征的指标体系。

evaluation of groundwater quality 地下水质评价

It refers to the evaluation after conducting analysis and research about the physicochemical property of the groundwater according to different purposes or applications.

根据不同目的和用途,对地下水的物理化学性质进行分析研究后,作出的评价。

evaporation-concentration process 蒸发浓缩作用

It refers to the phenomenon of increased salinity and mineralization in water caused by the concentration of water composition due to the evaporation of groundwater.

地下水因遭受蒸发,引起水中成分的浓缩,使水中盐分浓度增大,矿化度增高的现象。

evolution 演化

It is a process of system transition or bifurcation generated under the influence of internal and external factors. Evolution is a two-way process, either from simple to complex (progression) or from complex to simple (regression), and is not necessarily about the survival of the fittest.

系统在内外部因素的影响下所产生的跃迁或分岔的过程。演化是一个双向的过程,既可能由简单到复杂(进化),也可能由复杂到简单(退化),不一定优胜劣汰。

evolutionary economics 演化经济学

It refers to a vigorous and promising new field in the research of modern western economics. Compared with the static equilibrium analysis of neoclassical economics, evolutionary economics emphasises the study on "change", the important position of time and history in economic evolution, and institutional change.

现代西方经济学研究的一个富有生命力和发展前景的新领域,与新古典经济学的静态均衡分析相比,演化经济学注重对"变化"的研究,强调时间与历史在经济演化中的重要地位,强调制度变迁。

evolutionary game theory 演化博弈理论

On the basis of finite theory assumptions, it analyses the resource allocation behaviours of gamers and conducts strategy selection for the game being played, on the basis of finite theory assumptions, with the latest achievements in ecology, sociology, psychology and economics, and under the precondition that the game agents are assumed to have bounded rationality; Besides, this theory analyses game equilibrium of gamers with bounded rationality.

以有限理论假设为基础,结合生态学、社会学、心理学及经济学的最新发展成果,在假定博弈的主体具有有限理性的前提下,分析博弈者的资源配置行为及对所处的博弈进行策略选择,该理论分析的是有限理性博弈者的博弈均衡问题。

exclusive economic zone 专属经济区

It is an area adjacent to and beyond the territorial sea, extending as far as 200 nautical miles measured from the baseline that is used for calculating the breadth of the territorial sea. With a particular legal system, it is a special sea area between the territorial sea and high sea. Coastal states have sovereignty over natural resources, enjoy rights to construct and exploit artificial islands, installations and structures, exercise jurisdiction over marine scientific research and marine protection in the exclusive economic zone. Other nations enjoy the freedom to navigate in, fly over and to lay down submarine cables and pipelines in the exclusive economic zone.

在领海以外并连接领海,其宽度自领海基线量起不超过200海里的具有特定法律制度的海域。专属经济区介于领海和公海之间的一种特殊海域。在专属经济区内沿海国享有自然资源主权,建造和使用人工岛、设施与结构的权利,有海洋科研、海洋保护的管辖权。其他国享有航行、飞越、铺设海底电缆和管道的自由。

exclusive fishing zone 专属渔区

It is an area for coastal states to exercise exclusive fishing rights and fishery exclusive jurisdiction. Foreign fishermen shall not be engaged in fishing activities within this area, unless there is an international agreement or approval in accordance with the relevant laws and regulations.

沿海国行使专属捕鱼权和渔业专属管辖权的区域,该区域内,除非有国际协议或根据有关法律规章取得许可,外国渔民不得从事捕鱼活动。

exhaustibility of natural resources 自然资源耗竭性

It refers to the apparent consumption or zero state of resources reserves caused by the process when natural resources are being exploited and utilized. It may also refer to the changes of natural resources in position, shape, or form of existence as a result of exploitation and utilization.

自然资源在被开发或利用过程中导致明显消耗或资源蕴藏量为零的过程状态或改变其位置、形态、存在形式等。

experimental marine biology 实验海洋生物学

It is a discipline which deals with the morphogenesis, physiological, ecological and genetic phenomena of marine life through experiments on in vivo and in vitro tissues and cells in order to explore the laws of biological life activities.

对海洋生物的形态发生、生理、生态、遗传等生命现象,用活体和离体的组织、细胞进行实验研究,以探索生物的生命活动规律的学科。

exploitation area of mineral resources 矿产资源利用区

It is a sea area which is designated for the exploration and exploitation of mineral resources, including oil and gas regions, solid mining areas, etc.

为勘探、开采矿产资源需要划定的海域,包括油气区和固体矿产区等。

exploitation of coastal geothermal 海滨地热开采

It refers to the exploitation activity of coastal geothermal resources, such as a hydrothermal vent.

对海滨地热资源的开采活动,如海底热泉。

exploration of coastal geothermal resources 海滨地热资源勘查

It refers to the geological exploitation activity of coastal geothermal resources.

海滨地热资源的地质勘查活动。

exploration of marine natural gas 海洋天然气开采

It refers to the exploration of natural gas conducted in the sea or on the beach.

在海上和海滩进行的天然气开采活动。

export volume of marine products 海洋产品出口额

It refers to the total amount of domestic marine products for export. According to specifications in China, all export cargoes are counted by Free on Board (FOB) price in respective categories of aquatic products, crude oil and ships. The unit of measurement is Million Dollars (MD).

从我国国境出口的海洋产品的总金额。我国规定出口货物按离岸价格统计,按照水产品、原油、船舶分类统计。计量单位:百万美元。

exports 出口

It refers to products and labour services of local origin sold to other countries.

国内生产而在国外销售的物品与劳务。

exposed part above sea 海上暴露部位

It refers to the parts of marine structures exposed above sea level.

海洋构造物暴露于海平面以上的部位。

extension service of marine technology 海洋技术推广服务

It refers to related technology activities, technology extensions and transfers for extending new marine technologies, products, and crafts directly to the market.

将海洋新技术、新产品、新工艺直接推向市场而进行的相关技术活动,以及技术推广和转让活动。

extensive growth 粗放式增长

It refers to economic growth achieved based mainly on the expansion of the quantity of inputs in order to increase the quantity of outputs, opposite to that of intensive growth. It is characterized by high input, high consumption, high pollution, low output, and low profit, for example, rashly launching new projects, overextending, repeated construction and developing the "large and all inclusive" and "small and all inclusive" projects. This growth mode focuses more on the number and speed of the economic growth, while neglecting its quality and structure, which will exceed the bearing capacity of resources and environment, thus being unsustainable.

主要依靠生产要素投入的增加实现的经济增长。通常以高投入、高消耗、高污染、低产出、低效益为特征。如上项目、铺摊子、重复建设,甚至搞"大而全"、"小而全"等。这种增长方式较多注重经济增长的数量和速度,忽视经济增长的质量和结构,会突破资源和环境的承载能力而无法持久。

exterior river 外流河

They are the rivers that flow into the sea. They are mostly perennial rivers with a larger amount of water, often forming a huge river system.

流入海洋的河流。大都为常流河,水量较大,常常形成庞大的水系。

external edge of continental shelf 大陆架外部界限

(1) If the distance from the baseline of territorial waters to the external edge of the continent is less than 200 nautical miles, and then it can be extended to 200 nautical miles.

(2) If the distance from the baseline of territorial waters to the external edge of the continent extends more than 200 nautical miles, then the fixed points of the outer limit of the continental shelf cannot exceed 350 miles off the baseline which is used to measure the width of territorial waters.

(3) Or the fixed point of the outer limit of the continental shelf cannot exceed 100 nautical miles beyond the 2,500 metres depth contour.

(1)从领海基线到大陆外缘的距离不足200海里的,可扩展至200海里。(2)从领海基线到大陆边缘超过200海里的,大陆架外部界线的各定点,不应超过从测量领海的宽度的基线量起350海里。(3)或不应超过2 500米等深线以外100海里。

external edge of exclusive economic zone 专属经济区外部界限

It refers to an outer edge of waters formed by the determination of the width in an exclusive economic zone according to Article 57 of the *United Convention on the Law of the Sea*.

按照《联合国海洋法公约》第五十七条规定,确定专属经济区宽度而形成的水域之外缘。

extraction of bromine from seawater 海水提溴

It is the technical process of extracting bromine from seawater.

从海水中生产溴的工艺过程。

extraction of lithium from seawater 海水提锂

It refers to the technical process of extracting lithium from seawater.

从海水中生产锂盐的工艺过程。

extraction of magnesium from seawater 海水提镁

It is the technical process of extracting magnesium or magnesium compound from seawater.

从海水中制取镁及镁化合物的工艺过程。

extraction of marine crude oil 海洋原油开采

It refers to the extraction of crude oil in the sea and on the beach.

在海上和海滩进行的原油开采活动。

extraction of mirabilite from seawater 海水提芒硝

It refers to the technical process of extracting mirabilite (sodium sulfate) from seawater.

从海水中生产芒硝(硫酸钠)的工艺过程。

extraction of potassium from seawater 海水提钾

It refers to the technical process of extracting sylvite from seawater.

从海水中生产钾盐的工艺过程。

extraction of uranium from seawater 海水提铀

It is the process of extracting uranium from seawater.

从海水中提取铀的工艺过程。

F

fan 海底扇

It refers to a fan-shaped deposit body extending from the leading edge of a submarine canyon at the foot of a continental slope with gentle rise and fall, gullies being distributed on the surface.

又称"深海扇",大陆坡坡麓海底峡谷前缘向外倾斜延伸的扇状堆积体,地形起伏小,表面多沟谷分布。

fan delta 扇形三角洲

It refers to a delta in the shape of a fan, which is formed in areas with shallow seawater, rivers with high sediment concentration and estuaries with many branches.

海水较浅、河流含沙量较多且河口区岔流也较多的自然条件下形成的三角洲,常见其外形似扇形。

fast ice 固定冰

It refers to ice frozen to the shore, islands, or parts under the ocean.

与海岸、岛屿或海底部分冻结在一起的冰。

fault coast 断层海岸

Comprised of fault structure, fault coast refers to the coast where the overall extension direction of the coastline is consistent with the direction of the fault line. The uplifted areas along the fault plane are mostly cliffs, while the descending areas are steep seas.

由断裂构造组成,海岸线延伸总方向与断层线走向相一致的海岸。沿断层面抬升的地块多呈悬崖峭壁,下滑的为陡深的海洋。

feature concentration 要素密集度

It is the relative intensity of consuming various production factors in the process of commodity production.

在商品生产过程中,消耗各种生产要素的相对强度。

Federal Reserve Bank 联邦储备银行

It is a privately-owned regional financial institution under the US Federal Reserve System, which is authorised to issue Federal Reserve notes, implement clearing and collection, act as fiscal agents and deal government bonds under the guidance of the Federal Open Market Committee. According to the *Federal Reserve Act* in 1913, America is divided into 12 federal reserve districts with each district having one privately-owned federal reserve bank named after the city where it is located.

美国联邦储备系统所属的私营区域性金融机构,有权发行联邦储备券,执行票据结算与托收,代理国库,并在联邦公开市场委员会指导下买卖政府债券。据1913年《联邦储备法》规定,全国划分为十二个联邦储备区,每区设立一家私营的联邦储备银行,并以所在城市命名。

feeding ground 索饵场

It refers to the waters where fish clusters for feeds. Estuaries or cold-warm current junctions are the main waters where fish clusters for feed due to high biomass of organic matter, nutrient salts and fish baits.

鱼类集群索饵的水域。河口湾、寒暖流交汇处等有机质、营养盐类丰富饵料生物量高的水域为鱼类集群索饵的主要场所。

fetch 风区

It refers to the range of water over which a given wind has blown continuously.

受状态相同的风持续作用的海域范围。

financial intermediaries 金融中介机构

They refer to financial institutions where savers can lend money to borrowers indirectly.

储蓄者可以借以间接地向借款者提供资金的金融机构。

financial services for marine foreign exchange 海洋金融外汇服务

They refer to financial activities of foreign exchange provided to marine production.

为海洋生产提供的金融外汇活动。

financial system 金融体系

It refers to a set of financial institutions which facilitate the match of one person's savings with another's investment.

经济中促使一个人的储蓄与另一个的投资相匹配的一组机构。

fine and microstratification of ocean 海洋细微结构

It is a phenomenon about stratification of marine elements when a vertical dimension is between molecular dissipative scale and 100 metres.

垂向尺度在分子耗散尺度至 100 米之间的海洋要素分层现象。

finity of natural resources 自然资源有限性

The total amount of one or one kind of natural resources is a finite constant within a certain time and space.

在一定的时间和空间范围内某一种或某一类自然资源的总量是一个有限的常量。

first law of thermodynamics 热力学第一定律

It refers to a law which indicates that energy of a thermodynamic system can be transformed from one form to another without changing the total amount of energy.

热力系内物质的能量可以传递,其形式可以转换,在转换和传递过程中各种形式能源的总量保持不变。

first-year ice 一年冰

It is sea ice evolved from found ice, with the time of no more than one winter and the thickness of 30 centimetres to 2 metres.

由初期冰发展而成的,时间不超过一个冬季,厚度30厘米至2米的海冰。

fiscal policy 财政政策

It refers to a policy on government spending, revenue and debt level in order to enhance the employment level, lessen economic fluctuation, prevent inflation and achieve stable growth or decisions made on government income and spending.

为促进就业水平提高,减轻经济波动,防止通货膨胀,实现稳定增长而对政府支出、税收和借债水平所进行的选择,或对政府收入和支出水平所作的决策。

fish shelter 鱼礁

Also called an *artificial reef*, it refers to a facility used to protect fishery resources.

又称"人工鱼礁",指保护渔业资源的一种设施。

Fisher Effect 费雪效应

It refers to an economic theory proposed by the US mathematician-economist Irving Fisher (1867—1947), which states that the real interest rate equals the nominal interest rate minus the expected inflation rate. Therefore, real interest rates remain unchanged if the nominal rates increase at the same rate as inflation. Or the relation between the nominal interest rate, real interest rate and expected inflation rate is: nominal interest rate = real interest rate + expected inflation rate.

名义利率对通货膨胀率所进行的一对一的调整。名义利率、实际利率与通货膨胀率三者之间的关系是:名义利率=实际利率+通货膨胀率。

Fisheries Law of the People's Republic of China《中华人民共和国渔业法》

It is a basic law for adjustment of all kinds of social relations generated in the process of the development, utilization, protection and enhancement of fishery resources in Chinese waters. Adopted at the 18th *Meeting of the Standing Committee of the Ninth National People's Congress* on October 31, 2000, it came into effect as of December 1, 2000. It was revised at the 11*th Meeting of the Standing Committee of the Tenth National People's Congress* on August 28, 2004.

调整人们在中国水域开发、利用、保护、增殖渔业资源过程中所产生的各种社会关系的基本法律。2000年10月31日第九届全国人民代表大会常务委员会第十八次会议通过,2000年12月1日起施行。2004年8月28日第十届全国人民代表大会常务委员会第十一次会议修正。

fisheries oceanography 渔业海洋学

It is a discipline which studies physical, chemical and sediment factors of seawater, and survival, migration and distribution relationships of marine life.

研究海水的物理、化学、底质因素和海洋生物的生存、洄游、集散关系的学科。

fisheries stock assessment 渔业资源评估

It is to study the growth and death laws of fish by using mathematical models established according to data on fish biological characteristics and on fishery statistics; it is also to study the effects of fishing on fishery resources quantities and qualities, and at the same time to estimate and forecast the amount of resources and catches, thus providing scientific basis for fishery policies and measures.

根据鱼类生物学特性资料和渔业统计资料建立数学模型,对鱼类的生长、死亡规律进行研究;考查捕捞对渔业资源数量和质量的影响,同时对资源量和渔获量做出估计和预报,为制定渔业政策和措施提供科学依据。

fisheries water 渔业水域

It refers to spawning grounds, feeding grounds, overwintering grounds and migration channels of fish and shrimps, along with farms of fish, shrimps and algae.

鱼虾类的产卵场、索饵场、越冬场、洄游通道和鱼虾藻类的养殖场。

fishery damaged by disaster 渔业受灾

It refers to the loss of aquatic fishing, aquaculture and processing production losses caused by disasters.

水产捕捞、养殖和加工生产等因灾害造成损失的现象。

fishery ports 渔业港口

They refer to man-made or natural ports designated to serve fishery production, moorage and shelter fishery ships, from wind load to unload fishery harvests and supplement materials for fishery, including special fishery docks, special fishery water areas and special anchorage sites for fishery ships in comprehensive ports.

专门为渔业生产服务、供渔业船舶停泊、避风、装卸渔获物、补充渔需物资的人工港口或者自然港湾,包括综合性港口中渔业专用的码头、渔业专用的水域和渔船专用的锚地。

fishery products 水产品

They are seawater or freshwater economic animals and plants, as well as their processed products.

海水或淡水经济动植物及其加工品。

fishery resources use and conservation area 渔业资源利用和养护区

It is the waters designated for exploitation and conservation of fishery resources and development of fishery production, including fishing ports and fishing facilities base construction areas, culture zones, breeding areas, fishing areas and important fishery species conservation areas.

为开发利用和养护渔业资源、发展渔业生产需要划定的海域,包括渔港和渔业设施基地建设区、养殖区、增殖区、捕捞区和重要渔业品种保护区。

fishery zone 渔区

In the modern ocean, it is a special area of jurisdiction, established by coastal states to exercise exclusive fishing rights or to implement conservation of fishery resources. In this area, coastal states enjoy sovereign rights to fish and other marine biological resources. It is generally divided into exclusive fishing zones and conservation areas.

现代海洋中,一种特别管辖的区域。沿海国为了行使专属捕鱼权,或为了实施养护渔业资源而建立的特别管辖区域。该区域内,沿海国享有对鱼类及其他海洋生物资源的主权权利,一般分为专属渔区和养护区。

fishing intensity 捕捞强度

It refers to the capacity to catch some economic marine animals within a unit of time and a unit of water area, which depends on the size and number of fishing boats and nets and the level of fishing technique.

单位时间、单位面积水域内采捕某种经济海洋动物的能力。取决于捕捞船只和网具的大小与数量,捕捞技术的高低。

fishing stress 捕捞压力

It refers to stress and threats imposed on the material circulation of marine ecological systems by fishing production.

通过捕捞生产输出物质对海洋生态系统物质循环的胁迫。

fjord 峡湾

Geologically, it is a long, narrow inlet with steep sides or cliffs, created by glacial erosion.

海水侵入陆地形成的狭窄海湾。

fjord coast 峡湾型海岸

It is a coast with numerous fjords, characterized by a deeply concave bay, many islands and peninsulas, and a highly tortuous coastline. It can be found in a mountainous coast on the west bank of high-latitude continent.

峡湾众多的海岸。港湾深凹,多岛屿和半岛,海岸高度曲折。见于高纬大陆西岸的山地海岸。

flag state 船旗国

A ship that flies the national flag of a certain state has the nationality of that state. That state is the flag state of the ship. There must exist a genuine link between the state and the ship. The state must effectively exercise its jurisdiction and control in administrative, technical and social matters over ships flying its flag.

Ships in the high seas only comply with the international law and the law of the flag state.

船舶悬挂某一国家的国旗即具有该国国籍,这个国家即该船的船旗国。船舶在公海上只服从国际法和船旗国的法律。

flammable ice 可燃冰

Also called *natural gas hydrate*, it refers to the crystalline material which is a solid similar to ice formed by water and natural gas molecules under high pressures and low temperatures, distributed in deep sea sediments.

又称"天然气水合物",分布于深海沉积物中,由天然气与水在高压低温条件下形成的类冰状的结晶物质。

flanking connection effect 旁侧关联效应

It refers to a department connection effect formed when many industries offer related service to a leading industry in the process of production.

主导产业在进行产生过程当中,有许多产业为其提供相关的服务而产生的部门关联效应。

flat coast 低平海岸

Developed by the coastal plain, it is a coast where the land slope descends gradually into the sea and the waves are not strong.

沿海平原或沉溺陆地发育的,岸坡低缓、波浪作用轻微的海岸。

flood 洪水

It is a generic term for an overflow of water when river level submerges river beach, which is equal to or larger than bankfull discharge.

河流水位超过河滩地面溢流的现象的总称。为平滩和大于平滩的流量。

flood peak 洪峰

It is the highest point of river level or water flow during one flood or the entire flood season.

一次洪水或整个汛期水位或流量过程中的最高点。

flood period 汛期

[*atmospheric science*] It is the period of a year during which a river rises significantly owing to the concentration of seasonal precipitation or melting of ice and snow in the basin.

[*hydraulic science and technology*] It is a period from the rise of river flood at the beginning to the total receding at the end, which occurs generally from April to October in China.

[大气科学名词]流域内由于季节性降水集中，或融冰、化雪导致河水在一年中显著上涨的时期。

[水利科技名词]江河洪水从开始涨至全回落的时期，在中国一般为 4—10 月。

flood tide 涨潮

It is a rising process of sea water from a low-tide to a high-tide level.

海水由低潮到高潮水位上升的过程。

fluctuation cycle of groundwater 地下水动态变化周期

It refers to the time interval of the regime cycles of groundwater.

地下水动态呈有规律循环变化的时间间隔。

following sea 顺浪

In boating, it refers to a wave direction that matches the heading of the boat, or the included angle between the propagation direction of a wave and a boat is less than 45°.

波浪传播方向与船的夹角小于 45°或波向与流向一致的波浪。

forbidden fishing zone 禁渔区

It refers a water area where all kinds of fishing practice or some fishing gears are forbidden in order to protect fishery resources and ecological environment.

为保护渔业资源和生态环境所划定的，禁止一切捕捞生产活动或某些渔具作业的水域。

forecast services of marine environment 海洋环境预报服务

It refers to the forecast services for various marine environmental elements.

对海洋环境各要素进行的预报服务活动。

fortnightly tide 半月潮

It refers to a tide occurring at intervals of two weeks caused by the combined effects of the gravitational forces exerted by the Moon and the Sun and the rotation of the Earth.

月亮和太阳的引潮力产生的半月分波。

forward connection effect 前向关联效应

It refers to the department connection effect occurring when various industries provide raw materials, fuels and production facilities for the leading industry before the production process.

主导产业在进行产生之前,有许多产业为其提供原料、燃料和生产设备等而产生的部门关联效应。

fouling organism 污损生物

It refers to the living organism is the organism growing on a ship bottom, buoy, platform and a surface or interior of all other facilities in the sea, which is generally harmful.

生长在船底、浮标、平台和海中一切其他设施表面或内部的生物。这类生物一般是有害的。

fractal 分形

It refers to the aggregation of part and whole that are in some way similar to each other, characterized by the pattern of non-integer dimensions filling the space.

局部和整体以某种方式相似的集合,具有以非整数维形式填充空间的形态特征。

fractal dimension 分形维数

It is an index for characterizing fractal patterns. Typically, in Euclidean geometry, it is 0 for sets describing points (0—dimensional sets); 1 for sets describing lines (1—dimensional sets having length only); 2 for sets describing surfaces (2—dimensional sets having length and width); and 3 for sets describing volumes (3—dimensional sets having length, width, and height). However, with regard to complicated fractals like coastline, Koch curve, Sierpinski sponge, etc. it is hard to depict with integers like 1, 2 or 3. Therefore, fractal dimension is used to measure the irregularity of complexity.

描述分形的特征量。通常欧几里得几何中,直线和曲线是1维的,平面或球面是2维的,具有长、宽、高的形体是3维的;然而对于分形如海岸线、科赫曲线、谢尔宾斯基海绵等的复杂性无法用维数1、2、3这样的整数来描述,因而用分数维来作为复杂形体不规则性的量度。

fractal enterprise 分形企业

It is a new production mode depicted with the concept of self-similarity from fractal theory. The self-similarity of a fractal enterprise includes: (1) the self-similarity of the organization structure of the enterprise, that is, to build the structure of the enterprise centering on the process-centered; (2) the self-similarity of the objective, that is, the objective of the unit is consistent with the objective of the enterprise.

借用分形理论中的自相似概念描述的一种新的生产方式。分形企业的自相似性包括:(1)企业组织结构的自相似,即以过程为中心建立企业的组织。(2)目标自相似,即单元的目标与企业的目标一致。

fractal particle 分形体

It refers to a system characterized by fractals.

具有分形特征的系统。

fractal unit 分形元

It is a relatively independent basic unit divided from fractal integrity, including the basic information and quality of future integrity, from which it can develop into fractal integrity.

从分形整体中划分出来的具有相对独立性的基本单位,包含未来整体的基本信息与素质,由它可以发育成分形整体。

fracture zone 断裂带

It refers to a linear zone extending outwards, formed by a series of steep valley walls perpendicular to a mid-ocean ridge and irregular structures, such as asymmetrical ridges, troughs and escarpments.

由一系列垂直于大洋中脊的陡峭谷壁或不对称山脊、沟槽、陡崖等不规则实体构成,向外延伸呈线状排列的地带。

free port 自由港

It is a port where imported goods are exempt from customs formalities and tax payment of duties. Its range is uncertain, some being the specific areas of port, some extending to the adjacent areas, which are generally referred to as a free zone. It is usually a place for cargo handling, processing, storage, etc. The opposite is the port of entry, which aimes to encourage and promote international trade.

对进口的国外货物,不需要办理报关手续和缴纳税款的港口。范围不定,有的仅为港口的特定区域,有的则扩展至港口邻近地区,统称自由港。通常可进行货物装卸、加工、贮存等,否则即为报关港,目的是鼓励和促进国际贸易。

freedom of the high seas 公海自由

The high seas do not belong to any country. The basis of the legal system of the high seas is that every country shares the high seas equally. Activities permitted on the high seas include navigation, fishing, laying of submarine cables and pipelines, overflight of aircraft, construction of artificial islands and other facilities, scientific research, etc. But when countries exercise the above freedom, they should take into account other countries' interests to exercise the same freedom and their rights under the terms of the relevant international convention.

公海不属于任何国家,所有国家均能平等地共同使用是公海法律制度的基础。公海自由包括航行自由、捕鱼自由、铺设海底电缆和管道自由、飞越自由、建造人工岛屿和其他设施自由以及科学研究自由,但各国行使这些自由时必须顾及其他国家行使公海自用的利益以及其他由有关国际公约规定的权利。

freight turnover quantity 货物周转量

It is the sum of products of the weight of each cargo transported by port ships and the transportation distance of this shipment. The unit of measurement is ten thousand tons per kilometre.

港口船舶实际运送的每批货物重量与该批货物的运送距离乘积之和。计量单位:万吨/千米。

freight volume 货运量

It is the weight of cargos actually carried by port ships. The unit of measurement is ten thousand tons.

港口船舶实际运送的货物重量。计量单位:万吨。

fresh water manufacturing for industrial water use 工业用水淡水制造

It refers to production activities which use various seawater desalination technologies to process seawater as water for industrial purposes.

利用各种淡化技术将海水处理为工业专用水的生产活动。

freshwater ecology 淡水生态学

Freshwater ecology is a science which studies the mutual relationships between biological organisms and freshwater environment.

研究生物有机体与淡水环境之间相互关系的学科。

frictional unemployment 摩擦性失业

It exists because both jobs and workers are heterogeneous, and a mismatch can result between the characteristics of supply and demand. Such a mismatch can be related to skills, payment, work time, location, attitude, taste, and a multitude of other factors.

由于工人寻找最适合自己嗜好和技能的工作需要一定的时间而引起的失业。

fringing reef 裙礁

Also called *coastal reef*. it is a type of coral reef which grows and develops along the edge of a continent or an island. The surface of the reef has a generally equivalent height to the low tide of sea surface, tilting towards the sea from the coast.

又称"岸礁"、"裾礁",珊瑚礁的一种,沿大陆或岛屿的边缘生长发育的珊瑚礁。礁石表面大致与低潮水位相当,从海岸逐渐向海倾斜。

frozen processing of marine aquatic products 海洋水产品冷冻加工

It refers to the processing of freezing aquatic animals and plants such as fish, shrimp, crustacea, shellfish and algae cultivated or caught in the ocean for keeping fresh.

为了保鲜,将海水养殖或捕捞的鱼类、虾类、甲壳类、贝类、藻类等水生动植物进行的冷冻加工的活动。

fugitive emission 无组织排放

It is the non-control emission to air, water or land.

向空气、水体或土地的非控制排放。

function 功能

It refers to the value or effect natural or social affairs have on human survival and social development.

自然或社会事务对人类生存和社会发展具有的价值与作用。

functional aggregation 功能聚集

It refers to the interconnected production aggregation achieve in one area, with coordination and distribution of responsibilities, to achieve the optimal profit by expanding the production scale, increasing the production capacity or the number of enterprises, and mutually adopting the methods of combination and specialisation.

通过扩大生产规模,增加生产能力或企业个数,相互采取聚集联合化与专业化方式,形成分工协作,达到效益最优,在区域内相互联系的生产聚集。

functional zones of marine special reserves 海洋特别保护区功能分区

It means to demarcate various zones of

specific leading functions which can exert optimum benefits in the marine special reserves for the purpose of resource preservation and reasonable utilisation based on the natural resource conditions, environmental conditions, geographical locations, exploitation and utilisation conditions of the sea area and islands, by taking into account the needs of sustainability of the regional economy and society.

根据海域及海岛的自然资源条件、环境状况、地理区位、开发利用现状,并考虑地区经济与社会持续发展的需要,在海洋特别保护区内划分各类具有特定主导功能,有利于资源保护与合理利用,能够发挥最佳效益的区域。

future marine industry 未来海洋产业

It refers to the marine production industry that may be very likely to be set up in the near future based on the analysis and prediction of the development of science and technology, including deep-sea mining industry, seawater direct utilization industry, ocean energy utilization industry and marine biopharmaceutical industry etc.

根据科学技术发展的分析和预测,不久的将来完全可能建立的海洋生产行业,如深海采矿业、海水直接利用业、海洋能利用业和海洋生物制药业等。

future value 未来值

It is the value of an asset at a specific date. It measures the nominal future sum of money that a given sum of money is "worth" at a specified time in the future assuming a certain interest rate; it is the present value multiplied by the accumulation function.

在现行利率既定时,现在货币量将带来的未来货币量。

fuzziness 模糊性

It refers to the unclear state of a matter which is caused by its uncertainty in nature, definition and quantity. It describes the uncertainty of type genus, whereas, the consequence is definite. Uncertainty can be measured by a degree of membership, which is used to indicate to what degree a matter belongs to a genus.

事物本身是不确定的,它在本质上没有明确的含义,在量上没有明确的界限,导致事物呈现"亦此亦彼"的状态,是事物类属的不确定性,但事件发生的结果是确定的。模糊的不确定性用隶属度来度量。隶属度表示事物多大程度属于某一类。

G

game theory 博弈论

Also called *theory of games*, it refers to a discipline based on mathematics studying the optimal solution to problems in conflict.

又称"对策论",是一门以数学为基础的,研究对抗冲突中最优解问题的科学。

gap 隘口

It refers to a low point or opening between hills or mountains or in a ridge or mountain.

海岭或海隆上发育狭长、陡峭的地带。

gas hydrate 天然气水合物

It is also called *flammable ice*. Distributed in deep sea sediments, natural gas hydrate, in which a large amount of methane is trapped within a crystal structure of water, is a clathrate compound formed by water and gas molecules under high pressures and low temperatures, thus forming a solid similar to ice.

又称"可燃冰",分布于深海沉积物中,由天然气与水在高压低温条件下形成的类似冰状的结晶物质。

GDP real Gross Domestic Product (real GDP) 实际国民生产总值

It refers to the market value of all final goods calculated using price from a certain year in previous time as a base period price.

用从前某一年作为基期价格计算出来的全部最终产品的市场价值。

general ocean circulation 海洋总环流

It refers to entire seawater flow patterns of global oceans, generally defined as average state of seawater movement on a large scale.

又称"海洋基本环流",全球海洋的海水流型的总体,通常指大范围内海水运动的平均状态。

generation by offshore wind power 海上风能发电

It refers to the production activities of converting wind energy at sea (including coastal zones and islands) into electric energy.

将海上(包括海岸带和海岛)的风能转化成电能的生产活动。

generation of electricity by seawater salinity gradient 海水盐差能发电

It refers to the production activity of converting seawater salinity gradient energy into electricity.

将海洋盐差能转化成电能的生产活动。

geographically disadvantaged state 地理不利国

They refer to coastal states which cannot own the national waters jurisdiction over corresponding areas and cannot fully exercise their marine rights due to their own geographic restrictions, such as the coastal states of a closed sea or semi-closed sea, or the coastal states which cannot claim the rights to their own exclusive economic zone or continental shelf. In accordance with the *United Nations Convention on the Law of the Sea* of 1982, geographically disadvantaged states have the right to get involved in the development of the remainder of the biological resources of exclusive economic zones of other coastal states in the same sub-regions or regions. In other respects, the convention also puts geographically disadvantaged states on a par with other coastal states or gives them a certain degree of concern.

因本身地理条件的限制不能拥有相应面积的国家管辖海域和充分行使海洋权利的沿海国家。如闭海或半闭海的沿岸国或不能主张有自己的专属经济区和大陆架的沿海国。1982年《联合国海洋法公约》规定,地理不利国有权参与开发同一分区域或区域的其他沿海国专属经济区生物资源的适当剩余部分。公约还在一些方面使地理不利国处于同其他沿海国平等的地位或得到一定的照顾。

geological data 地质资料

It refers to the original geological data such as text, diagrams, sound images, electromagnetic media which were formed in the geological work, outcome geological data and the physical geological data such as rock core, specimens, polished thin sections, samples, etc.

在地质工作中形成的文字、图表、声像、电磁介质等形式的原始地质资料、成果地质资料和岩矿心、各类标本、光薄片、样品等实物地质资料。

geological hazard 地质灾害

It refers to the natural or human-induced geological phenomena which can endanger the safety of human life and property, living environment and of national development, or destroy the environment for human survival and development. In brief, it refers to the hazards caused by the geologic process. Geological hazards can fall into two categories according to their development process. One is slow degeneration geological hazards which can cause havoc to human beings in a relatively slow process, such as seawater intrusion, coastal erosion and land subsidence; the other is abrupt geological hazards which can break out suddenly resulting in disaster, such as landslide, collapse, debris flow, earthquake, etc.

由于自然产生或人为诱发的对人民生命与财产安全、生活环境和国家建设事业造成危害或使人类生存与发展环境遭受破坏的地质现象。简言之,即地质作用造成的灾害。地质灾害按其发展过程可分为两类,一类为缓变性灾害,以较缓慢的作用过程对人类造成危害,如海水入侵、海岸侵蚀和地面沉降等;另一类为突发性灾害,骤然发作成灾,如滑坡、崩塌、泥石流、地震等。

geology exploration for coastal and abyssal mineral deposits 海滨和深海矿产地质勘查

It refers to the geological exploration activities for coastal and abyssal mineral deposits.

海滨和深海矿产的地质勘查活动。

geology exploration of marine petroleum and natural gas 海洋石油天然气地质勘查

It refers to the geological exploration activities for energies such as marine petroleum and natural gas.

海洋石油和天然气等能源的地质勘查活动。

geology research and exploration of marine engineering 海洋工程地质调查与勘查

It refers to geology research and exploration activities of marine engineering.

海洋工程的地质调查与勘查活动。

geology research and exploration of marine environment 海洋环境地质调查与勘查

It refers to geological research and the exploitation activities of marine environments.

海洋环境的地质调查与勘查活动。

geomorphic age 地貌年龄

Geomorphic age refers to the age or era of the formation of geomorphology. It can be divided into the relative age and absolute age (according to geological age), the former one of which represents the geological time when geomorphology was formed; and the latter one of which represents the number of years from its formation to the present. The relative age of geomorphology is often described based on geomorphology evolution phase.

地貌形成的年龄或时代。分相对年龄和绝对年龄(按地质年龄)两种,前者是表示地貌形成的地史时期;后者表示地貌形成距今的年数。地貌相对年龄常结合地貌发育阶段来描述。

geomorphic classification 地貌分类

It refers to the classification in accordance with the causes of a landform and its surface features. The causes of a landform are complicated with different surface morphology is different, therefore the geomorphy formed is diverse. According to morphology, the mountains, hills, plateaus, plains and basins we usually talk about are just a few of the most basic geomorphic types; according to exogenic force, a landform can be classified into fluvial landform, lacustrine landform, arid landform, Aeolian landform, Karst landform, glacial landform, periglacial landform, coastal landform, Danxia landform, Yadan landform, etc.; according to endogenic force, a landform can be classified into land structural landform, folded landform, fault landform, volcano and lava flow landform; according to the composition of the rock-soil, a landform can be classified into loess landform, limestone landform, granite landform, etc.

根据地貌的成因和表面特征而划分的类别。地貌成因复杂,表面形态千差万别,因此形成的地貌多种多样。按形态划分,我们平时所说的山地、丘陵、高原、平原、盆地等即是几种最基本的地貌类型;按外营力划分,通常分为流水地貌、湖成地貌、干燥地貌、风成地貌、喀斯特地貌、冰川地貌、冰缘地貌、海岸地貌、丹霞地貌、雅丹地貌、岩溶地貌等;按内营力划分,有大地构造地貌、褶曲构造地貌、断层构造地貌、火山与熔岩流地貌等;按岩土组成划分,有黄土地貌、石灰岩地貌、花岗岩地貌等。

geomorphologic landscape 地貌景观

It is a generic term of the different kinds of surface morphology which relate to each other with respect to the causes of formation. For example, Karst landscape, mountain landscape, lake landscape, etc., are one of the important bases for the classification of land types.

成因上彼此相关的各种地表形态的总称。例如:喀斯特景观、山地景观、湖泊景观灯等,是土地类型划分的重要依据之一。

geomorphic types 地貌类型

They are the classification of the morphological characteristics of a land surface which refer to the different morphological categories based on the differences between the causes of formation and morphology. The same types of geomorphology bear the same or similar characteristics, and different types of geomorphology have distinct feature difference.

陆地表面形态特征的归类。指以成因和形态的差异,划分的不同地貌类别。同类型地貌具有相同或相近的特征,不同类型间有明显的特征差异。

geomorphologic unit 地貌单元

It refers to the unit of a landform, which is divided according to the causes of formation and development process. It can be classified into various scales in accordance with its size. For example, a mountainous region can be divided into a bigger geomorphic unit, while a mountain valley is subordinate to it.

地貌按成因形态及发展过程划分的单位。按规模大小可分为若干等级,如山地可划分为一个较大的地貌单元,而山地河谷是山地中次一级的地貌单元。

geomorphologic shape 地貌形态

It shape refers to the exterior shape of geomorphology which can be described with the parameters of length, width, height, depth and the size of the slope.

地貌的外部形状。是用其长、宽、高或深以及坡度大小等参数来进行描述的。

geostrophic current 地转流

[Marine science and technology] It refers to an oceanic flow in which the pressure gradient force is balanced by the Coriols effect.

[Atmospheric science] It refers to an oceanic flow related sea level pressure gradient in oceanography.

[海洋科技]水平压强梯度力和科氏力平衡条件下的海流。

[大气科学]海洋学中与海水水平压强梯度相联系的一种海流。

geotechnology 地球工程

It is an engineering project that involves the application of physics, chemistry, organism idiosyncrasy reaction process by human beings to combat climate changes, thus effectively reducing or managing the risks posed by climate change.Geotechnology can approximately fall into two categories. One is to remove the carbon dioxide

(CDR), which means to remove the greenhouse gases in the atmosphere with large-scale technology or engineering in order to effectively curb the increase of the earth's temperature; the other is to manage the solar radiation (SRM), which means to reduce the absorption of the solar radiation in the earth's atmosphere with engineering technology, thus to offset the increase of the earth's temperature caused by greenhouse gases in the atmosphere.

通过人为对地球系统的物理、化学或生物特质反应过程等进行干预来应对气候变化，减少并有效管理气候变化带来的风险的工程项目。地球工程可大体分为两类，第一类是二氧化碳移除(CDR)，即通过大规模的技术或者工程减少大气中的温室气体的含量，从而有效减少地球增温；第二类为太阳辐射管理(SRM)，即通过工程技术减少地球大气中太阳辐射的吸收，从而抵消大气中温室气体导致的地球增温。

glacial erosion landform 冰蚀地貌

It refers to various landforms as a result of erosion to the bottom and the side of rocks by a ice body and the substance it carries when glaciers happen.

当冰川发生时，冰体及其携带的物质对底部及旁侧的岩石进行刨蚀之后而残留的各种地形地貌。

glacial landform 冰川地貌

It refers to a surface configuration created by the action of glaciers.

由冰川作用形成的地表形态。

global climate change 全球气候变化

It refers to the statistically significant changes of average climatic states or longer periodic climatic fluctuations (typically 10 years or longer) globally. The reasons may involve the internal process and external forces of nature or the continuous alternation of atmospheric components and soil exploitation by human beings.

在全球范围内，气候平均状态统计学意义上的巨大改变或者持续较长一段时间（典型的为10年或更长）的气候变动。气候变化的原因可能是自然的内部进程、外部强迫，或者是人为地持续对大气组成成分和土地利用的改变。

Global Climate Observing System (GCOS) 全球气候观测计划

It was co-founded by the *World Meteorological Organization* (WMO), the *Intergovernmental Oceanographic Commission* (IOC) of *UNESCO*, the *International Council for Science* (ICSU) and the *United Nations Environment Programme* (UNEP) in 1992, which aim to build coordination mechanisms between international observing plans and national observing systems of all countries by implementing development plans and offering technical assistance and policy guidance, etc. GCOS targets the whole climatic system, including various physical, chemical and biological processes and atmospheric, marine, hydrographic, ice and land processes.

由世界气象组织（WMO）、联合国教科文组织（UNESCO）的政府间海洋委员会（IOC）、国际科学联盟理事会（ICSU）、联合国环境规划署（UNEP）于1992年共同发起的，该计划主要是通过制订发展计划、提供技术帮助和政策指导等手段,在各种国际观测计划和各国观测系统之间建立起协调机制。它主要研究对象是整个气候系统,包括各种物理、化学、生物的过程,以及大气、海洋、水文、冰层和陆地的过程。

global earth observing system 全球地面观测系统

Composed of sensors, communication devices, storage systems and various kinds of computers installed all over the world, it is applied to observe the earth and its dynamic processes, predict some phenomena and monitor the implementation of environmental convention of all countries.

由设置在全球各地的传感器、通信设备、存储系统和各种计算机组合而成的,用于观测地球、了解地球的动态过程,以便对一些现象加以预报,监测各国执行环境公约的实际情况等。

Global Observing Information Network (GOIN) 全球观测信息网络

It is a cooperation agreement signed by America and Japan in 1993, which aims to exchange global satellite observation data through an information network, evaluate the advantages and disadvantages of current satellite observation technology systems, put forward improvement measures and provide service to the construction of global observation information network. Meanwhile, GOIN will become a significant component of global observing information network infrastructure and earth observation systems. Currently, it has played a key role in monitoring global changes, disaster investigations and environment changes.

全球观测信息网络(GOIN)是1993年美国和日本达成的合作协议。其目的是通过信息网络交换全球环境的卫星观测数据,评估现有卫星观测技术系统的优点和弱点并提出改进措施,为建设全球观测信息网络(GOIN)服务。同时,全球观测信息网络(GOIN)也将成为全球信息基础设施的一个组成部分和地球观测系统十分重要的组成部分。现在它已经在全球变化、灾害调查和环境监测等方面,发挥着十分重要的作用。

Global Ocean Observing System (GOOS) 全球海洋观测系统计划

It is a large international ocean observing plan put forward by the intergovernmental oceanographic commission in 1992 with the purpose of establishing a model for observing global coasts and oceans and analysing ocean changes.

政府间海洋学委员会等国际组织1992年提出的,对全球沿海和大洋要素进行长期观测建立模型,分析海洋变化的大型国际海洋观测计划。

global sea level change 全球性海平面变化

Broadly speaking, it refers to the changes of sea level caused by the alternation of a glacial period and an interglacial period in the Quaternary Period. In a narrow sense, it refers to the rising of global sea level caused by global warming and glacier melting of the greenhouse effect due to human activities in the latest 300 years since the industrial revolution.

广义上指第四纪时期因冰期与间冰期交替导致的海面变化。狭义主要指工业革命以来近300年的时间里,人类活动造成的温室效应使全球气候变暖、冰川融化导致的全球海平面上升。

Global Sea Level Observing System (GLOSS) 全球海平面观测系统

It is a plan designed by an intergovernmental oceanographic commission for observing the long-term global climatic changes of sea level. Since the India Ocean earthquake in 2004, the plan objective of has turned to collect real time data of sea levels. Currently, the project is upgrading more than 290 observation stations to transmit real time data to the new established national tsunami centres through satellites. Solar panels are being installed in the system to ensure the smooth operation of observation stations even under severe weather conditions.

一个政府间海洋学委员会的计划,其目的是测量研究海平面全球长期的气候变化。自2004年印度洋地震后,该计划的目的已经改变为收集海平面的实时数据。该项目目前正在升级290余台观测站,以使他们能够通过卫星向新设立的国家海啸中心传送实时数据。他们还在组装太阳能电池板,令观测站即使在恶劣天气下仍能够继续运作。

global warming 全球气候变暖

Burning of fossil energies such as coal and gas or deforestation can produce various greenhouse gases such as carbon dioxide in the atmosphere and cause the "greenhouse effect", whose long-time accumulation can lead to global warming. Global warming can lead to the real location of global precipitation, melting of glaciers and frozen soil, frequent occurrence of extreme climate, rising of sea levels, etc. which endanger both the balance of the natural ecological system, food supply and living environment of human beings.

煤炭、石油等化石能源燃烧或砍伐、焚烧森林时,会产生二氧化碳等多种温室气体,这些温室气体存在于大气层中,产生"温室效应",长时间的积累会导致全球气候变暖。全球变暖会使全球降水量重新分配,冰川和冻土消融,极端气候频繁,海平面上升等,既危害自然生态系统的平衡,更威胁人类的食物供应和居住环境。

Golden Rule level of capital 资本的黄金律水平

It is the value of capital stock which maximizes a steady state level of consumption.

使消费最大化的稳定状态的资本存量值。

government-purchases multiplier 政府购买乘数

It is the ratio of income change to government-purchase expenditure change causing such an income change.

收入变动对引起这种变动的政府购买支出变动的比率。

gradation of natural resources 自然资源层位性

The structural arrangement of a natural resource system and the internal composition of various resources have certain sequences, exhibiting obvious gradation. If we see the horizon of natural resources as a vertical profile, the mineral resources mainly exist in the lower layer of the land, the interior of the lithosphere; soil organisms and land water resources are located on the land surface, namely, biosphere and hydrosphere; climate resources is at the top of the vertical system, namely the atmosphere.

自然资源系统的结构排列和各类资源内部的组成，都具有一定的序列，表现为明显的层位性。如果我们把自然资源层位看成一个垂直的剖面，则矿产资源主要存在于土地的下层，岩石圈的内部；土壤生物与陆地水资源则位处土地的表层，即生物圈和水圈；气候资源则处于垂直系统的最上层，即大气圈。

gravimetric point 重力点

It is a point for determination of gravity value.

测定重力值的点。

green barrier 绿色壁垒

It is also called *ecology barrier* and *environment barrier*. Employed by some developed countries with technology advantages through formulating legislation and compulsory technical regulations, a green barrier refers to trade barrier such as access restriction to foreign commodities in international trade for the purpose of protecting the environment and human health.

又称"生态壁垒"、"环境壁垒"。在国际贸易领域，一些国家凭借其科技优势，以保护环境和人类健康为目的，通过立法或制订严格的强制性技术法规，对国外商品进行准入限制的贸易障碍。

green chemical industry 绿色化工

It is a new processing mode of manufacturing chemical products by applying environmental protection concepts from the source, implementing source reduction to realise the optimisation and integration of production processes, recycling and reutilizing waste to lower costs and consumption, reducing the emission of toxic wastes, and alleviating the adverse effect of a product's full life cycle on the environment. With the emergence of green chemical industry, the environmental pollution of the chemical industry can be controlled more effectively from the source.

在化工产品生产过程中，从工艺源头上就运用环保的理念，推行源消减，进行生产过程的优化集成，废物再利用与资源化，从而降低成本与消耗，减少废弃物的排放和毒性，减少产品全生命周期对环境不良影响的一种生产方式。绿色化工的兴起，使化学工业环境污染的治理由先污染后治理转向从源头上根治环境污染。

green economy 绿色经济

Formed in the late 20th century, it refers to an economic development state or mode which coordinates economic growth and resource environments. Environmental science, which emerged in the same period, provides theoretical and technological preparation for its development. Green economy is the revolution and sublation of traditional economic growth modes and the only way to improve economic development. Any disobedience of green economy development modes, ignorance of growth quality and benefits, blind pursuit of high-speed development at the expense of wasting resource and destroying the environment will lead to the drastic fluctuation of economic activities and hinder real economic development.

经济增长与资源环境相协调的一种经济发展状态或发展模式,形成于20世纪末。20世纪末兴起并发展起来的环境科学为它提供了理论和技术上的准备。"绿色经济"是对传统经济增长模式的变革和扬弃,是未来人类社会经济发展的必由之路。违背"绿色经济"发展要求的增长模式,忽视增长的质量和效益、不惜浪费资源和破坏环境、片面追求一时高速度的做法,都会造成经济活动的大起大落,不能实现真正的发展。

green energy resources 绿色能源

[resource technology] They refer to resources which produce none or little greenhouse gases or pollutants, mainly including new energy and renewable energy.

[ecology] They are chemical energies which are transformed from solar energy and stored in green plants through photosynthesis. They can be processed or utilized directly to replace non-renewable energy resources such as coal, oil, etc. Under the guidance of sustainable development philosophy, green energy resources prove to keep harmonious coexistence with nature resources and environment.

[资源科技] 是指温室气体和污染物零排放或排放很少的能源,主要是新能源和可再生能源。

[生态学] 绿色植物通过光合作用将太阳能转化并储存于体内的化学能。人们直接或加工利用这些化学能作为能源,代替煤、石油等不可再生的能源。在可持续发展的理念下,绿色能源体现了与环境友好相容的自然资源的开发利用原则。

green food 绿色食品

They are foods produced in a pollution-free ecological environment and manufactured or processed according to the whole-process standardization. Synthetic pesticides and chemical fertilizers are not allowed, although certain approved pesticides may be used, making it comply with food quality standards which are set by national government. Green food logo can be used with the accreditation of specialised agencies.

在无污染的生态环境中种植及全过程标准化生产或加工,严格控制其有毒有害物质含量,使之符合国家健康安全食品标准,并经专门机构认定,许可使用绿色食品标志的食品。

green Gross Domestic Product (green GDP) 绿色国内生产总值

It is an index of economic growth with the environmental cost, environmental pollution and ecological cost of that growth factored into the statistic caliber of the conventional GDP.

将经济发展中资源成本、环境污染损失成本、生态成本纳入国内生产总值统计口径所形成的绿化后的国内生产总值。

green manufacturing 绿色制造

It is a modern manufacturing mode which gives consideration to both environmental influences and resource consumption. The goal of green manufacturing is to minimise negative impacts on the environment, maximise resource utilization and optimize coordination of enterprise economic benefits and social benefits in the whole life cycle of product design, manufacturing, packaging, transport, utilisation and disposal.

综合考虑环境影响和资源消耗的现代制造模式，其目标是使得产品从设计、制造、包装、运输、使用到报废处理的整个生命周期中，对环境负面影响最小，资源利用率最高，并使企业经济效益和社会效益协调优化。

green tourism 绿色旅游

It refers to a kind of alternative tourism which is related to rural tourism and compatible with nature tourism with little or no ecological impact on the destination.

可选择旅游的一种，与乡村旅游有一定联系，具有自然旅游的环境兼容性，对目的地有很小或没有生态影响。

green transport 绿色运输

It refers to transport which is featured by energy resources conservation and exhaust emission reduction. It can be implemented in the following ways: selecting transport vehicles and routes appropriately and avoiding roundabout and repetitive transport to realise energy conservation and emission reduction; improving technology of internal combustion engines and utilizing clean fuels to increase efficiency; and preventing leakage during transport to avoid serious environmental damage to certain areas.

以节约能源、减少废气排放为特征的运输。其实施途径主要包括：合理选择运输工具和运输路线，克服迂回运输和重复运输，以实现节能减排的目标；改进内燃机技术和使用清洁燃料，以提高能效；防止运输过程中的泄漏，以免对局部地区造成严重的环境危害。

greenhouse climate 温室气候

It is an atmospheric condition in a greenhouse which is characterized by a relatively high daytime temperature. Due to the greater transparency of the short-wavelength than the long-wavelength inside the greenhouse, the short-wavelength infrared range of the spectrum can passes through the glass roof and walls and is absorbed by the floor, earth, and contents, making the greenhouse warmer at daytime.

温室内的大气状况，其特征是由于玻璃遮盖对入射的短波辐射的透明度比对温室内的长波辐射的透明度要大，因而导致温室白天温度较高。

greenhouse effect 温室效应

[*atmospheric science*] It is a warming phenomenon caused by different absorption pattern of lower atmosphere in long-wavelength radiation and short-wavelength radiation.

[*ecology*] It is a phenomenon where greenhouse gases in the atmosphere prevent heat dissipation from the Earth's surface by absorbing long-wave length radiation so that the Earth surface temperature rises.

[大气科学] 低层大气由于对长波和短波辐射的吸收特性不同而引起的增温现象。

[生态学] 大气中的温室气体通过对长波辐射的吸收而阻止地表热能耗散,从而导致地表温度增高的现象。

greenhouse gas(GHG) 温室气体

[environmental protection] It is a gas that destroys the normal infrared radiation relationship between the atmosphere and the ground surface, absorbs the infrared radiation emitted by the Earth, and prevents heat dissipation from the Earth so that the air temperature on the Earth rises perceptibly.

[marine science and technology] It is a trace gas in the Earth's atmosphere, which allows short-wave length solar radiation to pass through freely and absorbs long-wave radiation (infrared rays) from the ground surface and air, thus causing the surface layer temperature to rise. It includes CO_2, CH_4, N_2O, CFC, etc.

[ecology] It is a gas component in the atmosphere, which is produced naturally or artificially and can absorb long-wave length radiation. For example, H_2O, CO_2, N_2O, CH_4, O_3 and CFC are the main greenhouse gases in the Earth's atmosphere.

[环境保护] 破坏大气层与地面间红外线辐射正常关系,吸收地球释放出来的红外线辐射,阻止地球热量的散失,使地球发生可感觉到的气温升高的气体。

[海洋科技] 在地球大气中,能让太阳短波辐射自由通过,同时吸收地面和空气放出的长波辐射(红外线),从而造成近地层增温的微量气体。包括二氧化碳(CO_2)、甲烷(CH_4)、氧化亚氮(N_2O)、氯氟烃(CFC)等。

[生态学] 大气中由自然或人为产生的能够吸收长波辐射的气体成分。如水汽(H_2O)、二氧化碳(CO_2)、氧化亚氮(N_2O)、甲烷(CH_4)、臭氧(O_3)和氯氟烃(CFC)是地球大气中的主要温室气体。

grike 溶沟

It is a channel formed by the corrosion and erosion of surface runoff along the jointing, bedding surface and cracks of soluble rocks.

地表径流沿着可溶性岩石的节理、层面和裂隙不断进行溶蚀和侵蚀所形成的沟道。

gross area of salt pan 盐田总面积

It is an entire area occupied by a salt pan, including brine storage, evaporation, brine keeping and crystallisation areas, and areas of ditches, trenches, pools, ridges within a beach, and beach lumps.

盐田占有的全部面积,包括储卤、蒸发、保卤、结晶面积,滩内的沟、壕、池、埝、滩坨等面积。

Gross Ocean Product(GOP) 海洋生产总值

An abbreviation for Gross Marine Economic Product, it refers to the final achievement from marine economic activities of all coastal resident units within a certain period, calculated by market value. GOP is the total sum of added value from marine industries and marine-related industries.

海洋经济生产总值的简称,指按市场价格计算的沿海地区常住单位在一定时期内海洋经济活动的最终成果,是海洋产业和海洋相关产业增加值之和。

gross regional product 地区生产总值

It refers to the final products of all resident units in a region during a certain period of time. The unit of measurement is hundred million yuan.

所有常住单位在一定时期内生产活动的最终成果。计量单位:亿元。

gross value of industrial output 工业总产值

It refers to the total value of industrial final products and industrial labour service activities produced or supplied by an industrial enterprise which is manifested in the form of currency. The unit of measurement is hundred million yuan.

工业企业在本年内生产的以货币形式表现的工业最终产品和提供工业劳务活动的总价值量。计量单位:亿元。

ground fracture 地裂缝

It refers to large fractures on the earth's surface going in one or several directions due to drought, groundwater recession, land subsidence, seismic structure movement or slope instability, etc.

由于干旱、地下水位下降、地面下沉、地震构造运动或斜坡失稳等原因造成的地面沿一个或几个方向产生宽大裂缝的现象。

groundwater balance 地下水均衡

It refers to the quantitative relation of the contrast among the variation of the total recharge, the total consumption and the storage capacity of the groundwater in a certain region (aquifer) over a certain time period.

某一地区(含水层)在一定时间段内,地下水的总补给量与总消耗量及地下水贮存量的变化量之间的数量对比关系。

groundwater discharge 地下水排泄

It refers to the process of the groundwater discharging from one aquifer to the earth's surface or another aquifer in different ways.

地下水从含水层中以不同方式排泄于地表或另一个含水层中的过程。

groundwater divide 地下水分水岭

It is the boundary of the groundwater basin.

地下水流域的分界线。

groundwater information system 地下水信息系统

Comprised of a database, model base and method base of the groundwater, it refers to the management information system used to provide comprehensive information about the groundwater regime for the users.

地下水数据库、模型库、方法库组成,为用户提供地下水动态的全面综合信息的管理信息系统。

groundwater occurrence 地下水赋存条件

It is a generic term for features such as groundwater embedment and distribution, water-bearing medium, water-bearing structure, etc.

地下水埋藏和分布、含水介质和含水构造等条件的总称。

groundwater runoff 地下径流

It refers to the part of precipitation which infiltrates into the ground to become groundwater, then discharges into the rivers as spring water or infiltration water.

渗入地下成为地下水,并以泉水或渗透水的形式泄入河道的那部分降水。

groundwater system 地下水系统

It refers to the basic unit of groundwater and its combination including water quantity, water quality input, operation and water output of the groundwater.

具有水量、水质输入、运营和输出的地下水基本单元及其组合。

growth pole 增长极

It refers to the first developed area of preferential growth upon which one or a group of leading sector centres.

一个或一组主导部门集中而优先增长的先发达地区。

gulf 海湾

[*marine science and technology*] It refers to a body of water connected to an ocean or sea by an inlet of land or between continents and islands.

[*resource technology*] It refers to a body of water surrounded by land with an area no smaller than a semicircle area with the diameter of a harbour entrance width.

[海洋科技]海或洋伸入大陆或大陆与岛屿之间的一部分水域。

[资源科技]被陆地环绕且面积不小于以口门宽度为直径的半圆面积的海域。

Gulf of Mexico 墨西哥湾

It is the adjacent sea of the Atlantic Ocean on the southeast edge of North America, which is located between America, Mexico and Cuba.

位于美国、墨西哥和古巴之间,北美洲东南边缘的大西洋的附属海。

Gulf of Thailand 泰国湾

Located in the southwest of the South China Sea, it is the gulf with an opening to the South between the Indochina Peninsula and the Malay Peninsula.

位于南海西南部,中南半岛和马来半岛之间开口向南的海湾。

gully 冲沟

It resembles large ditches or small valleys, but are metres to tens of metres in depth and width. When the gully formation is in process, the water flow rate can be substantial, which causes the significant deep cutting action into soil.

一种较大的、有间歇性水流活动的长条状谷地,由切沟发展而来。

guyot 海底平顶山

It refers to a seamount that is generally round or oval with a flat top.

顶部平坦,一般呈圆形或椭圆形的海山。

H

Haihe River 海河

It is one of the major rivers of North China, with a total length of 1,090 kilometers. Its drainage area is 264,617 square kilometers, its mean annual runoff is 22.6 billion cubic metres and its mean annual runoff depth is 85 millimetres. It flows through Shanxi (Jin), Hebei (Ji), Henan (Yu), Tianjin (Jin) and other provinces and regions, and its main tributaries are the Yongding River, Daqing River, Ziya River, Nanyun River and Beiyun River.

中国华北地区主要的大河之一,长度1 090千米,流域面积264 617平方千米,平均年径流量226.00亿立方米,平均年径流深度85毫米,流经晋、冀、豫、津等省区,主要支流有永定河、大清河、子牙河、南运河和北运河。

Haining tide 海宁潮

Also called *Qiantang tide*, it refers to the tidal bore at the estuary of Qiantang River, Hangzhou Bay, Zhejiang Province.

又称"钱塘潮",指浙江省杭州湾钱塘江口的涌潮。

halmyrolysis 海解作用

Also called *submarine weathering*, it refers to a constant physical, chemical or biogeochemical changing process of various substances which comprise the sea floor in a marine environment.

又称"海底风化作用",海洋环境中,组成海底的各种物质不断发生物理的、化学的或生物地球化学的变化过程。

halmyrolysis 海底风化作用

Halmyrolysis refers to a physical, chemical or biogeochemical changing process of various substances which comprise the sea floor in a marine environment.

又称"海解作用",海洋环境中,组成海底的各种物质不断发生物理的、化学的或生物地球化学的变化过程。

Hainan Island Marine Economic Zone 海南岛海洋经济区

Hainan Island has 1,618 kilometres of coastline and 490 square kilometres of intertidal zones. The superior marine resources are tropic marine biological resources, island and ocean marine tourism resources, and oil and gas resources, but the marine economic basis is weak. The main development directions are as follows: to develop island recreational tourism, tropical scenery tourism, marine ecotourism; to develop the processing and utilisation of marine natural gas resources; to improve the functions of Haikou, Yangpu and eight harbours, and enhance the transportation capability connecting to the inland; to strength larval rearing and farming base construction and encourage offshore fishing in open waters.

海南岛本岛海岸线长1 618千米,滩涂面积约490平方千米。优势海洋资源是热带海洋生物资源、海岛及海洋旅游资源和油气资源。海洋经济基础较薄弱。主要发展方向为:发展海岛休闲度假旅游、热带风光旅游、海洋生态旅游;发展海洋天然气资源加工利用;完善海口、洋浦和八所港口功能,加强与内陆连接的运输能力;抓好苗种繁育和养殖基地建设,鼓励发展外海捕捞。

holding orders of marine ship 海洋船舶手持订单量

This term refers to valid contract orders of marine ships undelivered to ship owners within a reporting period. The unit of measurement is Dead Weight Tonnage (DWT).

报告期末尚未交付船东的全部有效海洋船舶合同订单。计量单位:载重吨。

harbor 港口

It is a place where ships or vessels can pass in and out, berth or moor for passengers' embarking or disembarking, cargo handling, lightering, cargo storage, etc. It has corresponding port facilities and it is made up of a certain range of water area and land area. A harbour can also be composed of one or many harbour districts.

具有船舶进出、停泊、靠泊、旅客上下、货物装卸、驳运、储存等功能,具有相应的码头设施,由一定范围的水域和陆域组成的区域。港口可以由一个或者多个港区组成。

harbor craft scheduling 港务船只调度

It refers to the scheduling and management activities which can ensure the normal operation of the harbour.

保证海港正常运行对港务船只的调度与管理活动。

harbor engineering construction 海港工程建筑

It refers to the construction of the wharf, harbour basin, navigation channel and navigation facilities of a harbour.

海港码头、港池、航道和导航设施的施工。

harbor handling equipment manufacturing 海港装卸设备制造

It refers to the manufacturing of hoisting machinery, conveying machinery, cargo handling machinery, operation vehicles with hoisting gear and handling devices and other accessories used in a harbour wharf.

海港码头所用起重机械、输送机械、装卸货物机械,装有升降或搬运装置的工作车等及其配件的制造。

harbor navigation channel special machinery manufacturing 海港航道专用机械制造

It refers to the manufacturing of the special equipment or facilities used for dredging, decontaminating and ice breaking of a harbor and a navigation channel.

用于海港、航道清淤、清污、破冰等专用设备的制造。

harbor property management 海港物业管理

It refers to the service activities related to professional repairing, maintenance, and management of a harbour, wharf, warehouse and other areas.

对海港、码头、仓库等区域进行专业化维修、养护、管理等相关服务活动。

harbor storage services 海港仓储服务

They refer to professional freight yards, warehouses and activities of storage services-based logistics and distribution companies (centers) in a harbor.

在海港内专业货场、仓库及以仓储服务为主的物流、配送公司(中心)的活动。

harbour works 海港工程

It refers to permanent man-made structures adjacent to the seacoast. When delimiting the territorial sea, they are regarded as a part to determine the baseline of the territorial sea and other sea areas. "The outermost permanent harbour works which form an integral part of the harbour system are regarded as forming part of the coast." (Article 11 of UNCLOS)

毗连海岸的永久人工构筑物,在海域划界中,视为确定领海及其他海域界限的基线的一部分。"构成海港体系组成部分的最外部永久海港工程,为海岸的一部分"(《联合国海洋法公约》第十一条)。

hard systems 硬系统

Also called *well-structured systems*, they refer to systems with a clear mechanism that can be depicted by explicit mathematical models. Hard systems are actually engineering systems, for which there are better quantitative research methods that can calculate the system behaviour and optimal results.

又称良结构系统,是指机理清楚,能用明确的数学模型描述的系统。硬系统实际是一个工程系统,对于硬系统已有较好的定量研究方法,可以计算出系统行为和最优的结果。

hazard geology 灾害地质

It refers to geological factors which can endanger human life and property, namely various geological conditions that are likely to cause disasters, including some geological bodies and geological processes.

对人类生命财产能够造成危害的地质因素,即有可能成灾的各种地质条件,包括某些地质体和地质作用。

heat conduction 海水热传导

It refers to the molecule transformation or turbulence conduction phenomena occurring between adjacent and different temperature stratifications in seawater, due to the irregular movement of the molecule or fluid parcel.

海水中由于分子或流体块的不规则运动所产生的在相邻不同温度层间的分子变换或湍流传导现象。

heat island effect 热岛效应

It is one of the typical features of urban climates and a microclimate phenomenon indicating that the temperature of metropolitan areas is higher than that of the surrounding rural areas due to the changes of urban radiation, existence of waste heat from industries and humans, and reduction of evaporation heat consumption.

城市气候的主要特征之一。由于城市中辐射状况的改变,工业余热和生活余热的存在,蒸发耗热的减少,而形成的城市市区温度高于郊区温度的一种小气候现象。

Heilongjiang River 黑龙江

It is among the large Asian rivers and the world's tenth longest river, with the central part forming the border river between Russia and China. This river is formed in Luogu Village to the west of Mo River by the confluence of its two major affluents, the Ergune River in the south from the western slope of the Greater Khingan Mountains in Heilongjiang and the Shilka River in the north from the eastern foothill of Khentii Mountains in northern Mongolia. It is 4,350 kilometers long with a basin area of 1.843 million square kilometers, receiving major tributaries including Zeya River, Songhua River, Ussuri River and others along the course and finally flowing into Strait of Tartary in Russia.

黑龙江是亚洲大河之一,世界第十大河。黑龙江中段为中苏界江,源头有两支:南支额尔古拉河,源出黑龙江大兴安岭西坡,北支石勒喀河,源出蒙古北部肯特山东麓。两源在漠河以西的洛古村汇合后称黑龙江,全长 4 350 千米,流域面积 184.3 万平方千米,沿途接纳结雅河、松花江、乌苏里江等大支流,在苏联境内注入鞑靼海峡。

hemipelagic sediment 半深海沉积

Also called *continental slope sediment*, it refers to pelagic sediment in the continental slope or in waters with the depth between 200 and 2,000 metres. The continental slope sediment is characterized by transitional nature between a continent and ocean basin. The granularity of the sediments is smaller than that in shallow waters, and the sediments are mainly made up of by mud, followed by sand. The continental slope sediment is mainly composed of landslide sediment, sediment of sea floor canyon bottom and sediment of turbidity current at the canyon outlet.

又称"陆坡沉积",大陆坡区或水深在 200~2 000 米范围内的海底沉积。陆坡沉积物具有大陆和洋盆之间的过渡性质。沉积物粒度小于浅海,以泥为主,其次为砂。陆坡沉积物的成因主要有滑坡沉积物、海底峡谷底部沉积物和峡谷出口的浊流沉积物等。

heterogeneity 异质性

It is the variation degree (or strength) of a resource (or a property) in time and space that has a decisive effect on the existence of a species or a higher-level biological organization in an area (landscape or ecological system).

在一个区域里(景观或生态系统)对一个种或者更高级的生物组织的存在起决定作用的资源(或某种性状)在空间或时间上的变异程度(或强度)。

hetero-organization 他组织

It is formed of when an external environment exerts a decisive influence or under the effect of a command given by an independent subsystem, i.e. a controller.

在外界环境施加决定性影响的情况下或由一个独立的子系统,即控制者施加指令的作用下形成的。

hidden dune 暗沙

Also called *submerged reef*, it refers to the underwater reef which is not exposed even at low tide. It can be formed by organisms (such as a coral reef) or the extension parts of volcanic rocks or continental rocks under water. Normally 10 metres lower than the sea surface, a submerged reef is often isolated in the sea or close to the coast, which is unsafe for navigation.

又称"暗礁",低潮时仍不露出海面的水下礁石。可以由生物体组成(如珊瑚礁),也可以由火山岩礁或大陆岩石在水下的延伸部分组成。暗礁离水面一般不到10米,往往孤立海中或靠近海岸,对安全航行不利。

high destructive delta 高破坏性三角洲

In the regions where the marine action is stronger than the fluvial action, river sediment will soon be dominated by wave action or tidal action. The deposition thickness is small in the frontal edge of the delta, characterized by sand ridges of the beach and the deposition of the tidal channel. This slowly growing delta is called a high-destructive delta. According to the relative role it plays in the ocean, high-destructive deltas can be classified as wave-dominated deltas or tide-dominated deltas.

在海洋作用强于河流作用的地区,河流入海的泥沙很快地被波浪或潮汐作用控制,三角洲前缘沉积厚度较小,有海滩沙脊及潮道沉积。此类生长速度较慢的三角洲称为高破坏性三角洲。按海洋作用的相对重要性,高破坏性三角洲又分为浪控三角洲和潮控三角洲。

high energy marine environment 海洋高能环境

It refers to an area in which coastal sea water moves strenuously and sea-floor landform becomes unstable.

沿岸海水运动剧烈,海底地形地貌很不稳定的区域。

high sea 公海

It refer to the open waters of an ocean or a sea beyond the limits of the territorial jurisdiction of any and all states outside the internal waters, territorial waters, exclusive economic zone of the coastal states, and archipelagic waters of the archipelagic states.

沿海国内水、领海、专属经济区和群岛国的群岛水域以外不受任何国家主权管辖和支配的全部海域。

high-energy coast 高能海岸

It refers to a high wave energy coast on which the mean breaking wave is higher than 50 centimetres. Characterized by its open shape, it is constantly exposed to strong and steady winds and frontal surfaces, and it is also influenced by high wave energy action.

平均破波高度大于50厘米的高波能海岸。此类海岸外形开敞,经常性地暴露于强而稳定的地带风和锋面下,接受高波浪能量作用。

high-tech park 高科技园区

It is also called a *science & technology industrial park* or a *science park*. Generally relying on prestigious universities or scientific institutions, a high-tech park creates favourable conditions in a certain area in order to attract professors, scholars and researchers to initiate high-tech enterprises. The results of the science and research is directly converted into products in the market to bring about good economic benefits.

又称"科技工业园"、"科学城"、"高新技术园区"。一般依托名牌大学或科研机构,在一定区域内创造良好的条件,吸引教授、学者和研究人员兴办高技术企业,将科研成果直接转化为进入市场的产品,以获得良好的经济效益。

higher and low lunitidal interval 高低潮间隙

It refers to the time interval from the moon passing overhead to the next high or low tide at a given location of the sea.

海洋某点从月中天时刻到该点出现高低潮时的时间间隔。

highest groundwater level 地下水最高水位

It refers to the maximum value of groundwater level in a certain observation time period.

在某一观测时段内,地下水水位的最高值。

highly constructive deltas 高建设性三角洲

They refer to the highly growing deltas formed in regions where the fluvial action is stronger than the marine action. These deltas can be classified as bird-foot deltas or arcuate deltas based on the differences between the extension mode and the shoreline shape of the deltas. The former one can be represented by the American Mississippi River Delta. The latter one is exemplified by the Chinese Yellow River Delta.

在河流作用强于海洋作用的地区形成的高增长的三角洲。按三角洲扩展方式和滨线形态的不同,高建设性三角洲可以分为鸟足状三角洲和扇形三角洲。前者如美国的密西西比河三角洲,后者如中国的黄河三角洲。

hole 海底洼地

It refers to a pint-sized submarine concave ground.

小型的海底凹地。

horizontal region economic cooperation 水平区域经济合作

It refers to collaborative activities between two parties whose economic development is roughly at the same level, goods produced by them being generally at the same stage in the product life cycle, and the technical contents of production factors provided being basically identical.

合作双方经济发展水平大致相当,企业生产商品在产品生命循环阶段上所处层次大体相同,双方提供生产要素的技术含量基本一致的协作活动。

horst 地垒

It refers to the raised part between two faults which bear the same topographic features, and run approximately parallel but have opposite trends.

两个地貌性质相同,且走向大致平行,但倾向相反的断层之间的上升部分。

hot spring 海底热泉

It refers to a hot spring gushing out of a rock fracture in the seafloor.

海底岩石裂隙中喷涌出的热水泉。

human capital 人力资本

Derived from human resources, it refers to a capital form that is specifically and mainly centred on human resources and represented by the exploitation and utilisation of human abilities, including education, work experience, health and nutrition conditions of the population, etc.

在人力资源的基础上派生出来的，具体是指以人力资源为核心，主要以人的能力的利用、开发为表现的，包括教育、工作经验以及人口的健康营养状况等因素在内的资本形式。

human resources 劳动力资源

They refer to the set of individuals who make up the workforce of an organization, business sector, or economy.

又称"人力资源"、"劳动资源"，指某种范围内的人口总体所具有的劳动能力的总和。

hydraulic engineering 水利工程

It is the project constructed to control, harness, allocate, protect, develop and utilize surface water and ground water in nature to bring benefits for human being.

对自然界的地表水和地下水进行控制、治理、调配、保护，开发利用，以达到除害兴利的目的而修建的工程。

hydrobiology 水生生物学

It is a discipline which studies life phenomena and biological processes in water, as well as its interrelation with environmental factors.

研究水域环境中的生命现象和生物学过程及其与环境因子间相互关系的学科。

hydrobiont 水生生物

Hydrobiont refers to animals and plants living, as a whole or in part, in various waters, including freshwater organisms and marine organisms.

全部或部分生活在各种水域中的动物和植物。包括淡水生物和海洋生物。

hydrocole 水生动物

It is heterotrophic organisms living in the water. They can not make food themselves, and they acquire nutrition from plants, other animals and organic residues.

在水中生活的异养生物。它们自身不能制造食物，营养靠摄食植物、其他动物和有机残体获得。

hydrographic control point 海控点

The abbreviation of marine hydrographic measurement and control point, it refers to the encryption control points located near the coast and on islands laid between the first-grade and fourth-grade national geodetic control networks.

海洋测量控制点的简称，指在国家一等到四等大地控制网点间布设的位于海岸附近和海岛上的加密控制点。

hydrographic survey 海洋测量

A hydrographic survey is a generic term for the survey of sea and adjacent land, which includes marine geodetic surveys, water-depth surveys, submarine topographic surveys, ocean engineering surveys, marine gravity and magnetism surveys as well as other special surveys and investigations.

对海洋区域及邻近陆地的各种测量工作的统称。包括海洋大地测量、水深测量、海底地形测量、海洋工程测量、海洋重力、磁力测量及其他专题测量与调查。

hydrographic survey and charting 海洋测绘

It is a generic term for hydrographic surveying and charting, which includes measuring and surveying adjacent lands, rivers and lakes, obtaining basic geological information of the sea, composing charts of all kinds, and compiling navigation data, etc.

海洋测量与海图制作的总称。包括对其邻近陆地和江河湖泊进行测量和调查,获取的海洋基础地理信息,制作各类海图和编制航行资料等。

hydrogeologic unit 水文地质单元

It is a groundwater system with uniform recharge boundary and recharge, runoff and discharge conditions.

具有统一补给边界和补给、径流、排泄条件的地下水系统。

hydrogeological condition 水文地质条件

It is a generic term for the conditions of groundwater deposit, distribution, recharge, runoff and discharge, water quality, quantity and other geological conditions of their formation, etc.

地下水埋藏、分布、补给、径流和排泄条件,水质和水量及其形成地质条件等的总称。

hydrogeological division 水文地质分区

It is the division of a research area based on the different purposes and the difference in hydrogeological conditions.

针对不同目的将研究区按水文地质条件的差异性而划分的若干个块段。

hydrography and nautical cartography services 海洋测绘服务

It refers to all service activities conducted for hydrography, nautical cartography, and marine geological information system engineering, etc.

为海洋测绘、海图编制、海洋地理信息系统工程等进行的服务活动。

hydrologic cycle 水文循环

It describes the circular movement of water on, above and below the surface of the earth. The water moves from the surface to the atmosphere, then to the runoff, accumulated in the soil or waters, by the physical processes of evaporation, condensation, precipitation, and infiltration, then evaporates from the surface again and the same circulation run in cycles.

地球上的水从地表蒸发,凝结成云,降水到径流,积累到土中或水域,再次蒸发,进行周而复始的循环过程。

hydro project 水利工程

It is the project constructed to control, harness, allocate, protect, develop and utilize surface water and ground water in nature to bring benefits for human beings.

对自然界的地表水和地下水进行控制、治理、调配、保护,开发利用,以达到除害兴利的目的而修建的工程。

hydrology 水文

It refers to all sorts of changes and movements of water in nature, such as the distribution and intensity of rain, the water level of rivers and lakes, the water velocity and flow rate of rivers, sediment and salt content in currents, etc. It is necessary to know various conditions of a particular area before planning a water conservancy project so as to guarantee a complete and thorough plan.

水在自然界的各种变化和运动情况。雨水的分布和大小,江河湖泊水面的高低,江河水流的快慢和流量多少,以及水流中泥沙、盐类含量等现象,都是水文现象。计划一个水利工程前,必须了解这个区域里的各种情况才能有周密的根治计划。

hydrophytes 水生植物

They are plants of which at least part of their life is spent in aquatic environments. They are also referred to as aquatic plants or aquatic flora.

至少有一部分生命阶段是在水中度过的植物。

hydropower resources 水能资源

They are energy resources stored in rivers, lakes and seas in the form of potential energy, kinetic energy, etc.

以势能、动能等形式存在于江河湖海中水体的能量资源。

iceberg deposit 冰海沉积

It refers to an ice deposit at the bottom of the ocean carried in an ice tongue, iceberg or ice shelf floating on the edge of a coast. It is widely distributed at the bottom of the ocean, such as the modern sea ice deposit zone of 370-1,300 kilometres wide surrounding the South Pole.

漂浮于海岸边缘的冰舌、冰山、冰棚中所挟带的冰碛物在海洋底部的沉积。较广泛分布于海底,如围绕南极有一个宽达370~1 300千米的现代冰海沉积带。

ichthyology 鱼类学

It is a discipline which studies fish classification, morphology, physiology, ecology, phylogeny and geographical distribution.

研究鱼类的分类、形态、生理、生态、系统发育和地理分布的学科。

ill-structured systems 不良结构系统

Also called *soft systems*, soft systems are the ones with unclear mechanisms and cannot be described in a specific mathematical model.

又称"软系统",是指机理不清,很难用明确的数学模型描述的系统,如社会系统和生物系统。

implantation of marine forest tree 海洋林木种植

It refers to the implantation of forest trees in an offshore area and a tidal mud flat.

在近海和海涂进行的林木种植活动。

imports 进口

They refer to goods or services which are manufactured abroad while sold in domestic market.

国外生产而在国内销售的物品与劳务。

incineration at sea 海上焚烧

It refers to the burning of waste materials or other substances deliberately on the incineration facilities at sea excluding incidental actions which happen during a normal operation on vessels, platforms or other man-made structures.

以热摧毁为目的,在海上焚烧设施上,故意焚烧废弃物或者其他物质的行为,但船舶、平台或者其他人工构造正常操作中所附带发生的行为除外。

income of sea rental by the provincial government 海域出让金

It refers to the income received by the provincial government for renting use right of state-owned sea area to entities or individuals in accordance with the law for a term of years. A term of years means the maximum number of years for the right to use sea area by tendering or auction. The law to be followed is *Law of the P.R.C on the Administration of the Use of Sea Areas*.

省级政府依法将国家所有的海域使

用权在一定年限内出让给单位或个人所收取的价款。所谓一定年限是指通过招标或者拍卖的海域使用权最高年限。所谓依法是依据《中华人民共和国海域使用管理法》的规定。

increase and protect fishery resources fees 渔业资源增殖保护费

Departments of fishery administration under the people's governments at or above the county level should work out overall plans and take measures to increase fishery resources in fishery waters under their jurisdiction. They can levy fees for fishery resources multiplication and conservation on units and individuals which reap benefits from fishery resources. The fees will be embarked for increasing and protecting fishery resources.

县级以上人民政府渔业行政主管部门应当对其管理的渔业水域统一规划，采取措施，增殖渔业资源。县级以上人民政府渔业行政主管部门可以向收益的单位和个人征收渔业资源增殖保护费，专门用于增殖和保护渔业资源。

index line 标志线

It refers to a line connected by several index points.

由标志点连接而成的线。

index of labour cost 劳动费指数

It refers to the average labour cost of product per unit weight. The larger the index is, the higher the possibility of transferring from a minimum freight area to a low-price labour cost area and vice versa.

每单位重量产品的平均劳动费。如果劳动费用指数大，那么从最小运费区位移向廉价劳动费区位的可能性就大；否则，这种可能性就小。

index point 标志点

It refers to a clearly marked point, the coordinate of which can be measured to calculate the coordinate of boundary point.

具有明显标志并可通过对其坐标的测量推算界址点坐标的点。

index system for marine ecological monitoring 海洋生态监测指标体系

It refers to a qualitative and quantitative monitoring index system, which is induced from the perspective of ecology to directly analyse and assess the ecological quality condition and changing trend of marine natural reserves, based on ecological principles and characteristics of oceanography and marine organisms.

应用生态学原理，结合海洋学和海洋生物学的特点，从生态学角度归纳出的能够直接分析和评估海洋自然保护区生态质量状况及其变化趋势的定性和定量监测指标系统。

Indian Ocean 印度洋

It is the world's third largest ocean stretching between Asia, Oceania, Africa and Antarctica.

位于亚洲、大洋洲、非洲和南极洲之间的世界第三大洋。

indicated species 指示种

Indicator species are the iconic species of marine biological communities appearing in a certain state and in a certain sea area.

海洋生物群落在一定海域一定状态出现的标志性的物种。

industrial ecology 产业生态学

It refers to a discipline that studies the whole metabolic process in which the natural resources are extracted from the earth, transformed into commodities which can be bought and sold to meet the needs of humanity and then become waste in social production activities, which involve organization and management system, dynamics mechanism responsible for production, consumption and regulating behaviors, method of cybernetics and interrelation with life supporting system. It is concerned with the shifting of industrial process from linear (open loop) systems, in which resource and capital investments move through the system to become waste, to a closed loop system where wastes can become inputs for new processes.

一门研究社会生产活动中自然资源从源、流到汇的全代谢过程,组织管理体制以及生产、消费、调控行为的动力学机制、控制论方法及其与生命支持系统相互关系的系统学科。

industrial layout 产业布局

It refers to the distribution and combination of various industries of national economy in space.

国民经济各产业在空间上的分布和组合现象。

industrial layout orientation 产业布局指向

The industrial layout orientation refers to the tendency of choosing an industrial location under the action of various factors and layout mechanisms. The main industrial layout orientation types include energy, raw material sources, consumption area, labour force, transportation junction, hi-tech, etc.

在各种因素和布局机制的作用下,一个产业区位选择的趋向。主要产业布局指向类型有能源指向、原料地指向、消费地指向、劳动力指向、交通运输枢纽指向、高科技指向等。

industrial layout policy 产业布局政策

It refers to a policy which regulates the allocation of essential productive factors in a geographic space.

调节生产要素在地理空间上的配置政策。

industrial organization policy 产业组织政策

It refers to policies which regulate and control resource allocation structures within an industry to solve the contradiction between economy of scale and competitive assets.

调控一个产业内的资源配置结构的政策,以解决规模经济与竞争资产的矛盾。

industrial policy 产业政策

It refers to the sum of various policies the government has taken to be directly or indirectly involved in economic activities of industry and business production. With various industries of regional economy as target, the aim of these policies is to realize certain economic and social goals by protecting, supporting, adjusting and improving various industries.

政府为了实现一定的经济和社会目

标,以区域经济各产业为对象,通过对各产业的保护、扶植、调整和完善,直接或间接参与产业或企业的生产经济活动的各种政策的总和。

industrial structure policy 产业结构政策

It refers to the policy of a government regulating resources allocation among industries and its connectivity, involving two questions which are structure coordination and structure evolution respectively.

政府调节资源在产业间配置的构成及其关联性的政策,涉及结构协调和结构进化两个问题。

industrial trade cooperation 工业贸易合作

It encompasses the productive cooperation and processing trade cooperation of two parties in a manufacturing field.

又称"工贸使用",它包括合作双方在制造业领域的生产合作和加工贸易合作两个方面。

industry chain 产业链

It is a path and process through which one or several resources are continuously transferred through several industrial levels towards a downstream industry until reaching the consumers. It is also a comprehensive reflection of industrial levels, industrial correlation, resources processing depths and demand satisfaction degrees. It can reduce the cost of industrial aggregation and increase the added value of products through research on industrial rules, focus on correlation and continuity of industries, further development from multiple levels and extension of industrial chains.

一种或几种资源通过若干产业层次不断向下游产业转移直至到达消费者的路径与过程,是产业层次、产业关联程度、资源加工深度以及满足需求程度的综合反映。通过研究产业规律,注重产业之间的相关性、承接性和多层次的深度开发,延长产业链,可以降低产业聚集成本,提高产品附加价值。

infiltration coefficient of precipitation 降水入渗系数

It refers to the ratio of the amount of groundwater infiltrated and recharged by precipitation to gross precipitation per unit area in a given place.

一个地区单位面积上降水入渗补给地下水的量与总降水量的比值。

inflation 通货膨胀

It refers to a persistent increase in the level of consumer prices or a persistent decline in the purchasing power of money, caused by an increase in available currency and credit beyond the proportion of available goods and services.

在纸币流通条件下,货币的发行量超过商品流通中所需要的货币量而引起货币价值下降、物价持续而普遍上涨的现象。纸币、含金量低的铸币、信用货币过度发行都会导致通货膨胀。

influence coefficient 影响力系数

It is the degree of production demand influence on various sectors of the national economy where one sector adds one unit for final use.

某一部门增加一个单位最终使用时,对国民经济各部门所产生的生产需求波及程度。

information efficiency 信息有效

It reflects in a rational way all available information about the description of asset prices.

以理性方式反映所有可获得的信息的有关资产价格的描述。

information of geographical names 地名信息

They are generic terms for textual and numerical information which can reflect the proper names of places and their attributes.

反映地名及其属性的文字和数字信息的总称。

infrastructure 基础设施

It is a public service system necessary for the normal operation of social and economical activities in a region, including economic infrastructure and social infrastructure.

用于保证区域社会经济活动正常进行的公共服务系统,包括经济基础设施和社会基础设施。

inhabited island 有居民海岛

It is an island belonging to the registered residence of household registration management.

属于居民户籍管理的住址登记地的海岛。

inner waters 内水

They refer to water areas on the landward side of a territorial water baseline of a coastal state, including lakes, rivers and their estuaries, inner waters, ports, gulfs, straits and other water areas which are located within territorial waters.

又称"内陆海"、"封闭海"、"内海"。沿海国领海基线陆地一侧的水域。包括湖泊、河流及其河口、内海、港口、港湾、海峡以及其他位于领海以内的水域。

inshore 内滨

It refers to a zone extending from a low-tide level toward the sea to the outer edge of a breaker zone.

低潮线向海一侧,直至破浪带外界的地带。

inshore fishing 近海捕捞

It refers to the hunting of various natural aquatic species, which occurs within a sea area of an exclusive economic zone and continental shelf.

在专属经济区、大陆架以内海域从事的对各种天然水生动植物的捕捞活动。

institutional competition 制度竞争

It is the competition between rules or rule systems selected by the main forces of an institution. It is a process of institution search and discovery, institution learning, institution imitation and institution innovation, as well as the discovery of a more suitable institution.

制度主体所选择的规则或规则体系之间的竞争,它是一个制度搜寻和发现的过程,也是一个制度学习、制度模仿和制度创新,以及发现更适宜制度的过程。

institutional equilibrium 制度均衡

It is an evenly stable or relatively static state of various powers in a whole institutional system at a specific time. In other words, a state of mutual adaptation and coordination rather than incompatibility exists between any specific institutions in the institutional system. Meanwhile, both individuals and organizations can achieve their respective targets to the utmost extent through selecting their own best behaviours in the behaviour space allowed by the institutional system.

整个制度系统中的各种力量在特定时间上所达到的某种势均力敌的稳定或相对静止的状态,即制度系统中任何具体制度之间都不存在互斥关系,而是处于相互间适应协调的状态,同时,每个个人和组织都能在制度系统所允许的行为空间中通过选择自己的最佳行为而同时最大限度地实现各自的目标。

institutional evolution 制度演化

It is a repeated dynamic process in which an institution transforms from a game equilibrium to disequilibrium, and then to another game equilibrium.

制度从一个博弈均衡到非均衡,再到另一个博弈均衡的一个不断反复的动态过程。

institution 制度

It refers to mechanisms and structures of social order which are used to normalize or give incentive to the behaviors important to the main body of an institution, thus making a society more stable. An institution is generated by repeated games among the main forces of different institutions based on their own interests.

不同制度主体之间基于自身利益进行多次重复博弈而产生的、用以规范或激励制度主体的行为,给集体或社会带来稳定的认知性及标准化结构。

institution core 制度核

It refers to elements such as ethics, values, cultural orientation, truth, goodness, and beauty, etc. forming the most basic meaning of an institution in a certain historical period.

在一定历史时期,形成制度的最基本意义的道德、价值观、文化取向、真、善、美等要素。

insular plain 岛架平原

It refers to an island shelf which has a large geographic entity of flat and vast topography as the main body of the island shelf, with a mean slope of less than 0°10′.

岛架上地形平坦、广阔的大型地理实体,为岛架的主体,平均坡度一般小于0°10′。

Integrated Global Ocean Station System(IGOSS) 全球海洋站系统

Currently called the Integrated Global Ocean Station Net, it is a global real time ocean observing system composed of large numbers of national facilities (data observing, communication and processing equipment) and coordinated by an intergovernmental oceanographic commission whose headquarters is located in Paris, France. Founded in 1967, the conference held in 1968 made decisions as follows: to establish various manned, unmanned, stable and movable ocean buoy observing stations to realise the automation and long-distance telemetry and carry out ocean observation via satellite. Temperature and depth test plans and ocean pollution monitoring test programs were conducted in 1972 and 1975 respectively. Due to the complexity of the legal status of ocean data acquisition systems and marine laws, IGOSS is developing slowly.

现名全球联合海洋台站网,是计划建立的一个全球性的实时海洋观测服务系统,由大量的国家设备(资料观测、通信和处理设备)所组成,由政府间海洋学委员会进行协调,总部设在法国巴黎。1967年开始组建,1968年开会并作出决议:建立各种有人、无人、固定、可动的海洋浮标观测站,使海洋观测进入自动化、远距离遥测阶段,并利用人造卫星进行海洋观测。1972年开始执行温深测量试验计划,1975年开始执行海洋污染监测试验计划。由于海洋资料获取系统的法律地位与海洋法问题交织在一起,所以该系统发展比较缓慢。

integration of natural resources 自然资源整体性

All kinds of natural resources do not exist in isolation, but are mutually related and restricted to form a complex resource system.

各类自然资源之间不是孤立存在的,而是相互联系,相互制约而组成一个复杂的资源系统。

integration of the land and sea 海陆一体化

It means to integrate a sea system and land system which are originally relatively isolated into a new unified whole with the application of system theory and synergetics theory, in accordance with the internal connection between the two geographic units of sea and land, in order to achieve a more effective allocation of the sea and land resources by integrated planning, linkage development, and the combination and integrated management of industry chains.

根据海、陆两个地理单元的内在联系,运用系统论和协同论的思想,通过统一规划、联动开发、产业链的组接和综合管理,把本来相对孤立的海陆系统,整合为一个新的统一整体,实现海陆资源的更有效配置。

intensive agriculture 集约农业

It refers to an agricultural production system characterized by the high use of inputs such as capital, labour and chemical fertilizers or by crop rotation to increase average yield per unit land area.

在单位面积上投入大量的劳力、资本、肥料等,或实施轮作以提高单位面积平均收获量的农业。

intensive cultivation 集约养殖

[*fisheries science*] It is a way of cultivating aquatic economic animals with high density of aquatic fry, high inputs of material and energy and fine management per unit water area.

[*marine technology*] It is a way of high density cultivation characterized by relying on advanced instrument and equipment, to achieve high output and high economic benefits.

[水产学]单位水体苗种密度高、物质和能量投入多、管理精细的一种水产经济动物养殖方式。

[海洋科技]采用先进仪器设备和

管理技术,实施高密度、高产量、高经济效益的养殖方法。

intensive growth 集约式增长

It refers to an economic growth relying on increase of production efficiency, characterized by high technology, low input, low consumption, low pollution, high output, high benefit and high added value with economy structure to be optimized gradually. Technological progress is the basis and source for the growth while constant adjustment of ideology and system (such as political and legal system, economic system and economic structure) is the prerequisite.

主要依靠生产效率的提高实现的经济增长。通常以高技术为依托,以低投入、低消耗、低污染、高产出、高效益、高附加值和经济结构不断趋向优化为特征。这种增长方式以技术进步为基础和源泉,以制度(政治和法律制度、经济体制、经济结构等)和思想意识的不断调整为必要条件。

intensive utilization of natural resources 自然资源集约利用

It is a way of resource utilisation which makes intensive investment in labour, capital, technology and other factors of production to get higher output and economic benefits.

集中投入较多的劳动、资金、技术和其他生产要素,以获取更多产出和经济效益的资源利用方式。

intercontinental sea 陆间海

It generally refers to a wide and deep sea, which is located between continents, such as the Mediterranean Sea and the Caribbean Sea.

位于大陆之间的海,面积和深度都较大,如地中海和加勒比海。

interference 干涉

It refers to abnormal changes of amplitude when two waves of the same or similar frequency superimpose on each other.

海洋上频率相同或相近的波叠加后,产生振幅异常变化的现象。

Intergovernmental Panel on Climate Changing(IPCC) 联合国政府间气候变化委员会

It is an intergovernmental body jointly established by the World Meteorological Organization (WMO) and the United Nations Environment Programme (UNEP) in 1988, whose mission is to evaluate the current state of scientific knowledge on climate change, potential socio-economic impacts by climate change and possible countermeasures of adapting and slowing down climate change.

联合国政府间气候变化委员会是世界气象组织(WMO)及联合国环境规划署(UNEP)于1988年联合建立的政府间机构。其主要任务是对气候变化科学知识的现状,气候变化对社会、经济的潜在影响以及如何适应和减缓气候变化的可能对策进行评估。

interior river 内流河

Also called *Inland River*, it is a river flowing into inland deserts or lakes.

又称"内陆河",河水流没于大陆内部沙漠或内陆湖泊的河流。

intermittent stream 季节性河流

It is a stream that flows during rainy season and possibly dries up in drought season with its water source supplied by surface runoffs.

水源主要由地表径流补给、雨季期间出现水流而旱季可能干枯的河流。

intermountain basins 山间盆地

Also called *basins*, they refer to the low-lying and relatively closed area with a flat bottom between seamounts or sea knolls.

又称"盆地",指海山或海丘之间地形低洼、相对封闭的区域,底部地形平坦。

internal tide 内潮

It is generated as the surface tides move stratified water up and down sloping topography, which produces a wave in the ocean interior.

海水内界面处的潮波现象。

internal unit of sea lot 宗海内部单元

It refers to the waters zoned within a sea lot according to the way the sea is used.

指宗海内部按用海方式划分的海域。

internal wave 内波

They are gravity waves that oscillate on the pycnocline within a fluid medium.

流体内部密度跃层界面上的波动。

International Association for the Physical Science of the Ocean (IAPSO) 国际海洋物理科学协会

It is an association subordinate to the International Union of Geodesy and Geophysics (IUGG) and its predecessor is the International Association of Physical Oceanography (IAPO) which was established in 1931 and changed to the present name in 1967. The Association is supported by committees such as marine geophysics, marine chemistry, physical oceanography, Tides and the Mean Sea Level, etc., and China is one of its member states.

国际大地测量学和地球物理学联合会(IUGG)下属的一个协会,前身为1931年成立的国际物理海洋学协会(IAPO),1967年改用现名,协会下设海洋地球物理学、海洋化学、物理海洋、潮汐与平均海平面等委员会,中国是其成员国之一。

International Chamber of Shipping 国际航运协会

It is an international civil shipping organization, whose purpose is to safeguard the right interests of ship owners from different countries, which operate fleets based on the principles of free trade. The International Chamber of Shipping was established in London in 1921.

国际性民间航运组织,其宗旨是维护在自由贸易原则基础上经营船队的各国船舶所有人的权益。成立于1921年,会址设在伦敦。

International Law of the Sea 国际海洋法

It is a generic term for the legal system regarding various marine regions which aims to adjust the principles and rules in respect of exploitation, utilization and management of the ocean among different countries. It has not only the common characteristics of international law, but also the special concept and legal system different from other departments of law within international law.

关于各种海洋区域的法律制度,调整国家之间在开发、利用和管理海洋方面相互关系的原则与规则的总称。既具有国际法的一般特征,又具有与国际法中的其他部门法不同的特殊概念和法律制度。

International Maritime Organization (IMO) 国际海事组织

Originally called *the Inter-Governmental Maritime Consultative Organization*, established on January 13th, 1959, it was changed to the present name of International Maritime Organization on May 22nd in 1982. With its headquarters in London, the IMO is a specialized agency of the United Nations which deals with marine safety matters and develops marine technology. The IMO's purpose and mission are up to enhance marine safety, improve efficiency of shipping, prevent and control marine pollution caused by marine vessels, and deal with legal matters related to the above.

1959 年 1 月 13 日成立的政府间海事协商组织,1982 年 5 月 22 日改名为国际海事组织。总部设在伦敦,是联合国处理海上安全事务和发展海运技术方面的专门机构之一。其宗旨和任务是促进海上安全、提高船舶航行效率、防止和控制船舶对海洋污染等以及处理与上述事项有关的法律问题。

international nautical mile 国际海里

Also called *a nautical mile*, it (symbol n miles) is a unit of distance used in sea navigation. It is about one minute of arc of latitude measured along any meridian, or about one minute of arc of longitude at the equator. As the meridian is an ellipse, its curvature at different latitudes varies, therefore, one minute of arc of latitude measured along any meridian is also different. One nautical mile is equal to 1,852 metres.

又称"海里",航海上计量距离的单位,符号为 n mile。它等于地球椭圆子午线上纬度 1 分(1 度等于 60 分,1 圆围为 360 度)所对应的弧长。由于地球子午圈是一个椭圆,它的不同纬度的曲率是不同的,因此,纬度 1 分所对应的弧长也是不相等的。1 海里 = 1 852 米。

international rivers 国际水道

They refer to rivers which connect to high seas, flowing through two or more countries and are open for merchant ships from different countries on the basis of riparian state agreements. In a broad sense, international waters encompass the rivers which flow through the territory of two or more countries, including boundary rivers and rivers of many countries; in a narrow sense, international waters refer to the rivers of a special class which flow through several countries, high seas and allow free navigation of merchant ships from different countries.

又称为"国际河流",同公海相通、流经两个或两个以上国家并经沿岸国协议对各国商船开放的河流。广义上包括所有流经两个或几个国家领土的河流,即包括界河和多国河流在内;狭义上则专指特殊一类的河流,指流经几个国家而可流到公海并由各国商船在平时自由航行的河流。

international seafloor 国际海底区域

They refer to a seabed, ocean floor and subsoil beyond the jurisdiction of the state, namely, the deep seafloor and its subsoil outside the Exclusive Economic Zone of the surrounding countries and the continental shelf.

国家管辖范围以外的海床和洋底及底土,即各国专属经济区和大陆架以外的深海海底及其底土。

International Seabed Authority 国际海底管理局

It is an organization established by the *Law of the Sea Convention* to manage the resources in the international seabed area in 1994. The state parties of the *Law of the Sea Convention* are all members of the International Seabed Authority and the principal organs of the Authority are the Assembly, the Council, the secretariat and the Enterprise.

根据1982年《联合国海洋法公约》,于1994年成立的管理国际海底区域资源的组织。《联合国海洋法公约》缔约国都是该管理局的成员,主要机构有大会、理事会、秘书处和企业部。

International Whaling Commission (IWC) 国际捕鲸委员会

It is an international body set up by the terms of the *International Convention for the Regulation of Whaling* (ICRW) in Washington, D.C., United States, on December 2nd 1946. The purpose and mission of the IWC are as follows: to investigate the number of whales, to work on measures to hunt and protect the whale stocks of the Pacific Ocean (for example, to determine the protected and non-protected species of whales; to prescribe open and closed seasons and areas for whaling and marine sanctuary, to designate the whaling time and tools, etc.), and to implement strict international supervision towards the whaling industry.

1946年12月2日根据《国际捕鲸公约》在华盛顿成立,其宗旨和任务是:调查鲸的数量;制定捕捞和保护太平洋鲸藏量的措施,如确定鲸的保护品种和非保护品种、开放期和禁捕期、开放水域和禁捕水域、捕捞时间和工具等;对捕鲸业进行严格的国际监督。

internet services 互联网服务

They cover internet management and transmission of data and images for marine production, management and service.

为海洋生产、管理、服务提供的互联网管理,数据、图像传送服务。

interregional trade 区域贸易

It refers to the commodity transactions between different regions.

一个地区与其他地区进行商品交易的活动。

intersection of coastal boundary 海岸界点

It refers to the point of intersection between the boundary point of an administrative area of land at country level closest to the coast and the coastal line.

陆域县级行政区域界线向海一侧最靠近海岸的界点或其延长线与海岸线的交点。

intertidal benthos 潮间带生物

It refers to organisms that live on the surface of an intertidal zone and in the bottom sediments of a water body.

生活在潮间带地表的植物和底表与底内的动物。

intertidal mudflat culture 滩涂养殖

It is the production activity of cultivating marine economic animals and plants by making use of a soft muddy or sandy muddy mudflat in an intertidal zone after flattening the mudflat and building dikes and dams for it.

利用潮间带软泥质或沙泥质滩涂，加以平整、筑堤、建坝，养殖海水经济动植物的生产活动。

intertidal zone 潮间带

Also called *foreshore*, *seashore* and sometimes referred to as the littoral zone, it is the area that is above water at low tide and under water at high tide (in other words, the area between tide marks). In China, it specifically refers to areas between coastlines and zero-metre of a nautical chart.

又称"滩涂"、"海滩"。高潮线与低潮线之间的地带，相当前滨。我国具体指海岸线与海图零米线之间的地带。

intervention on the high seas 公海干预

It refers to the rights of coastal states for adopting necessary measures on the high seas in order to avoid or mitigate the damage of serious pollution accidents to the coast and waters.

沿海国在公海上采取必要措施，防止或减轻公海上发生严重污染事故对其海岸和水域污染损害的权利。

investigation of groundwater withdrawals 地下水开采量调查

It refers to the periodic or aperiodic investigation or statistical analysis of the actual groundwater exploitation used by city, industry and farmland irrigation in order to obtain the statistics of the actual groundwater exploitation in the studied area.

为了取得研究区内地下水实际开采量的数据，对城市、工业、农田灌溉等地下水实际开采量进行定期或不定期的调查和统计工作。

investigation of tourism resources 旅游资源调查

It refers to the study and recording of individual tourism resource according to the classification standard of tourism resources.

按照旅游资源分类标准，对旅游资源单体进行的研究和记录。

investment 投资

It is the behaviour of economic subjects' which invest their capital at home or abroad in a certain economic and social field in order to obtain economic benefits.

经济主体为了获取经济利益而把资本投向国内或国外某项经济社会领域的行为。

investment environment 投资环境

It is a generic term for all external factors, which can influence an enterprise's production and operation activities as well as their results in an area where capital is invested.

存在于受资区域内,能够影响企业生产经营活动过程及其结果的一切企业外部因素的总称。

investment in infrastructure 基础设施投资

It refers to the workload of infrastructure construction activities providing the society with common production conditions in monetary items, including water industry, railway transportation, highway transportation, pipeline transportation, air transportation, postal communication, public service, environmental protection, etc in the tertiary industry. The unit of measurement is hundred million yuan.

以货币表现的向社会提供作为共同生产条件的基础设施建设活动的工作量。包括第二产业中的电力、煤气及水的生产和供应业;第三产业中的水利业、铁路运输、公路运输、管道运输、航空运输、邮电通信、公共服务和环境保护等。计量单位:亿元。

investment volume of fixed assets 固定资产投资额

They refer to the workload of constructing or purchasing fixed assets which are manifested by currency and it is a comprehensive index to reflect the investment scale, rate, proportional relation and direction for use in terms of economic type. The investment in fixed assets of the whole society can be categorized as state-owned investment, collective investment, individual investment, joint operation, joint-stock system, foreign business, Hong Kong, Macao, Taiwan investment or others. According to the Channel of Management, the investment of fixed assets of the whole society can be divided into capital construction, transformation and renovation, investment in real estate development and investment in other fixed assets. The unit of measurement is hundred million yuan.

以货币表现的建造和购置固定资产活动的工作量,它是反映固定资产投资规模、速度、比例关系和使用方向的综合性指标。全社会固定资产投资按经济类型可分为国有、集体、个体、联营、股份制、外商、港澳台商、其他等。按照管理渠道,全社会固定资产投资总额分为基本建设、更新改造、房地产开发投资和其他固定资产投资四个部分。计量单位:亿元。

island 海岛

It refers to a naturally formed land area that is surrounded by water and remains above water when the high tide comes. It includes inhabited islands and uninhabited islands.

四面环(海)水并在高潮时高于水面的自然形成的陆地区域,包括有居民海岛和无居民海岛。

island arc 岛弧

It is a type of arc-shaped archipelago situated between the mainland and the ocean basin.

大陆与海洋盆地之间呈弧形分布的群岛。

island chain 岛链

It is a generic term of island arcs in a chain-like conformation.

对链状排列的诸岛弧之总称。

island harbor 岛港

It is the harbour situated between the sea-coast and islands.

岛屿沿岸或与岛屿之间的港口。

island protection 海岛保护

It refers to the protection of an island and the ecological system of its surrounding seas, the protection of uninhabited islands and the protection of islands for special use.

海岛及其周边海域生态系统保护、无居民海岛自然资源保护和特殊用途海岛保护。

island protection special fund 海岛保护专项资金

It refers to a special fund used for island protection, ecological restoration and scientific research activities.

用于海岛的保护、生态修复和科学研究活动的专项基金。

island statistics investigation system 海岛统计调查制度

It is the groundwork for the development, protection and management of an island. Knowing and grasping the resources, environment, economy and management of islands of our country in a timely and accurate way can help to provide guidance on accelerating the development and construction to islands, and strengthen the management of them. Besides, it is beneficial for the government and administrative departments at all levels to provide macro-control and strategic guidance to the development of islands and oceans; and it is also beneficial for the sustainable exploitation and utilization of islands in our country and the protection of them.

海岛统计调查是海岛开发、保护与管理的基础性工作,及时、准确了解和掌握我国海岛的资源、环境、经济、管理等情况,是加速海岛开发建设、加强海岛管理的向导,有利于各级政府、管理部门对我国海岛和海洋发展的宏观调控和战略指导,有利于我国海岛的可持续开发利用和保护。

islands for special use 特殊用途海岛

They include an island wherein a territorial sea base point is located, an island for national defense, an island in a nature reserve, etc.

包括领海基点所在的海岛、国防用途海岛、自然保护区内的海岛等。

isobathymetric line 等深线

It refers to a curve linking all points of equal depth, which can reflect the fluctuation of the ocean floor. The basic isobathymetric lines are: 0 metres, 5 metres, 10 metres, 20 metres, 30 metres, 50 metres, 100 metres, 200 metres, 500 metres, 1000 metres, 2000 metres, etc.

深度相等的各点连成的曲线,用以显示海底表面的起伏。基本等深线为:0米、2米、5米、10米、20米、30米、50米、100米、200米、500米、1000米、2000米等。

isohaline 等盐线

It refers to a line linking points of equal salinity on a map of salinity distribution.

在盐度分布图上,表征盐度相等各点的连线。

isotherm 等温线

It refers to a line linking points of equal temperature on a map of ocean water temperature distribution.

在海洋水温分布图上,水温相等各点的连线。

J

Japan Sea 日本海

It is a marginal sea in the northwest of the Pacific Ocean located between the Japanese archipelago and the Asian continent.

位于日本群岛和亚洲大陆之间,太平洋西北部的边缘海。

jetty 突堤

It is a general term for a breakwater or dock artificially built with one end connected with the coast and the other end projecting from the land out into the sea. When the waves are mainly from the open sea on one side and the other side has a natural landform for shelter, a single jetty is generally used; when the coast is open without a natural shelter, the form of two jetties getting encircling, also called *a double-jetty form*, is mostly used, to shape port water area with good sheltering.

人工构筑的一端与岸相连,另一端伸向海中防波堤或码头的统称为突堤。当波浪主要来自外海一侧,另一侧由天然地形掩护时,一般采用单突堤;在没有天然掩护的开敞海岸,大多采用两条突堤合围的形式,又称双突堤,以形成掩护良好的港口水域。

Joint Global Ocean Flux Study (JGOFS) 全球联合海洋通量研究

Presided by the Scientific Committee of Oceanic Research between 1990 and 2004, JGOFS is an international ocean science research program which studies the atmosphere, ocean surface and interior ocean, quarterly and interannual fluxes of carbon and its influences on climate changes.

1990—2004 年由海洋研究科学委员会主持的,研究大气、大洋表层和大洋内部区域,季度至年际碳通量及其对气候变化影响的国际海洋科学研究计划。

joint implemented(JI) 联合履行机制

It is one of the three flexibility mechanisms set out in the Kyoto Protocol to help countries with binding greenhouse gas emissions targets (so-called Annex I countries) meet their obligations. JI is described in Article 6 of the Kyoto Protocol. Under Article 6, any Annex I country can invest in an emission reduction project (referred to as a "Joint Implementation Project") in any other Annex I country as an alternative to reducing emissions domestically. In this way countries can lower the costs of complying with their Kyoto targets by investing in projects that reduce greenhouse gas emissions in an Annex I country where reducing emissions may be cheaper, and then using the resulting Emission Reduction Units (ERUs) towards their commitment goal.

发达国家之间通过项目级合作,所实现的温室气体减排抵消额实现温室气体减排抵消,可以转让给另一发达国家缔约方,但是同时必须在转让方的允许排放限额上扣减相应的额度。

joint production 合作生产

It refers to production activity by enterprises in various areas to work together for products, that is, cooperative enterprises take on the production of some products or part of the operation in a general project and finally jointly complete the entire project.

不同地区的企业共同完成某项产品的生产活动,通常表现为合作企业各自承担总项目中部分产品或部分工序的生产,最后共同完成全部项目。

K

Karst 喀斯特

It is a geological phenomenon which is formed by the water dissolution of soluble rocks.

水对可溶岩的溶蚀作用所产生的地质现象。

key reserves 重点保护区

They include areas involving national maritime rights and interests and national defense security, such as territorial sea baseline points, military uses, etc., areas for rare and endangered marine species, economic species and their habitats, and areas where natural landscapes, natural ecosystems and historical sites with certain representativeness, typicality and special protection values are taken as the main targets of protection.

包括领海基点、军事用途等涉及国家海洋权益和国防安全的区域，珍稀濒危海洋生物物种、经济生物物种及其栖息地，以及将具有一定代表性、典型性和特殊保护价值的自然景观、自然生态系统和历史遗迹作为主要保护对象的区域。

keystone species 关键种

It is a species that is critical or crucial in the food web.

食物网中处于关键环节起到控制作用的物种。

Kombu (*Laminaria japonica*) 昆布

Also called laminaria japonica, it is an edible kelp from the Laminariaceae of Heterokontophyta and is widely eaten in East Asia. Its frond is brown, flat and belt-like with a maximum length of 7 meters. The tree branch-shaped holdfast at its base can attach it to the seafloor rocks or the bottom parts of vessels in a low-temperature ocean.

又称"海带"，褐藻门，海带科。藻体褐色，扁平呈带状，最长可达 7 米，基部有固着器树状分枝，用以附着海底岩石，在船舶船底也常见附着，生长于水温较低的海中。

Korea Strait 朝鲜海峡

Located between the southeast of Korea Peninsular and Tsushima Island, it is an important passage connecting the Yellow Sea with the Sea of Japan.

位于朝鲜半岛东南与对马岛之间，沟通日本海和黄海的重要通道。

Kyoto Protocol《京都议定书》

It is an additional agreement of the UN climate convention, which was approved at the third meeting of the Conference of the Parties on *Framework Convention on Climate Change* which was held in Kyoto of Japan in December 1997 and came into effect on 16th, February 2005. The protocol sets standards on carbon emissions for all countries, namely: from 2008 to 2012, the carbon emissions of most industrial countries should be reduced by 5.2% compared with that in 1990 on average.

1997年12月在日本京都举办的《气候框架公约》第3次缔约方大会上通过的联合国气候公约的附加协议，2005年2月16日开始强制生效。该议定书为各国的二氧化碳排放量规定了标准,即:在2008年至2012年间,全球主要工业国家的工业二氧化碳排放量比1990年的排放量平均要低5.2%。

L

labor-force participation rate 劳动参工率

It refers to the percentage of labour force in an adult population.

劳动力占成年人口的百分比。

labor force 劳动力

It refers to the population with labour ability, including both the employed and the unemployed.

既包括就业者又包括失业者的具有劳动能力的人口。

labor-intensive industry 劳动密集型产业

Contrary to a technology-intensive industry, it is an industry in which labour input is higher in the percentage of production factors allocation. The main characteristic is that the consumption of materialized labour is lower than that of living labour, the scientific and technological content and additional value of product is low. With the improvement of science and technology and organic composition of capital, a labour-intensive industry will be gradually replaced by a capital-intensive industry.

与"技术密集型产业"相对,在生产要素的配置比例中,劳动力投入比重较高的产业。基本特点是:物化劳动消耗比重较低而活劳动消耗比重较高,产品的科技含量和附加值低。随着科学技术的进步和资本有机构成的提高,劳动密集型产业将逐渐被资金技术密集型产业所取代。

lagoon 潟湖

It is a body of water separated from the sea by elongated, shallow sandy deposits (barrier islands or sand spits). A lagoon often has one or more narrow tidal channels connected to the sea.

以狭长、低平的沙质堆积体(堡岛或沙嘴)与海分隔的水域称为潟湖。潟湖常有一个或一个以上狭窄的潮汐通道与大海相通。

land 土地

It is the solid surface of the earth that is not permanetly covered by water, including all attributes of the biosphere under and above it in vertical directions, and it is a substance system formed by near surface climate, geomorphology, surface geology, hydrology, soil, animals and plants, as well as the interactive of results between past and current human activities.

地球陆地表面具有一定范围的地段,包括垂直于它上下的生物圈的所有属性,是由近地表气候、地貌、表层地质、水文、土壤、动植物以及过去和现在人类活动的结果相互作用而形成的物质系统。

land area of development zone 开发区土地面积

It covers a land area of development zones of different kinds and levels in coastal cities. Development zones in coastal cities refer to development areas of various types which are established on ports or along coastal lines, including economic (technical) development zones, bonded areas, Hi-Tech Industrial Parks, border economic cooperation zones, etc. The unit of measurement is square kilometres.

沿海城市各级各类开发区的土地面积。沿海城市开发区是指沿海城市依托港口或海岸线建设的各类开发区,包括经济(技术)开发区、保税区、高新技术产业园区、边境经济合作区等,下同。计量单位:平方千米。

land carrying capacity 土地承载力

It is the ratio of land resource productivity under certain conditions and per capita consumption standards under a certain living standard.

在一定条件下土地资源生产力与一定生活水平下的人均消费标准之比。

land potential 土地潜力

It is the capacity of a land unit to provide sustainable benefits for agriculture, forestry, animal husbandry, tourism, etc., depending on the limitation of natural elements at a given operation and management level.

在一定经营管理水平下,由自然要素的限制性所决定的,某一土地单元对农业、林业、牧业和旅游业等几种土地利用大类提供持续效益的能力。

land quality 土地质量

It reflects how well land functions satisfy human needs.

土地功能满足人类需要的优劣程度。

land resources 土地资源

They are areas of land for human use in current and foreseeable technical economic conditions.

在当前和可预见未来的技术经济条件下,可为人类利用的土地。

land subsidence 地面沉降

It is the motion of the earth's surface shifting downward as the soil layer consolidates and densifies further, due to the fact that the excessive groundwater (or oil or gas) on a large scale results in the descending of the level of water (oil, atmospheric pressure).

由于大范围过量抽汲地下水(或油、气)引起水位(或油、气压)下降,土层进一步固结压密而造成的地面向下沉落。

land-based aquaculture 陆基养殖

It refers to aquaculture facilities and mode constructed on land.

以陆地为基础建造的养殖设施和养殖模式。

landform 地貌

It refers to the various forms of the earth surface (including the seabed), caused by the interaction of the endogenic forces and exogenic forces.

地球表面(包括海底)的各种形态。由内营力和外营力相互作用而形成。

landlocked state 内陆国

It is a country which is entirely enclosed by land territory of other countries and therefore, has no coastal outlet.

被其他国家的陆地领土所包围,因而没有出海口的国家。

landslide 滑坡

Also known as a landslip, it is a geological phenomenon that includes a wide range of ground movements, such as rock falls, deep failure of slopes and shallow debris flows. Landslides can occur in offshore, coastal and onshore environments. Although the action of gravity is the primary driving force for a landslide to occur, there are other contributing factors affecting the original slope stability, including both natural disaster and human activities.

斜坡上的部分岩(土)体在自然或人为因素的影响下,沿某一明显的界面发生剪切坡滑向坡下运动的现象。

land-sourced pollutants 陆源污染源

They refer to places or facilities where pollutants are emitted from land to sea, cause or may cause marine environmental pollution.

从陆地向海域排放污染物,造成或者可能造成海洋环境污染的场所、设施等。

La Niña 拉尼娜

It is a coupled ocean-atmosphere phenomenon that is the counterpart of El Niño. During a period of La Niña, the sea surface temperature across the equatorial Eastern Central Pacific Ocean will be lower than normal by 3-5℃.

与厄尔尼诺相反的现象,即赤道东太平洋海温较常年偏低。

large department system 大部门体制

It refers to a system in which the departments of similar businesses and comparable functions can be merged and centralized, to be managed by a larger department, thus creating a service-oriented government from the aspects of system, organizational structure and personnel; and integrating and streamlining government system structure.

又称"大部分体制",是指在政府的部门设置中,把业务相似、职能相近的部门进行合并,相对集中,由一个大部门进行管理,从体制、机构与人员方面落实服务型政府的转变,整合和梳理体制结构。

large marine ecosystem (LME) 大海洋生态系统

It refers to a marine ecosystem with an area of over 200,000 square kilometers, characterized by its distinct bathymetry, hydrography and productivity. The wide coastal areas close to the continent are fit for the reproduction, growth and feeding of the marine populations.

面积超过 200 000 平方千米,具有独特水文、海底地形、生产力,以及适于种群繁殖、生长和取食的接近大陆宽广海域的生态系统。

law and regulation of sea 海洋法规

It is a law and regulation made by legislature and governments for marine management.

立法机关和政府制定的管理海洋的法律和规章制度。

Law of One Price 一价定律

Also know as the purchasing power parity (PPT), it is an economic concept which posits that "a good must sell for the same price in all locations".

The purchasing power parity (PPT)

is an equivalent coefficient between currencies calculated based on different price levels in different countries, which allow us to make a reasonable comparison among gross domestic products of different countries in economics. There is a large deviation between theoretical exchange rate and real exchange rate.

又称"购买力评价",经济学的一个著名假说,该定律说明同样的产品在同一时间在不同地点不能以不同的价格出售。购买力平价是一种根据各国不同的价格水平计算出来的货币之间的等值系数,使我们能够在经济学上对各国的国内生产总值进行合理比较,这种理论汇率与实际汇率可能有很大的差距。

law of property 物权法

It, as an important part of the civil law, is the sum total of laws and regulations which regulate people (natural person, legal person, other organization, or country in special circumstances)'s dominance relationship over property.

民法的重要组成部分,调整人(自然人,法人,其他组织,特殊情况下可以是国家)对于物的支配关系的法律规范的总和。

Law of the People's Republic of China on Evaluation of Environmental Effects《中华人民共和国环境影响评价法》

Short for Environmental Protection Law, it was formulated in order to focus on the environmental impact, control pollution, protect the ecological environment and take timely measures to reduce potential trouble fundamentally, globally and from the perspective of source of development. Its significance lies in the finding of a more reasonable and better environmental management mechanism and mobilizing all aspects of social force, thus forming a new mechanism to protect environment together which involve government approval, unified supervision and management of the environmental protection administration, accountability taken by concerned departments for the environmental impact produced by planning, and public participation.

简称《环评法》,是为了从根本上、全局上和发展的源头上注重环境影响、控制污染、保护生态环境,及时采取措施,减少后患。规划环境影响评价最重要的意义,就是找到了一种比较合理的环境管理机制,充分调动社会各方面的力量,可以形成政府审批,环境保护行政主管部门统一监督管理,有关部门对规划产生的环境影响负责,公众参与,共同保护环境的新机制。

Law of the People's Republic of china on the Administration of the Use of Sea Areas《中华人民共和国海域使用管理法》

It is a comprehensive law for regulating in the exclusive activities relating to the use of sea areas within the sea surface, water volume, seabed and subsoil of the inland waters and territorial seas of the People's Republic of China. Adopted at the 24th Meeting of the Standing Committee of the Ninth National People's Congress of the

PRC on October 27, 2001, it came into effect as of January 1, 2002.

规范在中国内海和领海的水面、水体、海床和底土从事排他性用海活动的综合性法律。2001年10月27日中华人民共和国第九届全国人民代表大会常务委员会第二十四次会议通过,2002年1月1日起施行。

Law of the People's Republic of China on the Protection of Offshore Islands《中华人民共和国海岛保护法》

It was enacted with a view to protecting the ecosystems of islands and their surrounding waters, reasonably developing and utilizing the natural resources of islands, safeguarding national marine rights and interests, and promoting sustainable economic and social development. Adopted at the 12th Meeting of the Standing Committee of the 11th National People's Congress of the PRC on December 26, 2009, it came into effect as of March 1, 2010.

为了保护海岛及其周边海域生态系统,合理开发利用海岛自然资源,维护国家海洋权益,促进经济社会可持续发展而制定的法规。2009年12月26日中华人民共和国第十一届全国人民代表大会常务委员会第十二次会议通过,2010年3月1日起施行。

Law of the People's Republic of China on the Territorial Sea and the Contiguous Zone《中华人民共和国领海及毗连区法》

It is a law for regulating all activities in China's territorial sea and the contiguous zone. Adopted at the 24th Meeting of the Standing Committee of the Seventh National People's Congress on February 25, 1992, it was promulgated by Order No. 55 of the President of the People's Republic of China on February 25, 1992, and took effect as of the date of promulgation.

规范和调整在中国领海和毗连区从事一切活动的法律。1992年2月25日第七届全国人民代表大会常务委员会第二十四次会议通过,1992年2月25日中华人民共和国主席令第五十五号公布施行。

law of the sea 海洋法

It is a body of law establishing the legal status of marine areas and jurisdiction of each nation over the marine area, which is a legal marine document concerning navigation, resources exploitation, scientific research, environmental protection, marine management, etc.

确立各类海洋区域的法律地位,及各国在各类海洋区域内的权限。是从事航行、资源开发、科学研究、环境保护、海洋管理等方面具有法律性的海洋文件。

Law on the Exclusive Economic Zone and the Continental Shelf of the People's Republic of China《中华人民共和国专属经济区和大陆架法》

It is a law for regulating in all activities in China's exclusive economic zone and on its continental shelf. Adopted at the 3rd Meeting of the Standing Committee of the Ninth National People's Congress of the PRC on June 26, 1998, it was effective as of the date of promulgation.

规范和调整在中国专属经济区和大陆架上从事一切活动的法律。1998年6月26日中华人民共和国第九届全国人民代表大会常务委员会第三次会议通过并施行。

laws and regulations for marine environmental protection 海洋环境保护法规

It is a generic term for international laws, conventions and regulations for protecting the marine environment from pollution.

以保护海洋环境不受污染为目的的国际法、公约和规则的统称。

layout orientation 布局指向

It refers to an industrial layout under the joint action of various factors and a layout mechanism, which generally reflects the tendency of a particular area.

在各种因素和布局机制共同作用下的产业布局,往往反映出对某一类地域的倾向。

leading industry 主导产业

It refers to an industrial sector which plays a leading role in a region with a relatively high degree of specialization and a longer industrial chain.

一个区域有带动作用的、专业化程度比较高的产业部门,具有较长的产业链条。

legal system management of resources 资源法制管理

It is to use legal means to adjust the various irregularities that may occur in the process of human development, utilisation, protection and management of resources.

用法制的手段调整人类在资源开发、利用、保护和管理过程中可能发生的各种不规范行为。

length of berth 码头泊位长度

It refers to the actual length of an area which is used for berthing ships to load and unload cargoes and passengers, including the length of berths of various fixed and floating quays. The unit of measurement is metres.

用于停系靠船舶,进行装卸货物和上下旅客地段的实际长度。包括固定和浮动的各种形式码头的泊位长度。计量单位:米。

length of nature coastline 自然岸线长度

It is the length of an undeveloped coastline formed by the interaction of ocean and land. The unit of measurement is kilometres.

未经开发的由海陆相互作用形成的岸线的长度。计量单位:千米。

levee 堤

It refers to the water retaining structure built along the banks of rivers, lakes and seas. The structures built on both sides of the rivers can be called "river embankments" or "river levees"; the structures built on the coast of the sea can be called "sea dikes" or "seawalls".

沿江、河、湖、海的边岸修建的挡水建筑物,建在江、河两岸的,称"江堤"或"河堤";建在海边的,称"海堤"或"海塘"。

Liao River 辽河

It is a principal river of two branches

in northeast China. The eastern branch (Dong Liao He) originates in Saha ridge of Jilin Province and the western branch (Xi Liao He) derives from Zipen Mountain. The two branches meet in Changtu county and flow to the Bohai Sea through Yingkou. It is 1430 kilometres long with a basin area of 164 thousand square kilometres.

中国东北地区南部大河。有东西两源,东辽河源出吉林省萨哈岭,西辽河源出辽宁省自盆山,在昌图县汇合,经营口入渤海。长1430千米,流域面积16.4万平方千米。

license of sea area use 海域使用权证

It refers to a permit for the use of a state-owned sea area issued to any entity or individual through legal procedures by the State Council or local people's government. In other words, when the application for the use of sea areas is approved or obtained through tendering or auction, any entity or individual should go through registration formalities and be registered by the State Council or local people's government in accordance with laws and regulations.

单位或者个人依法取得对国家所有的特定海域使用权的法律凭证,即单位或个人使用海域的申请被批准或者通过招标、拍卖方式取得海域使用权后,应当办理登记手续,由国务院或地方人民政府依照法律规定登记注册,并向海域使用权人颁发海域使用权证书。

license system for fishing 捕捞许可证制度

It is clearly stipulated by the *Fisheries Law of the People's Republic of China* and other laws and regulations. It is the obligation and legal right of fishing businesses and individuals for applying for fishing permits to environmental protection administration departments and fish according to the stipulation of the fishing permits.

捕捞许可证制度是《中华人民共和国渔业法》等法律法规明确规定的,捕捞单位或个人向环境保护行政主管部门申请领取捕捞许可证,并按照许可证的规定进行捕捞,是法律规定的义务和权利。

liman coast 溺谷型海岸

It refers to an estuary coast formed by the submergence of a coastal valley when crust descends or a sea level rises, where a traverse sandbank, sandspit or rear lagoon tend to come into being, typically represented by the north share of the Black Sea.

地壳下降或海平面上升使海水淹没沿海谷地而形成的河口湾型海岸,在湾口处常形成横贯的沙坝、沙嘴和后方的潟湖。以黑海北岸最典型。

linearity 线性

It indicates a proportional relationship between two variables, which can be graphically represented as a straight line in rectangular coordinates. It stands for regular and smooth motion in space and time. In a linear system, the behaviour or property of the whole is equal to the sum of the parts.

量与量之间成比例关系,用直角坐标形象地画出来是一条直线。在空间与时间上代表规则与光滑的运动。在线性系统中,整体的行为或性质等于部分之和。

liquefaction of sand 砂土液化

It describes a process and phenomenon whereby saturated sand substantially loses shear strength and changes from a solid into a liquid state.

饱和砂土的抗剪强度趋于零,由固体状态转化为液体状态的过程和现象。

liquidity 流动性

It refers to a measure of the extent to which a person or organization has cash to meet immediate and short-term obligations, or assets that can be quickly converted to do this.

一种资产兑换为经济中交换媒介的容易程度。

littoral current 沿岸流

It is a current flowing along a coast with no direct relation with the waves.

沿海岸流动的与海浪无直接关系的海流。

littoral placer 滨海砂矿

It refers to minerals formed by heavy minerals when they concentrate in loose seabed sediments under the action of inland rivers, waves, tides, tidal currents and ocean currents in coastal areas. It includes two types, one is a metal placer, the other is a non-metal placer.

又称"海底砂矿"。滨海地带内流河、波浪、潮汐、潮流和海流等作用,使重矿物富集于海底松散沉积物中而形成的矿产,包括海滨金属砂矿和非金属砂矿两种类型。

littoral sanatorium 滨海疗养院

It refers to medical facilities established in coastal areas for long-term or choric illnesses, focusing more on rehabilitation than on treatment.

滨海地区建立的以疗养、康复为主,以治疗为辅的医疗机构。

littoral sediment 近岸沉积

It refers to the offshore sediment of which the water depth is between 0 and 20 metres.

又称"滨海沉积",指近岸水深 0~20 米范围内的沉积。

littoral wetland 滨海湿地

It refers to waters and its coastal wet zones with a depth of less than six metres at low tide mark. It includes permanent waters with a depth of less than six meters, intertidal zones (or flooding zones), coastal lowland, etc.

低潮时水深浅于 6 米的水域及其沿岸浸湿地带,包括水深不超过 6 米的永久性水域、潮间带(或洪泛地带)和沿海低地等。

living resources 生物资源

They refer to genetic resources, organisms, population, biological systems and any components included, which have a practical or potential value and use for human beings.

对人类具有实际的或潜在的价值与用途的遗传资源、生物体、种群或生态系统及其中的任何组分的总称。

locational factor 区位因素

It refers to the aggregation of various influencing factors which determine that engaging economic activities in a given area or several areas of the same kind will obtain more benefits than doing same business activities in other areas.

在特定的地点或在某几个同类地点进行经济活动比在其他地区进行同种经济活动可能获得更多利益的各种影响因素的集合。

longitudinal coast 太平洋型海岸

It is a coast with a general direction of the coastline approximately parallel to that of a geological tectonic line.

又称"纵向海岸",海岸线延伸的总方向与地质构造线的走向近似平行的海岸。

longshore bar 沿岸沙坝

It refers to ridges of sand or gravel deposit formed under the action of coastal currents, which is distributed above a high tide line.

分布在高潮线以上,由沿岸流作用形成的垄状砂质或沙砾质堆积体。

low-carbon city 低碳城市

It refers to a city which advocates a low carbon mode of economic development, constructs green transportation system and building, changes the consumption concept of the residents, and innovates the low-carbon technology in order to achieve a dynamic balance between carbon emission and carbon treatment.

在城市空间经济社会发展过程中,倡导低碳经济发展模式,实施绿色交通和建筑,转变居民消费观念,创新低碳技术,从而实现碳排放与碳处理动态平衡的城市。

low-carbon concept stock 低碳概念股

It refers to the stock of listed company whose business is energy conservation and environmental protection at the stock market.

证券市场里以节能环保为题材的上市公司。

low-carbon economy 低碳经济

It is a global economic development model used to reduce carbon emissions. It was initially found in the UK Energy White Paper, *Our Energy Future-Creating a Low Carbon Economy*, which pointed out that the global temperature rise should be maintained not to exceed 2℃. To achieve this, the global greenhouse gas emission which will peak in the future 10 to 15 years, should be cut in half in 2050. Therefore, a low-carbon economy model should be established, and the essence of which is to increase energy efficiency and improve the energy structure; the core of which is the innovation of energy technology and policy.

减少碳化物排放的全球经济发展模式。最早见诸政府文件是在2003年的英国能源白皮书《我们能源的未来:创建低碳经济》,白皮书提出:努力维持全球温度升高不超过2℃。这就要求全球温室气体排放量在未来10~15年内达到峰值,到2050年则削减一半,为此,需要建立低碳经济模式,其本质是提高能源效率和改善能源结构,核心是能源技术创新和政策创新。

low-carbon energy 低碳能源

It refers to energy of high efficiency, low consumption, low pollution and low carbon emission, which includes nuclear energy, clean coal, and renewable energy sources, among which renewable energy sources includes solar energy, wind energy, hydroenergy, marine energy, geothermal energy, biomass energy, etc.

高能效、低能耗、低污染、低碳排放的能源,包括可再生能源、核能和清洁煤,其中可再生能源包括太阳能、风能、水能、海洋能、地热能及生物质能等。

low-carbon production 低碳产业

It is based on low energy consumption and low pollution, and it is the carrier of the development of the low-carbon economy. Its development will stimulate the transformation development of the current high-carbon industry, create new opportunities for industrial development, form a new economic growth point and promote rapid economic development. The core of low-carbon production development is to adjust the industrial structure, including not only the structural adjustment of the large industrial system which comprises industry, agriculture, and service industry, but also the internal structural adjustment of the three main industries.

以低能耗、低污染为基础的产业,是低碳经济发展的载体。其发展将带动现有高碳产业的转型发展,催生新的产业发展机会,形成新的经济增长点,促进经济高速发展。发展低碳产业的核心是调整产业结构,不仅包括工业、农业、服务业构成的大产业体系的结构调整,还包括三大产业内部的结构调整。

low-carbon society 低碳社会

It refers to a society of low carbon emissions and a balanced ecological system. In this society, human behaviour is more environmental-friendly, and people can live in harmony with nature.

一个碳排放量低、生态系统平衡、人类的行为方式更加环保、人与自然和谐相处的社会。

low-carbon technology 低碳技术

It is an important symbol of the core power of a nation's core competitiveness and it is the fundamental solution to the ever-growing problem of the ecological environment and energy resources. Low-carbon technology covers many fields like oil, chemical engineering, electric power, transportation, architecture, metallurgy, all so on including the clean and efficient utilization of coal, the high-value conversion of oil and gas resources and coal-bed gas, development of renewable energy sources and new energy, energy conservation transformation of the traditional technology, and the utilization and sequestration of CO_2.

低碳技术是国家核心竞争力的一个重要标志,是解决日益严重的生态环境和资源能源问题的根本出路,低碳技术广泛涉及石油、化工、电力、交通、建筑、冶金等多个领域,包括煤的清洁高效利用、油气资源和煤层气的高附加值转化、可再生能源与新能源开发、传统技术的节能改造以及二氧化碳捕获、利用与封存等。

low coast 低平海岸

Developed by the coastal plain, the flat coast is a coast where the land descends gradually into the sea and the waves are not strong.

沿海平原或沉溺陆地发育的,岸坡低缓、波浪作用轻微的海岸。

low-energy coast 低能海岸

It refers to the coast which is protected from the action of strong waves by coastal cape. The average height of its breaking wave is less than 10 centimetres.

受海岬等保护而免受强浪作用,平均破波高小于10厘米的海岸。

low tidal highland 低潮高地

It refers to a piece of naturally formed land surrounded by water that is exposed at low tide but submerged at high tide.

在低潮时四面环水并高于水面但在高潮时没入水中的自然形成的陆地。

low water line 低潮线

It refers to a line occurring on the coast or the beach where the low tide water surface intersects with the coastline when the sea water recedes to the low tide water level.

海水后退至低水位时,低潮水面与岸边相接处沿岸或海滩上出现的痕迹线。

lowest groundwater level 地下水最低水位

It refers to the minimum value of groundwater level in a certain observation time period.

在某一观测时段内,地下水水位的最低值。

M

magnesium chloride manufacturing 氯化镁制造

It is a production activity of extracting magnesium chloride from bittern and subsurface brine.

以苦卤、地下卤水为原料提取的氯化镁的生产活动。

magnesium fertilizer manufacturing 镁肥制造

It refers to the process of manufacturing magnesium by utilizing marine chemical products such as magnesium chloride.

利用氯化镁等海洋化工产品制造镁肥的生产活动。

main business cost 主营业务成本

It is the actual cost arising from the principal business of enterprises. The unit of measurement is ten thousand yuan.

企业经营主要业务发生的实际成本。计量单位:万元。

main business income 主营业务收入

It refers to the total income as a result of daily activities such as the sale of goods, provision of services by enterprises. The unit of measurement is ten thousand yuan.

企业在销售商品、提供劳务等日常活动中所产生的收入总额。计量单位:万元。

main business tax and extra charges 主营业务税金及附加

They refer to taxes levied on the principal business of enterprises, including business tax, consumption tax, city maintenance and construction tax, resource tax, land value-added tax and education surcharges. The unit of measurement is ten thousand yuan.

企业经营主要业务应负担的营业税、消费税、城市维护建设税、资源税、土地增值税、教育费附加。计量单位:万元。

main stream 干流

Also known as a trunk stream, it refers to a stream or a river which is formed by the confluence of several tributaries and flows directly into the sea or inland lakes.

又称"主流",直接流入海洋或内陆湖泊的河流,通常由若干条支流汇集而成。

malacology 软体动物学

Malacology is a discipline which deals with the study of classification, morphology, reproduction, development, ecology, physiology, biochemistry, geographic distribution of mollusk and their relationship with human beings.

研究软体动物的分类、形态、繁殖、发育、生态、生理、生化、地理分布及其与人类关系的学科。

manganese nodule 锰结核

Also called *polymetallic nodule*, polymetallic nodule refers to the nodule-like manganese, iron oxide and oxide deposits formed in the sediments on ocean floors.

又称"多金属结核",自生于海底表层沉积物中的呈结核状的铁锰氢氧化物和氧化物矿床。

management for marine public safety 海洋公共安全管理

It refers to the management by police working for marine inspections, harbours, customs, etc.

海上稽查、海港、海关等人民警察的管理活动。

management of coastal sand mining 海滨砂矿业管理

It refers to the management activities for coastal placer mining-related affairs conducted by the social groups of relevant industry.

有关行业社会团体对海滨砂矿业相关事务的管理活动。

management of marine disaster effect 海洋灾害管理

It refers to the work concerning disaster investigation, assessment, disaster statistical statement and disaster release about marine disaster.

对灾情调查、评估,灾情统计、信息与报告,灾情发布等的全面管理。

management of marine lighthouse and navigation aid 海上灯塔航标管理

It refers to maintenance and management activities held to ensure the normal use of a marine lighthouse and navigation aid.

为保证海上灯塔航标正常使用进行的维护、管理活动。

management of marine oil and gas 海洋油气业管理

It refers to the management of transactions of marine oil and gas by the relevant industry authority.

有关行业主管机构对海洋油气事务的管理活动。

management of marine shipping industry 海洋船舶工业管理

It refers to the management of marine shipping industries by the relevant industry authority.

有关行业主管机构对海洋船舶工业经营的管理活动。

management of marine transportation 海洋交通运输业管理

It refers to management activities for marine transportation by government departments at all levels.

各级政府部门对海洋运输相关事务的管理活动。

management of marine waste disposal 海洋倾废治理

It refers to the disposing and treatment of the sea waters for accepting waste within a special sea area.

对用于接纳废弃物的特殊海域的海水进行处理、处置的活动。

management of sea salt and marine chemistry industry 海盐及海洋化工业管理

It refers to the management activities of sea salt and marine chemistry industries by the relevant industry authority.

有关行业主管机构对海盐及海洋化工经营的管理活动。

management on sea area use 海域使用管理

It refers to any controlled activity by the country on exclusive use of a specific sea area continuously over three months to ensure scientific and reasonable utilization of sea area resources for the sake of protecting marine resources and ecological environment.

国家为了保护海洋资源和生态环境,确保海域资源的科学、合理利用而对持续使用特定海域三个月以上的排他性用海活动所采取的控制行为。

mangrove 红树林

It refers to the saline marsh community composed of evergreen trees and shrubs represented by mangrove plants, growing on the silt beach in the tropics and subtropics.

热带、亚热带淤泥质海滩,以红树科植物为代表的常绿乔灌木组成的盐生沼泽群落。

mangrove coast 红树林海岸

It refers to the tropical and subtropical biological coasts formed by mangroves. Mangrove lives on muddy coasts and estuarine tidal flats with an abundance of organic matters, whose crowns float above sea level and roots are exposed in low tide. The dense mangrove helps slow the tidal speed, reduce wave energy, promote silt disposition, obstruct against silt into the port and reduce silting in the port. Mangrove coasts are widely distributed in mid or low latitude coastal countries in Southeast Asia, American continent, Africa, and others while mangrove coasts in China are widespread in coastal areas of Hainan, Guangdong, Guangxi, Fujian and Taiwan.

热带和亚热带地区由红树林组成的生物海岸称为红树林海岸。红树林生长在富含有机质的淤泥质海岸和河口浅滩上,高潮时树冠飘荡在水面上,低潮时露出根部。茂密的红树林有降低潮流流速,削减波浪能量和促使泥沙淤积的作用,同时也有阻挡泥沙进入港口,减少港口淤积的效果。红树林海岸广泛分布于东南亚及美洲、非洲等中低纬度的沿海国家,中国的红树林沿岸主要分布在海南、广东、广西、福建和台湾沿海地区。

man-land relationship 人-地关系

It is an abbreviation for the relationship between human and geographical environment, which is typically defined as the relationship between human society and activities and the natural environment, and is broadly explained as the relationship between the survival and development of human society or human activities and the geographical environment.

人们对人类与地理环境之间关系的一种简称。对它的经典解释是人类社会及其活动与自然环境之间的关系,非经典解释认为人-地关系是指人类社会生存与发展或人类活动与地理环境(广义的)的关系。

man-sea relationship 人-海关系

A man-sea relationship refers to the mutually interactive relationship between human activities and seas (various factor structures such as resources, environment, disasters, etc.). It is a natural part of a man-land relationship, which reflects the effect and influence of seas on human society, expresses the understanding of the ocean and highlights the mutual response and feedback to the interaction of man and sea.

人类活动与海洋(资源、环境、灾害等各种要素结构)之间的互感互动关系。它是人地关系的天然组成部分,一方面反映海洋对人类社会的影响与作用;另一方面表达了人类对海洋的认识与把握,突出人-海相互作用过程中的彼此响应和反馈。

manufacture of marine amusement rides 海上游乐设备制造

It refers to the manufacture of marine amusement rides and amusement equipments.

海上游乐设备和游艺器材的制造。

manufacture of marine oil storage devices 海洋石油储油装置制造

It is the manufacturing of special metal containers for marine liquefied petroleum gas.

海洋石油液化气等专用金属容器的制造。

manufacture of submarine cable 海底电缆制造

It refers to the manufacture of multi-purpose submarine cables used in electricity distribution, power transmission, communication, lighting, etc.

用于电力输配、电能传送、通信传播及照明等各种用途的海底电缆的制造。

manufacture of submarine fiber cable 海底光缆制造

It refers to the manufacture of a submarine fiber cable which can convert electricity signals into light signals in order to transmit sound, words, images and other information.

将电的信号变成光的信号,进行声音、文字、图像等信息传输的海底光缆的制造。

manufacture of submarine transmission pipeline 海底运输管道制造

It refers to the manufacturing activities of submarine pipelines to transport oil, natural gas and other fluid products.

用于输送石油、天然气及其他流体产品的海底管道的生产活动。

manufacturing for marine wave-energy prime mover 海洋波浪能原动机制造

It refers to the manufacturing of a prime mover for generating electricity by using marine wave energy.

利用波浪能源发电的原动机的制造。

manufacturing of machinery for marine aquatic products processing 海洋水产品加工机械制造

It refers to the manufacturing of machinery for fish processing, shrimp processing, shellfish processing, algae and kelp processing, sea blubber processing, etc.

包括鱼类处理及加工机械、虾类加工机械、贝类加工机械、藻类和海带加工机械、海蜇加工机械等的制造。

manufacturing of marine aquaculture drug 海洋水产养殖药品制造

It is the manufacturing of drugs for marine aquaculture.

海洋水产养殖用药品的制造。

manufacturing of marine aquaculture feedstuff 海洋水产养殖饲料制造

It is the manufacturing of feedstuff for marine aquaculture.

海洋水产养殖用饲料的制造。

manufacturing of marine aquaculture feed 海洋水产饲料制造

It is the manufacturing of feed with marine aquatic products such as fishbone, shrimp and shellfish.

用鱼骨、虾、贝等海洋水产品生产饲料的加工活动。

manufacturing of marine biological drugs 海洋生物药品制造

It is the manufacturing of bio-chemical drugs by biotechnological means, using marine animals and plants as the raw materials, such as polysaccharide sulphate.

以海洋动植物为原料,利用生物技术生产生物化学药品的生产活动,如藻酸双酯钠。

manufacturing of marine canned aquatic product 海洋水产品罐头制造

It refers to the manufacture of hard-and-soft-packing cans of marine aquatic products and the manufacture of marine aquatic products aboard a ship.

海洋水产品的硬包装和软包装罐头制造,以及在船舶上从事海洋水产品罐头的加工活动。

manufacturing of marine fishery machinery and accessories 海洋渔业机械与配件制造

It is the manufacture and repair of special equipment and accessories for mariculture, marine fishing, etc.

海水养殖、海洋捕捞等专用设备与配件的制造及修理。

manufacturing of marine fishery navigation equipments 海洋渔业导航设备制造

It refers to the manufacturing of devices for marine fishery navigation, guidance and location.

海洋渔业导航、制导、定位仪器仪表的制造。

manufacturing of marine fishing ropes and nets 海洋渔绳渔网制造

It refers to the manufacturing of fishing ropes, rigging, cables and rope wires, including metal and nonmetal fishing wires, fishing nets, etc.

各种渔用绳、索具、缆绳、合股线的生产活动,包括金属和非金属渔用丝、绳、渔网等。

manufacturing of marine fishing supplies 海上垂钓用品制造

It refers to the manufacturing of various tools or supplies specially used for marine fishing, like the fishing pole, other fishing supplies or auxiliary supplies.

海上垂钓专用的各种用具及用品制造,如钓鱼竿及其他钓鱼用具和辅助用品等。

manufacturing of marine hydrology instruments 海洋水文专用仪器制造

It is the manufacturing of special instruments, apparatus and similar devices for marine hydrology, such as hydrographic measuring instruments.

海洋水文专用仪器和仪表及类似装置的制造,如水道测量仪器等。

manufacturing of marine petrochemical products 海洋石油化工产品制造

It refers to the manufacturing of organic intermediate, hydrocarbon and its halogenated derivatives, etc., using marine petroleum as a raw material.

以海洋石油为原料,制造有机中间体、烃类及其卤化衍生物等的生产活动。

manufacturing of marine proprietary Chinese medicines 海洋中成药制造

It refers to the manufacturing of traditional Chinese medicines directly used for prevention and treatment of human diseases, with marine animals and plants as raw material.

以海洋动植物为原料直接用于人体疾病防治的传统药的制造。

manufacturing of matching equipment for offshore production 海洋石油生产配套设备制造

It is the manufacturing of protection products for marine petroleum platforms, special marine petroleum cables and pipes, etc.

包括海洋石油平台保护用品、海洋石油专用缆索和管材等的制造。

manufacturing of metal packing containers for marine aquatic products 海洋水产品金属包装容器制造

It is the manufacturing of metal packaging containers and accessories for transportation and packaging of marine aquatic products.

主要为海洋水产品运输或包装而制作的金属包装容器及附件的制造。

manufacturing of ocean thermal and salinity energy prime mover 海洋温盐差能原动机制造

It is the manufacturing of prime movers to generate electricity with thermal energy and salinity energy.

利用温差、盐差能源发电的原动机的制造。

manufacturing of offshore oil exploration instrument 海洋石油勘探专用仪器制造

It refers to the manufacturing of special geophysical instruments, apparatus and similar devices for geological exploration, drilling and production of marine petroleum and natural gas, such as offshore oil drilling and logging instruments.

海洋石油和天然气的地质勘探、钻采等地球物理专用仪器、仪表及类似装置的制造,如海洋石油钻探测井仪器等。

manufacturing of refrigeration equipments for marine aquatic products 海洋水产品专用制冷设备制造

It is the manufacturing of refrigeration equipments for marine aquatic products.

海洋水产品冷冻冷藏设备制造。

manufacturing of special equipment for marine mineral exploitation 海洋矿产勘探专用设备制造

It is the manufacturing of special equipment for geological exploration of marine mineral resources, such as geophysical rigs.

海洋矿产资源的地质勘探专用设备的制造,如物探钻机等。

manufacturing of special equipment for offshore oil exploration 海洋石油勘探专用设备制造

It is the manufacturing of special equipment for geological exploration of marine petroleum and natural gas.

海洋石油和天然气的地质勘探专用设备的制造。

manufacturing of special instrument for marine mineral exploitation 海洋矿产勘探专用仪器制造

It is the manufacturing of special instruments, apparatus and similar devices of geophysics for geological exploration, drilling and production, etc. of marine mineral resources, such as petrophysical properties measuring instruments and its accessories.

海洋矿产资源的地质勘探、钻采等地球物理专用仪器、仪表及类似装置的制造,如岩矿物理性质测量仪及配套附件等。

manufacturing of special instruments for marine meteorology 海洋气象专用仪器制造

It refers to the manufacturing of special instruments, apparatus and similar devices for marine meteorology.

海洋气象专用仪器和仪表及类似装置的制造。

manufacturing of coating material for marine vessel 海洋船舶专用涂料制造

It refers to the manufacturing of coating materials, auxiliary materials, etc. for marine vessel.

包括海洋船舶专用涂料及其辅助材料等的制造。

manufacturing of deck machinery 海洋船舶甲板机械制造

It is the manufacturing of cargo loading and unloading machines, anchor winch steering engine, control actuators, mooring winches, boat hoists, etc. for marine vessels.

包括海洋船舶用货物装卸机械、起锚机、舵机、系泊绞车、起艇机等的制造。

manufacturing of electrical equipment for marine vessel 海洋船舶电气设备制造

It includes the manufacturing of electric engines, cables, power transformation equipment, apparatus, control equipment, etc. for marine vessels.

包括海洋船舶的电机、电缆、变电设备、仪表、控制设备等的制造。

manufacturing of engines and propulsion equipment for marine vessel 海洋船舶发动机和推进设备制造

It refers to the manufacturing of combustion engine, steam turbine, boiler, shafting, propeller and corollary equipment.

包括海洋船舶用内燃机、汽轮机、锅

炉、轴系、推进器及其配套设施的制造。

manufacturing of equipment for automatic control system for marine electric power 海洋电力自动控制系统装置制造

It refers to the manufacturing of equipment such as industrial control systems, apparatus, and regulating and controlling devices during the process of marine electric power generation.

海洋电力生产过程中工业控制系统、仪表和调节控制等装置的制造。

manufacturing of equipment for marine environmental pollution control 海洋环境污染防治专用设备制造

It refers to the manufacturing of special equipment for marine environmental pollution control, such as water pollution control equipment and waste disposal treatment equipment.

海洋环境污染防治的专用设备的制造,如水污染防治设备、废弃物处理设备等。

manufacturing of fixed offshore mooring devices 海洋固定停泊装置制造

It refers to the manufacturing of fixed offshore mooring devices or non-navigation oriented ships such as lightships, firefighting vessels, crane ships, drilling ships and oil platforms.

海上固定停泊或不以航行为主的船只或装置的制造,如灯船、消防船、起重船、钻井船和石油平台等。

manufacturing of marine chemical commodity products 海洋化工日用产品制造

It refers to the manufacturing of daily chemical products using marine chemical products as raw materials, such as the manufacturing of soaps and syndetergent using alkali and potassium chloride as raw materials; the manufacture of cosmetics and oral clearing products with carrageenan as an addictive.

以海洋化工产品为原料制造日用化学产品的生产活动,如以碱、氯化钾为原料制造肥皂及合成洗涤剂;以卡拉胶为添加剂制造化妆品、口腔清洗用品等。

manufacturing of marine chemical pesticide 海洋化学农药制造

It is the manufacturing of chemical pesticide such as naled, using marine chemical products, such as bromine, as raw materials.

利用海洋化工产品(如溴素)为原料生产化学农药的活动,如二溴磷原药。

manufacturing of marine floating device 海洋浮式装置制造

It refers to the manufacturing of marine floating devices, such as dredgers, floating cranes, floating docks, cofferdams, mooring platforms and inflatable rafts.

海洋浮式装置的制造,如挖泥船、浮吊、浮船坞、隔离舱、泊位平台、充气筏等。

manufacturing of marine functional food 海洋保健营养品制造

It is an activity of extracting effective

components from marine organisms for manufacturing marine functional food, such as deep-sea fish oil and spirulina.

从海洋生物中提取有效成分,加工生产制造海洋保健营养品的活动,如深海鱼油、螺旋藻等。

manufacturing of marine navigation and communication equipment 海洋船舶导航、通信设备制造

It is the manufacturing of compasses, radars, communication equipment and underwater sound equipment for marine vessels.

包括海洋船舶用罗经、雷达、通信设备、水声设备的制造。

manufacturing of marine non-metal ship 海洋非金属船舶制造

It is an activity of manufacturing marine ships with non-metal materials such as timber, glass fibre and reinforced plastics.

以木材、玻璃纤维、增强塑料等非金属材料制造海洋船舶的活动。

manufacturing of marine tidal current energy prime mover 海洋潮流能原动机制造

It is the manufacturing of a prime mover for generating electricity from marine tidal current energy.

利用潮流能源发电的原动机的制造。

manufacturing of marine tidal energy prime mover 海洋潮汐能原动机制造

It is the manufacturing of a prime mover for generating electricity from marine tidal energy.

利用潮汐能源发电的原动机的制造。

manufacturing of instrument for application of marine electric power supply 海洋电力供应用仪表制造

It is the manufacturing of metering apparatus, automatic regulating and controlling devices, applied in marine electric power supplies.

海洋电力供应过程中使用的计量仪表、自动调节和控制仪器及装置的制造。

manufacturing of special equipment for marine mining 海洋采矿专用设备制造

It is the manufacturing of special equipment for coastal and sea-floor mining.

海滨和海底矿产开采的专用设备的制造。

manufacturing of special equipment for marine geology 海洋地质专用仪器制造

It is the manufacturing of special equipment for marine geological surveys.

海洋地质调查专用仪器的制造。

manufacturing of special instrument and apparatus for marine environmental monitoring 海洋环境监测专用仪器仪表制造

It refers to the manufacturing of special instruments and apparatus for monitoring a marine environment, such as test instruments for water quality and remote supervision systems for water pollution.

对海洋环境进行监测的专用仪器仪表的制造,如水质测试仪、水质污染遥测系统等。

manufacturing of special instruments for marine navigation 海洋航海专用仪器制造

It refers to the manufacturing of instruments and apparatus as well as similar devices for marine navigation, guidance and surveys, such as bearing compasses.

航海的导航、制导、测量仪器和仪表及类似装置的制造,如定向罗盘等。

manufacturing of special materials for treatment of marine environmental pollution 海洋环境污染处理专用材料制造

It includes the manufacturing of chemical agents and materials for pollution treatment during production activities in sea water utilization, marine chemistry, marine biomedicine, etc., such as water treatment medicament (flocculant, sludge dewatering agent and antifouling composition), and water treatment material, such as filler, filter material for biofilter and membrane material.

海水利用、海洋化工、海洋生物医药等生产过程中的污水处理所用的化学药剂及材料的制造,如水处理药剂(絮凝剂、污泥脱水剂、防垢剂);以及水处理材料,如填料、生物滤池用滤料、膜材料。

manufacturing of special products for marine chemistry 海洋化工专用产品制造

It is the manufacturing of special chemistry products using marine chemical products as raw materials, such as the manufacturing of extinguishants and fire retardants using magnesium chloride, bromine, etc. as raw materials, as well as the manufacturing of water treatment chemical products using magnesium hydrate as a raw material.

以海洋化工产品为原料,制造专用化学用品的生产活动,如以氯化镁、溴素为原料制造灭火剂、阻燃剂;以氢氧化镁为原料制造水处理化学用品等。

map of hazardous geology 灾害地质图

A map of hazardous geology is a professional map showing the intensity, scale and spatiotemporal distribution of geological hazards' disaster-inducing conditions (disaster-causing geological bodies and geological phenomena) and disaster-inducing factors.

表示地质灾害的诱灾条件(致灾地质体与地质现象)和诱灾因素强度、规模及其时空分布规律的专业图件。

marginal efficiency of capital (MEC) 资本的边际效率

The Marginal Efficiency of Capital (MEC) is the rate of discount which would make the sum of the present value of the returns expected from the capital asset during its life just equal to its supply price or replacement capital. Investment is dependent on the rate of interest and MEC.

资本的边际效率是一种贴现率,这种贴现率正好使一项资本物品的使用期内各预期收益的现值之和等于这项资本品的供给价格或者重置资本。投资由利率和资本边际效率决定。

marginal product of capital (MPK) 资本的边际产量

It is the additional output resulting, ceteris paribus, from the use of an additional unit of capital. The formula can be expressed as follows: $MPK = F(K+1, L) - F(K, L)$.

在劳动量不变的条件下,经济体因多用一单位资本而得到的额外产出量: $MPK = F(K+1, L) - F(K, L)$。

marginal basin 边缘盆地

A marginal basin is distributed inside a marginal sea which has an oceanic crust and transitional crustal structure.

分布于边缘海内的具有大洋型或过渡型地壳的构造盆地。

marginal propensity to consume (MPC) 边际消费倾向

In economics, it is a metric that quantifies induced consumption, the concept that the increase in personal consumer spending (consumption) occurs with an increase in disposable income (income after taxes and transfers). The proportion of disposable income which individuals spend on consumption is known as propensity to consume.

增加的消费与增加的收入之比率,也就是增加的一单位的收入中用于增加消费部分的比率。

marginal sea 边缘海

It refers to a semi-closed sea adjacent to a continent on one side and separated from the ocean by a peninsular and island arc on the other side.

又称"陆缘海",位于大陆边缘,一侧以大陆为界,另一侧为半岛、岛弧与大洋分隔的海域。

marginal sea basin 边缘海盆地

Also called *a back-arc basin*, it refers to a deep water basin with an oceanic crust structure on the side of a trench-arc system, located behind the island arc.

又称"弧后盆地",是指沟-弧体系陆侧具有洋壳结构的深水盆地,位于岛弧后方。

mariculture 海上养殖

It refers to the production activities of sea farming under a low water line of spring tides, which includes bottom sowing proliferation, raft culture, cage culture, etc.

在大潮低潮线以下从事海水养殖的生产活动。包括:底播增殖、浮筏养殖、网箱养殖等方式。

mariculture area 海水养殖面积

It refers to the water surface area in an intertidal zone, a shallow sea or bay where economic aquatic animals and plants such as fish, shrimp, crab, shellfish, alga, etc. are artificially cultivated. All of the above-mentioned areas should be counted into mariculture areas no matter whether there is a harvest within the reporting period. Some intertidal areas and water surface, where no or few fry is released and only general management is conducted, are counted out. The unit of measurement is hectares.

利用滩涂、浅海、港湾进行鱼、虾、蟹、贝、藻等海水经济动植物的人工养殖的水面面积。在报告期内无论是否收获其产品,均应统计在海水养殖面积中,但有些滩涂、水面不投放苗种或投放少量苗种,只进行一般管理的不统计为养殖面积。计量单位:公顷。

mariculture production 海水养殖产量

It refers to the amount of aquatic products caught in the farming seawaters where products are cultivated from artificially-released fry or naturally-adopted fry under artificial feeding and management. The unit of measurement is ten thousand tons.

从人工投放苗种或天然纳苗并进行人工饲养管理的海水养殖水域中捕捞的水产品产量。计量单位:万吨。

marine abrasion landform 海蚀地貌

As a type of coastal landform, it refers to an erosion form generated under the destructive action of constant pounding, scouring, grinding on the coast by waves, tides and sediments and gravels carried in waves and tides. The commonly seen marine abrasion landforms are sea cliffs, abrasion platforms, wave cut notches, blowholes, sea arches, sea stacks, etc. It is characterized by obvious changes, with some parts moving forward and some parts moving backward.

属于海岸地貌的一种类型,在波浪、潮流及其所携带的泥沙、砾石不断地冲击、冲刷、研磨破坏海岸的作用下形成的侵蚀形态。常见的有海崖、海蚀平台、海蚀穴和海蚀窗、海蚀拱桥和海蚀柱等。特点是变化明显,有的地方前进,有的地方后退。

marine accessories manufacturing 海洋饰品制造

It refers to the manufacturing of various accessories, such as pearl accessories, using marine animals and plants as raw material.

以海洋动植物为原料,经加工制作各种装饰品的活动,如珍珠饰品等。

marine accident 海难事故

It refers to a disaster occurring when mechanical equipments, cargoes, or personnel of the marine ships or offshore production platforms encounter marine natural disasters or accidents.

海洋船舶或海上生产平台的机械设备、所载货物及人员因遭遇海上自然灾害或意外事故而造成的灾难。

marine accidents 海事

It generally refers to all the accidents associated with marine transportation and operation. It may also refer to accidents which happen to ships or offshore structures during navigation, operation or mooring.

泛指一切与海上运输、作业有关的事故。船舶、水上结构物在航行、作业或停泊时所发生的事故。

marine adult higher education 海洋成人高等教育

It refers to activities for adult higher education with marine professions.

各种设有海洋专业的成人高等教育活动。

marine advisory services 涉海咨询服务

They refer to the offering of consultation, planning and designing services related to marine economy and technology to marine institutions.

为海洋企事业单位提供与海洋社会经济和海洋专业技术有关的咨询、策划、设计等活动。

marine agricultural services 海洋农业服务

They refer to the operational management services for agricultural production in a tidal mud flat.

为海涂农业生产提供的经营管理服务活动。

marine algae 海藻

They are simple and single organisms formed by basal cells which normally grow on the sea floor or some solid structures. Its main characteristics are: no vascular tissues, no phenomenon of differentiation among roots, stems and leaves; no flowers, fruits or seeds; no protective cell layers for reproductive organs; spores or gametes normally produced directly from single cells and no embryos.

海产藻类的统称，通常固定于海底或某种固体结构上，是基础细胞所构成的单株或一长串的简单植物。主要特征为：无维管束组织，没有真正根、茎、叶的分化现象；不开花，无果实和种子；生殖器官无特化的保护组织，常直接由单一细胞产生孢子或配子以及无胚胎形成。

marine amusement park services 海上游乐园服务

It refers to the marine amusement service activities which are equipped with large recreational facilities.

配有大型娱乐设施的海上游乐服务活动。

marine animal and plant protection 海洋动植物保护

It includes protection activities such as feeding, cultivating, reproducing, etc. of marine endangered wild animals and plants, and the management of marine habitats.

对海洋野生濒危动植物的饲养、培育、繁殖等保护活动，以及对海洋动植物栖息地的管理活动。

marine application software services 海洋应用软件服务

It refers to services of software designing, programming, analysing and testing for marine production and management, such as finance, taxation, industrial control and information search.

为海洋生产、管理提供应用软件设计、编制、分析及测试方面的服务，包括财务、税务、工业控制和信息检索等。

marine aquatic products retail 海洋水产品零售

It refers to the specific retail of marine aquatic products.

专门经营海洋水产品的零售活动。

marine archives 海洋档案馆

It refers to management and services activities to all kinds of marine archives.

对各类海洋档案文件进行的管理和服务活动。

marine benthos 海洋底栖生物

It refers to a great variety of organisms that live at the marine base surface or in the sediments of the ocean. It includes phytobenthos and zoobenthos. Phytobenthos are mainly macroalgae and microalgae while zoobenthos include representatives of all classification systems. Benthos can be divided into epibenthos, endobenthos and hyperbenthos based on the relationship between organisms and substrates.

底栖生物是由生活在海洋基底表面或沉积物中的各种生物所组成,种类繁多。底栖生物包括底栖植物和底栖动物。底栖植物主要是一些大型藻类和单细胞藻类,底栖动物则包括各分类系统的代表。按照生物和底质的关系,底栖生物可分为底上、底内和底游3种类型。

marine berth and dock engineering construction 海洋船台船坞工程建筑

It refers to the construction of berths and docks for marine shipping industries.

海洋船舶工业船台船坞的建筑。

marine bioacoustics 海洋生物声学

It is a scientific discipline which studies acoustic behaviours, characteristics, etc. of marine biology.

研究海洋生物的声学行为与特性等的学科。

marine biological chart 海洋生物图

It is a special chart which reflects the distribution of marine organisms, including marine plankton charts, benthos charts, fish charts, bird charts, etc.

反映海洋生物分布情况的专题海图,包括海洋浮游生物图、底栖生物图、鱼类图、鸟类图等。

marine biological resources 海洋生物资源

They refer to marine organisms that can reproduce and update lives continuously with exploitation and utilization value.

海洋中具有生命的能自行繁衍和不断更新的且具有开发利用价值的生物。

marine biology 海洋生物学

It is a scientific discipline which studies life phenomenon as well as its occurrences and development rules in the ocean.

研究海洋中一切生命现象及其发生、发展规律的学科。

marine biomedical research 海洋生物医学研究

It refers to research and experimental development related to marine biomedicine.

与海洋生物医药有关的研究与试验发展活动。

marine biooptics 海洋生物光学

It is a scientific discipline about biological processes influenced by optics. In other words, it is about the optical process during which the upper layer of the ocean is influenced by biological processes.

研究受光学影响的生物过程的学科,即对海洋上层受生物过程影响的光学过程研究。

marine biotechnology 海洋生物技术

It refers to the high technology about producing useful substances or the orderly improvement of marine biological genetic traits with the application of metabolic processes of marine organisms or organisms, based on the principles and methods of marine biology and engineering.

运用海洋生物学与工程学的原理和方法,利用海洋生物或生物代谢过程,生产有用物质或定向改良海洋生物遗传特性所形成的高技术。

marine biotic community structure 海洋生物群落结构

It refers to characteristics such as species composition, special pattern and temporal dynamics, etc. of marine biotic communities.

海洋生物群落的物种组成、空间格局和时间动态等特征。

marine cave 海蚀洞

It refers to a cave formed by a marine abrasion action on weak parts of a sea cliff (soft rock, fractures and intersections of fractures).

海蚀作用在海蚀崖软弱处(软岩、裂隙或裂隙交会处)形成的洞穴。

marine chemistry 海洋化学

It is a scientific discipline about chemical composition, chemical properties and the chemical process of marine substances and flux on all kinds of interfaces, and about chemical problems during seawater resources exploitation.

研究海洋中各种物质的化学组成、化学性质、化学过程、各种界面上的通量以及海水资源开发利用中的化学问题的学科。

marine chemistry industry 海洋化工业

It is an industry of manufacturing chemical products such as sodium hydrate, soda ash and potassium fertilizer with chemicals extracted from sea water, such as marine salt, bromine, potassium and magnesium as raw materials.

以海水中提取的海盐、溴、钾、镁等化学物质为原料,加工生产烧碱、纯碱和钾肥等化工产品的生产行业。

marine city 海上城市

It refers to a city built on an offshore artificial island or floating city, having the function of new type of cities.

建在海上人工岛上或漂浮式的城市,具有新型城市功能。

marine climate 海洋性气候

Also called *coastal climate*, it refers to a mild, humid and cloudy climate in coastal areas and islands where the climate is much more influenced by the ocean than by continent.

又称"滨海气候"、"海岸带气候"。指陆地沿海和海岛气候受大陆影响小,受海洋影响大,气候温和、湿润、多云。

marine computer system management 海洋计算机系统管理

Marine computer system management includes services of designing, integrating, assembling, etc. of computer systems for marine production, management and services.

为海洋生产、管理、服务提供计算机系统的设计、集成、安装等方面的服务。

marine conductivity 海水导电性

Seawater contains salt and it has good conductivity. The marine conductivity can be indicated by seawater electrical conductivity. The electrical conductivity per cubic metre seawater is called *specific conductance*, which is a physical quantity, and also an important physicochemical property of seawater.

海水含有盐分,具有良好的导电性,

海水的导电性可以用海水电导率来表示。1立方米的海水的电导称为"电导率",是表示海水物理性能的一个物理量,也是海水的一个重要的物理化学性质。

marine container manufacturing 海洋集装箱制造

It refers to the manufacturing of exclusive containers for marine transportation.

海洋交通运输专用集装箱的制造。

marine convective mixing 海水对流混合

It refers to an exchange of the upper and lower layers seawater. When the density of the upper layer seawater is higher than that of the lower layer seawater, the upper layer seawater sinks and the lower layer of seawater rises due to the action of gravity. The existence of pycnocline in the seawater will hinder the convective mixing, while the aggravation of the convective mixing will eventually destroy the existence of various clines and layers.

海水上下的交换作用。当上层海水的密度大于下层海水的密度时,在重力作用下,上层海水下沉,下层海水则上升。由于海水中存在着密度跃层,使对流混合受到阻碍,然而对流混合的加剧,最终将破坏各类跃层的存在。

marine crafts manufacturing 海洋工艺品制造

Marine crafts manufacturing refers to the manufacturing of a variety of crafts through art processing, such as shell carving, coconut shell carving and shell carving pictures are using marine shells, agad, etc. as raw materials.

以海洋贝壳和海滨植物种木等为原料,经艺术加工而制成的各种工艺品的生产活动,如贝壳雕、椰壳工艺品、贝雕画。

marine crops planting 海洋农作物种植

It refers to the crops planning enclosed within a tidal mud flat, including grains, oil crops, legumes, cotton, etc.

在海涂范围以内,以围垦形式进行的农作物种植,包括谷物、油料作物、豆类、棉花等。

marine crude oil processing and petroleum products manufacturing 海洋原油加工及石油制品制造

It is an activity of extracting liquid or gaseous fuel from marine crude oil and processing marine petroleum products.

从海洋石油中提炼液态或气态燃料,以及海洋石油制品的生产活动。

marine cryology 海冰学

Marine cryology refers to the science which studies the generation and disappearance process, distribution, and movement of sea ice, snow and their types, physical property and chemical property.

研究海洋中冰、雪的生消过程、分布、运动、类型及其物理、化学性质的学科。

marine cultural relics and cultural protection 海洋文物及文化保护

It refers to the protection and management of marine culture, such as marine folk art, folk custom and immovable cultural relics along the coast, which are of historic, cultural, artistic and scientific values and identified within the scope of cultural relics protection by the department concerned.

沿海地区具有历史、文化、艺术、科学价值并经有关部门鉴定列入文物保护范围的不可移动文物,以及海洋民间艺术、民俗等海洋文化的保护和管理活动。

marine culture 海洋文化

It refers to spiritual, behavioral, social as well as materialistic civilization created by human being throughout the history of human exploration and exploitation of the ocean. It is an important part of Chinese culture.

人类在探索、开发、利用海洋的历史长河中,创造的一种具有精神的、行为的、社会的和物质的文明生活内涵,是中华民族文化的重要组成部分。

marine delimitation 海洋划界

It refers to a maritime boundary delimited by coastal countries based on an agreement. The territorial sea is delimited by the midline among countries without agreement.

沿海国通过协议划定彼此间的海上边界。对领海而言,如果有关国家之间无协议,则通过中间线来实现。

marine deposit 海相沉积

Marine deposit, sediment in the ocean, is divided into littoral facies, continental facies, bathyal facies, abyssal deposit, etc., based on the depth of sea-floor sediment formation. The characteristic is that facies changes remain small and that it can maintain homogeneous lithologic characters on a relatively large scale. Main rocks are clastic rocks, claystones, ferruginous rocks, silicalites, etc.

在海洋中沉积的物质。按海底沉积物形成的深度不同可分为滨海相、陆栅相、次深海相、深海相沉积等。特点是相变不大,可以在较大范围内保持均一的岩性,主要岩石是碎屑岩、泥岩、铁质岩、硅质岩等。

marine depositional landform 海积地貌

It is a type of coastal landform. By the action of waves and tides, the sand and dirt at coastal areas may have occurrences of horizontal or longitudinal movement. When the movement is blocked or the momentum decreases, the sand and dirt will stop and accumulate to form a marine depositional landform, such as beaches, sand pits, sandbanks, etc.

海岸地貌的一种类型。是海岸地带的泥沙在波浪和潮流的作用下,发生横向或纵向的搬移运动,当其受到阻碍或动力减弱时,便停积下来形成的地貌。如海滩、沙嘴、沙坝等。

marine deposition plain 海积平原

It refers to a coastal plain which is flat and slightly leaning towards the sea, formed by the deposition of offshore materials under the action of the movement of tides, waves and winds.

在潮流、海浪和风的搬移作用下,近岸物质沉积形成的地表平坦,微向海倾的沿海平原。

marine disaster 海洋灾害

Marine disaster occurs when abnormal

or drastic changes take place within marine natural environment, leading to events that endanger the society, economy and life and property in the sea and offshore area. It includes storm surge disaster, strong wind and wave disaster, sea ice disaster, sea fog disaster, seafloor earthquake disaster, volcanic eruption disaster, etc. Man-made marine disasters are caused by changes in marine natural condition due to human activities.

海洋自然环境发生异常或激烈变化,导致在海上或海岸带发生的严重危害社会、经济和生命财产的事件。海洋灾害主要有:风暴潮灾害、大风大浪灾害、海冰灾害、海雾灾害和海底地震与火山喷发引起的灾害等。人为活动导致的海洋自然条件改变而引发的灾害,称为人为海洋灾害。

marine disaster prevention 海洋防灾

It refers to the preventive measures taken before a disaster to reduce and relieve damages by marine disasters, such as constructing disaster prevention projects and making disaster prevention plans.

为减少、减轻海洋灾害造成的损失,在灾前采取的预防性措施。如建设防灾工程、制定防灾预案等。

marine disaster warning and forecasting 海洋灾害预报和警告

It refers to the work concerning forecasting or warning for potential disaster such as storm tsunami, tsunami, surge, sea ice, sea level rise, marine climate anomaly, red tide, and marine geological disaster, and its time, location, range of influence, intensity, damage, etc.

对可能发生的灾害及时间、地点、影响范围、强度和损失程度等进行预报或发布警报的工作,包括风暴潮、海啸、巨浪、海冰、海平面上升、海洋气候异常、赤潮、海洋地质灾害等。

marine disaster reduction 海洋减灾

It includes marine disaster monitoring, forecasting, warning, prevention, disaster resisting, disaster relief as well as post-disaster recovery and reconstruction, in order to reduce and relieve damages from marine disasters.

包括对海洋灾害的监测、预报、警报、防灾、抗灾、救灾及灾后恢复重建等,以减少和减轻海洋灾害所造成的损失。

marine disposal zone 海洋倾废区

It refers to a special area exclusively for dumping waste according to scientific, reasonable, safe and economical principles and procedures stipulated by national marine administrative departments, through demonstration, selection and delimitation, with approval of the State Council.

国家海洋主管部门按一定程序,以科学、合理、安全和经济的原则,经论证、选划并经国务院批准公布的专门用于接纳废弃物的特殊海域。

marine drug resources 海洋药物资源

Marine drug resources include marine plants, animals and mineral resources for use in medicine.

海洋中可供药用的植物、动物、矿物资源。

marine ecological disaster 海洋生态灾害

A marine ecological disaster, which harms marine and coastal ecosystems, is caused by human factors or natural variation.

人为因素或自然变异导致的损害海洋和海岸生态系统的灾害。

marine ecology 海洋生态学

It is a scientific discipline about the rules of survival, development and extinction of marine organisms as well as its interrelationship with physicochemistry and biological environments.

研究海洋生物的生存、发展、消亡规律及其与理化、生物环境间相互关系的学科。

marine economic crops planting 海洋经济作物种植

It is about the planting of economic crops including vegetables, flowers, fruits, etc., enclosed within a tidal mud flat.

在海涂范围以内,以围垦形式进行的经济作物种植,包括蔬菜、花卉、水果等。

marine ecosystem 海洋生态系统

A marine ecosystem refers to a unified whole formed through continuous substance exchange and energy transmission between marine biological groups and environment of sea bottoms and water layer areas. It is a biological unit that is functionally stable and self-adjusting.

海洋生物群落与海底区和水层区环境之间进行不断物质交换与能量传递所形成的统一整体,具有相对稳定功能并能自我调控的生态单元。

marine ecosystem bearing capacity 海洋生态系统承载力

It refers to the maximum ecological serving capacity, especially the maximum supplying and bearing capacity, continuously supplied by marine ecosystems for human activities and biological survival under certain conditions.

一定条件下海洋生态系统为人类活动和生物生存所能持续提供的最大生态服务能力,特别是资源与环境的最大供容能力。

marine ecosystem dynamics 海洋生态系统动力学

It is a scientific discipline about dynamic changes of marine ecosystems driven by marine dynamic conditions.

研究海洋生态系统在海洋动力条件驱动下动态变化的学科。

marine ecosystem ecology 海洋生态系统生态学

It is a scientific discipline about the interaction between physical processes, chemical processes and biological processes of structures and functions within a marine ecosystem.

研究海洋生态系统内的结构及其功能的物理过程、化学过程和生物过程的相互作用和相互制约的学科。

marine ecosystem function 海洋生态系统功能

It refers to substance circulation, energy flowing, information transmission as well as its adjustment and control in a marine ecosystem.

海洋生态系统中的物质循环、能量流动、信息传递及其调控作用。

marine ecosystem health 海洋生态系统健康

This term means that a marine ecosystem is vibrant as time goes, able to maintain its organizational structure and autonomy, and easy to be recovered under an external threat.

海洋生态系统随着时间的进程有活力并且能维持其组织结构及自主性,在外界胁迫下容易恢复。

marine education 海洋教育

Marine education is about setting up institutions for marine professional education or for marine vocational training, in accordance with the relevant regulations of the country.

依照国家有关法规开办海洋专业教育机构或海洋职业培训机构的活动。

marine electric power industry 海洋电力业

It refers to electric production activity about the utilization of ocean energy and ocean wind along the coast.

在沿海地区利用海洋能、海洋风能进行的电力生产活动。

marine energy resources 海洋能源

It is a generic term for renewable natural resources of sea water, such as tidal energy, wave energy, ocean current energy, temperature-difference energy and osmotic energy.

海水所具有的潮汐能、波浪能、海(潮)流能、温差能和盐差能等可再生自然能源的总称。

marine engineering management 海洋工程管理

It refers to the organization and management of marine equipment, submarine engineering and other engineering activities, including auditing and supervising submarine cables, pipe laying, etc.

对海洋设施、海底工程和其他工程开发活动的组织管理,包括审核和监督海底电缆、管道的铺设等。

marine engineering management services 海洋工程管理服务

This term refers to services for marine engineering-related preparation, planning, fabrication costs, financing, budgeting, ground, bidding, consultation, supervision, etc.

与海洋工程有关的筹建、计划、造价、资金、预算、场地、招标、咨询、监理等服务活动。

marine engineering survey and design 海洋工程勘察设计

It refers to geological surveys and engineering designs before marine engineering constructions, such as surveys and designs for marine harbour engineering and marine oil engineering.

海洋工程建筑施工前的地质勘察和工程设计,如海港工程的勘察设计、海洋石油工程的勘察设计。

marine environment 海洋环境

The marine environment concerns all the seas and oceans on the earth. By region, it can be divided into estuaries, bays, coastal oceans, open seas and oceans. By marine environmental elements, it can be divided into sea water, sediments, marine organisms and the atmosphere above the sea.

地球上海和洋的总水域,按照海洋环境的区域性可分为河口、海湾、近海、外海和大洋等,按照海洋环境要素可分为海水、沉积物、海洋生物和海面上空大气等。

marine environmental capacity 海洋环境容量

This term refers to the maximum payload for accepting pollutants by a certain sea area, on the premise of making full use of marine self-purification capacities as well as avoiding marine pollution damage.

在充分利用海洋自净能力并且不造成海洋污染损害的前提下,某一海域所能接纳的污染物最大负荷量。

marine environmental element 海洋环境要素

It is a fundamental link among marine environmental systems and a basic unit for marine environmental structures.

海洋环境系统的基本环节,海洋环境结构的基本单元。

marine environmental element forecasting 海洋环境要素预报

It is an activity to forecast marine elements and surface meteorological elements, such as sea wave, seawater temperature, tide, current and suitability.

对海浪、海温、潮汐、潮流、适宜度等海洋环境要素和海面气象要素的预报活动。

marine environmental forecasting and prediction 海洋环境预报预测

It is about making qualitative and quantitative forecasting about future marine hydrological regimes and seawater movement conditions in a sea area, based on observation data and specific methods.

根据观测资料和特定的方法,对某海区未来的海洋水文状况和海水运动状况作出定性的或定量的预测。

marine environmental hydrodynamics 海洋环境流体动力学

It is a scientific discipline about hydrodynamic processes and its evolution rules in a marine environment.

研究海洋环境中的流体动力过程及其演变规律的学科。

marine environmental management 海洋环境管理

Marine environmental management refers to an administrative activity to prevent, relieve and control marine environmental damage and degradation by means of administration, law, economy, scientific technology for the sake of balance of marine ecosystems and sustainable development of marine resources environment.

以海洋生态环境平衡和海洋资源环境的可持续利用为目标,运用行政、法律、经济、科技技术等手段,防止、减轻和控制海洋环境损害或退化的行政行为。

marine environmental monitoring 海洋环境监测

It is about systematic observation and monitoring of marine environmental elements and indexes in a planned way.

对海洋环境要素或指标,有计划地进行系统的观测、监视。

marine environmental pollution damage 海洋环境污染损害

This damage occurs when substances or energy are pulled into a marine environment directly or indirectly, consequently harming marine biological resources and human health, hindering fisheries and other legal activities, damaging sea water utilisation quality, impairing environmental quality, etc.

直接或者间接地把物质或者能量引入海洋环境,产生损害海洋生物资源、危害人体健康、妨害渔业和海上其他合法活动、损害海水使用素质和减损环境质量等有害影响。

marine environmental protection 海洋环境保护

It includes activities to manage and regulate the marine environment, prevent and control marine pollution, protect and improve marine ecosystems by means of science and technology, administration, law, etc.

采用科技、行政、法律等手段,管理和整治海洋环境,预防和控制海洋污染,保护和改善海洋生态环境的一切活动。

marine environmental protection management 海洋环境保护管理

It is about the supervision and management of activities that will influence a marine environment within a sea area under the national jurisdiction or activities on land and along the coast; it is about the organization and management conducted for surveys, monitoring, observation, evaluation, etc. of a marine environment.

对一国管辖海域、沿海陆域内从事影响海洋环境活动的监督管理,以及对海洋环境的调查、监测、监视和评价等进行的组织管理活动。

marine environmental protection industry 海洋环境保护业

This industry includes activities concerning the protection, regulation and eco-repair of marine natural environments by means of marine environmental monitoring management, as well as the exploitation and application of environment-friendly marine technology and equipment.

通过海洋环境的监测管理、海洋环保技术与装备的开发应用而进行的海洋自然环境保护、治理和生态修复整治活动。

marine environmental protection technology 海洋环境保护技术

This technology is applied to solving marine environmental pollution and marine ecological damage as well as maintaining coordinated development between human beings and the environment.

解决海洋环境污染和海洋生态破坏,维持人类与环境协调发展的技术。

marine environmental quality 海洋环境质量

It refers to a requirement from human beings as to the quality level of the overall marine environment, the overall water, substrates and organisms, for the sake of being suitable for survival and reproduction as well as social and economic development.

人类从适宜生存和繁衍以及社会经济发展考虑,对海洋环境的总体或它的水质、底质以及生物等提出的优劣程度的要求。

marine environmental science 海洋环境科学

It is a scientific discipline about various pollutants, pollution sources, pollution dispersion mechanisms, causes and consequences of pollution, pollution control and prevention, etc. in a marine environment.

研究海洋环境中的各类污染物、污染源、污染扩散机制、污染的原因和后果、污染的控制和防治等的一门学科。

marine environmental self-purification capability 海洋自净能力

It refers to the capability of the ocean to accommodate and digest waste water or pollutants. The capability mainly depends on the ocean as a giant container, seawater movement as well as geological, chemical and biological process in sea water, all of which to some degree can dilute and dissipate damages by waste water and pollutants. Marine environmental self-purification capability is a great treasure to human beings.

海洋能容纳消化污水或污染物的能力。这种能力主要是依靠海洋本身的巨大容积和海水的流动,以及海水中的地质、化学和生物作用,它们都可以在不同程度上稀释、消散污水和污染物可能造成的种种危害。海洋自净能力是人类的一项宝贵资源。

Marine Environment Protection Law of the People's Republic of China 《中华人民共和国海洋环境保护法》

This Law was formulated in order to protect the marine environment and resources, prevent pollution damage, maintain ecological balance, safeguard human health and promote the development of marine programmes. On December 25, 1999, the 13th Meeting of the Standing Committee of the Ninth National People's Congress of the PRC adopted the revised version. It came into effect as of April 1, 2000.

中国为了保护海洋环境及资源,防止污染损害,保护生态平衡,保障人体健康,促进海洋事业发展而制定的法律。1999年12月25日中华人民共和国第九届全国人民代表大会常务委员会第十三次会议修订通过,2000年4月1日起施行。

marine equipment manufacturing industry 海洋设备制造业

It refers to the manufacturing of apparatus, devices, equipment, accessories, etc. for marine production and management.

为海洋生产与管理活动提供仪器、装置、设备以及配件等的制造活动。

marine erosion 海蚀作用

It is a generic term about the abrasion and dissolution among debris under the influence of ocean dynamic, chemistry and biology.

海洋动力、化学和生物作用下,岩屑之间的磨蚀作用和溶蚀作用的总称。

marine erosion plain 海蚀平原

It refers to a flat and seaward coastal plain formed by the erosion and destruction of waves, ocean currents, tides, etc. towards the coastal lands.

波浪、海流、潮流等对沿岸陆地侵蚀、破坏形成的地表平坦，微向海倾的沿海平原。

marine erosion terrace 海蚀阶地

It refers to a staircase-shaped landform formed by the lifting and subsidence of marine benches or sea beaches due to the rising and falling of sea levels and vertical tectonic movement caused by climate or geological structure actions.

气候或地质构造作用引起的海平面升降或垂直的构造运动，使海蚀台及海滩抬升或沉降形成的阶梯状地形。

marine facies sedimentation 海相沉积

Marine facies sedimentation, sediment in the ocean, is divided into littoral facies, continental facies, bathyal facies, abyssal deposit, etc., based on the depth of sea-floor sediment formation. The characteristic is that facies changes remain small and that it can maintain homogeneous lithologic characters on a relatively large scale. Main rocks are clastic rocks, claystones, ferruginous rocks, silicalites, etc.

在海洋中沉积的物质。按海底沉积物形成的深度不同可分为滨海相、陆栅相、次深海相、深海相沉积等。特点是相变不大，可以在较大范围内保持均一的岩性，主要岩石是碎屑岩、泥岩、铁质岩、硅质岩等。

marine financial and investment services 海洋金融投资服务

The term refers to financial activities of providing fund and credit for marine production.

为海洋生产提供资金、信贷的金融活动。

marine financial guarantee services 海洋金融担保服务

The term refers to financial guarantee services for marine production.

为海洋生产提供的金融担保活动。

marine fish oil extraction and related products manufacturing 海洋鱼油提取及制品的制造

It refers to an activity of extracting oil from marine fish or fish liver for products manufacturing.

从海鱼或鱼肝中提取油脂，并生产制品的活动。

marine fishery 海洋渔业

Marine fishery includes mariculture, marine fishing, marine fishery services and the processing of marine aquatic products.

包括海水养殖、海洋捕捞、海洋渔业服务以及海洋水产品加工等活动。

marine fishery resources 海洋渔业资源

Also called marine aquatic resources, they refer to animals and plants with an exploitation value within a sea area, including marine fish, crustacean, shellfish, large-sized algae, etc.

又称"海洋水产资源"，指海域中具有开发利用价值的动植物。包括海洋鱼类、甲壳类、贝类和大型藻类资源等。

marine fishing industry 海洋捕捞业

It is an industry for catching natural fishes and other aquatic economic animals and plants in the sea with fishing tackles, fishing boats and equipment.

利用各种渔具、渔船及设备，在海洋中捕获天然的鱼类和其他水生经济动、植物而形成的生产行业。

marine fishing production 海洋捕捞产量

It refers to fishing production within domestic seas of a nation, exclusive of deep-sea fishing. The unit of measurement is ten thousand tons.

本国国内海域捕捞产量，不包括远洋捕捞。计量单位：万吨。

marine food and beverage services 海洋餐饮服务

This term refers to food and beverage services, predominantly with aquatic products, which include services of dinner, fast food and other seafood specialities provided by hotels, restaurants, inns, dining halls and other dining places.

以海洋水产品为主的餐饮服务活动，包括宾馆、饭店、酒楼、餐厅及其他餐饮场所提供的正餐、快餐服务，以及其他海鲜风味小吃等服务。

marine forestry services 海洋林业服务

It refers to various supportive services for forestry production in a tidal mud flat.

为海涂林业生产提供的各种支持性服务活动。

marine foundation software services 海洋基础软件服务

Marine foundation software services are about the designing, programming, analyzing and testing of application software for marine production and management, inclusive of system software, database software, image processing, etc.

为海洋生产、管理提供应用软件设计、编制、分析及测试方面的服务，包括系统软件、数据库软件、图像处理等。

marine freight port 海洋货运港口

A marine freight port provides the following services: loading and uploading of port ship cargo, placement and piling-up of port cargo, piloting of port vessels, port-ship tugging services, maritime transport cargo packing, loading and uploading of containers, and berthing and material supply for freight ships.

包括港口船舶货物装卸服务、港口货物停放、堆存服务、港口船舶引航活动、港口拖船服务、海上运输货物打包、集装箱装拆服务、货运船舶停靠和物资供应服务。

marine functional zone 海洋功能区

It is an area delimited in terms of natural resources conditions, environmental conditions, geological zones, and the exploitation status of a sea area and an island, with consideration of sustainable economic and social development both nationally and regionally. It is the smallest unit in marine functional zoning.

根据海域及海岛的自然资源条件、环境状况、地理区位、开发利用现状，并考虑国家或地区经济与社会持续发展的需要，所划定的具有最佳功能的区域，是海洋功能区划最小的功能单元。

marine general higher education 海洋普通高等教育

It refers to a general higher education within marine professions.

各种设有海洋专业的普通高等教育活动。

marine geodesy 海洋大地测量学

A branch of geodesy, it refers to a discipline that studies the establishment of marine geodetic network and confirmation of the earth' shape and size and the shape and change of sea surface as well.

大地测量学的一个分支。研究建立海洋大地控制点网及确定地球形状和大小、海面形状与变化的科学。

marine geography 海洋地理学

Marine geography is a scientific study about spatial structure characteristics, the development law of marine geological environments, and the interaction between human activities and marine environments, as well as the exploitation and protection of marine resources, marine economy, territorial politics and management.

研究海洋地理环境的空间结构特点与发展变化规律,人类活动与海洋地理环境相互作用,以及海洋资源的开发利用与保护、海洋经济、疆域政治与管理的学科。

marine geology 海洋地质学

Marine geology is a scientific study about geological features and evolution history of the coast and ocean floor, including landforms, sedimentary processes, tectonic evolutions, as well as the formation and utilization of submarine mineral resources.

研究海岸、海底的地质特征及演变历史,包括地形地貌、沉积过程、构造演化和海底矿物资源的形成与利用的学科。

marine geophysics 海洋地球物理学

Marine geophysics is a scientific discipline about the physical properties of the parts of the earth covered by sea, as well as its relationship with the constitution and construction of the earth.

研究地球被海水覆盖部分的物理性质及其与地球组成、构造关系的学科。

marine heat of evaporation 海洋蒸发热

It refers to the calories needed when one gram of sea water evaporates into vapor of the same temperature. Evaporation heat of sea water is of slight difference to that of pure water.

使1克海水化为同温度的蒸汽时所需热量的卡数。海水的蒸发热与纯水的蒸发热相差甚微。

marine heat ratio 海水热比

It refers to the amount of heat needed to make the temperature of 1 gram of seawater rise by 1℃.

使1克海水温度增加1℃时所需要的热量。

marine high-tech industry 海洋高新技术产业

It refers to production and service industries formed on marine knowledge intensiveness, high technology and new technology.

由海洋知识密集型和高技术以及新技术而形成的生产和服务行业。

marine humanistic landscape 海洋人文景观

It refers to marine scenery and relics, created by human beings, with values of sightseeing, leisure, entertainment and tourism.

由人类创造的具有观光、休闲、娱乐、游览价值的海洋景物和遗迹。

marine hydrology 海洋水文学

It is a scientific discipline about the origin, distribution, circulation, movement, etc. of sea water.

研究海水起源、分布、循环、运动等变化规律的学科。

marine impact heap 海洋冲击堆

It is a heap with the shape of a circular cone which lies in the vent of a continental-slope submarine canyon, smaller than a submarine fan.

大陆坡麓海底峡谷出口处,比海底扇的规模小、呈圆锥形的堆积体。

marine information service industry 海洋信息服务业

It includes management and services for marine libraries and archives, services for marine publication, services for marine satellite remote sensing, services for marine communication and other marine information services.

包括海洋图书馆与档案馆的管理和服务、海洋出版服务、海洋卫星遥感服务、海洋电信服务、计算机服务以及其他海洋信息服务活动。

marine information technology 海洋信息技术

It refers to the technology of carrying out scientific management, statistical analysis and comprehensive services for marine information.

对海洋信息进行科学管理、统计分析及综合服务的技术。

marine international organization 海洋国际组织

It is an international marine body set up by the United Nations and other international organizations, such as the Intergovernmental Oceanographic Commission (IOC).

由联合国和其他国际组织所设立的国际性涉海机构,如政府间海洋学委员会(IOC)。

marine libraries 海洋图书馆

It refers to the management and services for special marine libraries, archives, and document libraries, including marine digital libraries.

专门的海洋类图书馆、资料馆、文献馆的管理和服务活动,包括海洋数字图书馆。

marine living resources 海洋生物资源

Marine living resources refer to marine organisms that can reproduce and update lives continuously with exploitation and utilization value.

海洋中具有生命的能自行繁衍和不断更新的且具有开发利用价值的生物。

marine management 海洋管理

It refers to the management activities of marine agencies at all levels taking law, policy, administration and economic measures.

各级涉海管理机构采用法律、政策、行政和经济手段进行的管理活动。

marine mechanic school education 海洋技工学校教育

It refers to the educational activities of

all mechanic schools with marine professions.

各种设有海洋专业的技工学校教育活动。

marine memorial museum 滨海纪念馆

It refers to a martyr cemetery, memorial hall, martyr memorial hall and so on in coastal areas.

滨海地区的烈士陵园、纪念堂和烈士纪念馆等。

marine metal ship manufacturing 海洋金属船舶制造

It is the manufacturing of marine ships with steel, aluminium, etc. as major materials.

以钢、铝等金属为主要材料制造海洋船舶的活动。

marine meteorological chart 海洋气候图

It refers to a special chart, which reflects meteorological formation factors, composition characteristics, changes as well as interrelationships within the ocean.

反映海洋范围内气候形成诸因素与组成的特征、变化及相互影响的专题海图。

marine meteorological services 海洋气象服务

Marine meteorological services are about the observation and services for marine meteorology.

海洋气象观测和服务等活动。

marine meteorology 海洋气象学

It is a scientific discipline about atmospheric phenomena and weather processes above the ocean and its adjacent area, as well as the interaction between the atmosphere and the ocean.

研究海洋和它的临近区域上空的大气现象,天气过程以及大气与海洋相互作用的学科。

marine microbial ecology 海洋微生物生态学

It is the scientific study of interaction between microbes and environment (including biotic and non-biotic environments) in a marine ecosystem.

研究海洋生态系统中微生物与环境(包括生物与非生物环境)之间相互作用的学科。

marine middle professional education 海洋中等专业教育

It refers to middle professional education activities by what is equipped with marine professions.

各种设有海洋专业的中等专业教育活动。

marine mining 海洋矿业

Marine mining is about the mining and beneficiation of coastal placers, gravels, geotherms, coal mines abyssal minerals, etc.

包括海滨砂矿、海滨土砂石、海滨地热与煤矿及深海矿物等的采选活动。

marine mixed layer 海洋混合层

A marine mixed layer is a layer in which active turbulence has homogenized some range of depths, using the mixing process of sea water.

有海水混合过程的水层。

marine natural landscape 海洋自然景观

Marine natural landscape is of value for sightseeing, leisure, recreation and tourism, which includes coastal landscape, island landscape, marine ecological landscape, submarine mountain landscape, etc.

具有观光、休闲、娱乐、游览价值的海洋天然景观。主要包括海岸景观、海岛景观、海洋生态景观、海底景观、水下山岳景观等。

marine natural monuments and nonliving resources protection 海洋自然遗迹和非生物资源保护

It refers to protection for nature reserves, such as marine geological relics, marine paleontological relics, marine nature landscape and marine nonliving resources.

包括海洋地质遗迹、海洋古生物遗迹、海洋自然景观、海洋非生物资源等自然保护区的保护活动。

marine nature reserves 海洋自然保护区

It refers to marine areas which are delimited for special protection and management in accordance with law for the sake of protecting marine natural environment and resources. These areas cover coast, estuary, gulf, island, coastal intertidal zone, coastal wetland and sea area within certain area, including the protected objects.

以海洋自然环境和资源保护为目的,依法把包括保护对象在内的一定面积的海岸、河口、海湾、岛屿、沿海滩涂、沿海湿地或海域划分出来,进行特殊保护和管理的区域。

marine news 海洋新闻

It refers to the interviewing and reporting of marine activities.

对海洋活动的采访报道活动。

marine organic chemistry products manufacturing 海洋有机化学产品制造

It refers to the manufacturing of hydrocarbon and halogenated derivatives, with marine oil chemical products (such as naphtha) as raw materials.

以海洋石油化工产品(如石脑油)为原料,制造烃类及其卤化衍生物的生产活动。

marine organic pollution 海洋有机污染

It refers to damages done by harmful organic matters accumulated in marine environment due to natural or human activities.

由于自然或人为活动等原因产生的有害有机物质在海洋环境中聚集所造成的危害。

marine organism 海洋生物

Marine organisms refer to various organisms in the ocean, including marine animals (invertebrates and vertebrates), marine plants, microbes, viruses, etc.

海洋中的各种生物,包括海洋动物、海洋植物、微生物及病毒等,其中海洋动物包括无脊椎动物和脊椎动物。

marine organism species protection 海洋生物物种保护

It refers to the protection of rare and endangered marine species.

对海洋珍稀、濒危生物物种的保护。

marine passenger port 海洋客运港口

A marine passenger port is a port where passenger service companies, passenger centres and passenger stations offer passenger services.

客运服务公司、客运中心、客运站提供的海港客运服务活动。

marine phycology 海藻学

It is the scientific study of algae in morphological structure, physiological function, reproductive mode, phylogenesis, ecology, classification, etc.

研究海藻的形态构造、生理功能、繁殖方式、系统发育、生态和分类等方面的学科。

marine physics 海洋物理学

It is the scientific study of marine physical property as well as change rule.

研究海洋的物理特性及其变化规律的学科。

marine pinnacle 海中岩峰

It refers to the columnar bedrock with a small area, which pokes relatively higher out of the sea.

面积很小，突出海面较高的柱状基岩。

marine plankton 海洋浮游生物

Marine plankton is a diverse group of organisms that float in water passively with ocean currents. The common characteristics are the lack of advanced locomotive organs, weak locomotivity or incapability of movement, and just floating with currents.

浮游生物指在水流运动的作用下，被动地漂浮在水层中的生物群。它们的共同特点是缺乏发达的运动器官，运动能力薄弱或完全没有运动能力，只随水流移动。

marine planning and planning management 海洋规划与规划管理

It is about the drafting and managing of marine basic law, regulation, policy and planning, including marine basic law, regulation and policies concerning coastal zones, islands, inner waters, territorial seas with contiguous zones, continental shelves, exclusive economic zones with seas under jurisdiction; the division of marine functional zones, marine exploitation planning, marine environmental protection and regulation planning, marine scientific and technological planning, marine economic development planning, etc.

对海洋基本法律、法规、政策和规划的起草、拟定及管理活动，包括：海岸带、海岛、内海、领海与毗连区、大陆架、专属经济区及其管辖海域的海洋基本法律、法规和政策；海洋功能区划、海洋开发规划、海洋环境保护与整治规划、海洋科技规划、海洋经济发展规划等。

marine plant 海洋植物

Marine plant, containing chlorophyll, is a kind of autotrophic organism that can carry on the photosynthesis to produce organic matter.

含有叶绿素，能进行光合作用生产有机物的自养型海洋生物。

marine plant and animal viewing service 海洋动植物观赏服务

It refers to a viewing service for marine animals and plants in coastal areas, such as aquariums and underwater worlds.

沿海地区海洋动植物观赏服务活动,如海洋馆、水族馆和海底世界。

marine policy 海洋政策

Marine policy is the code of conduct established by coastal countries for achieving development goal, strategy, policy, and development plan of marine cause and for handling foreign relations.

沿海国家为实现其海洋事业的发展目标、战略、方针,发展规划和处理涉外关系所制定的行动准则。

marine pollution 海洋污染

It is a phenomenon in which pollutants damage the marine ecosystem and lead to the degradation of seawater quality when they are discharged into the ocean through human activities.

人类活动排放的污染物进入海洋中,破坏海洋生态系统,引起海水质量下降的现象。

marine pollution control 海上污染治理

It refers to the management and governance of the pollutants discharged from marine ships, offshore oil platforms, etc.

对海洋船舶、海上石油平台等排放污染物的治理活动。

marine power-driven fishing vessels 海洋机动渔船

They refer to the vessels engaged in marine fisheries and auxiliary fisheries, powered by configuration machines. The unit of measurement is a vessel.

配置机器作为动力的从事海洋渔业生产和辅助渔业生产的船舶。计量单位:艘。

marine pressure 海水压强

The pressure (p) at some point in the sea refers to the weight of water column per unit area. The formula is $p = \rho g h$ (ρ for seawater intensity, g for acceleration of gravity, h for seawater thickness). The unit of measurement is Pa.

海洋中某一点的压强 p,指这一点单位面积上水柱的重量,等于海水厚度 h 与海水密度 ρ、重力加速度 g 的乘积,即 $p=\rho g h$。单位为 Pa(帕)。

marine protected area 海洋保护区

It is a designated area to protect rare and endangered marine species, ecological species and their habitats, marine natural landscapes, natural ecological systems and historical sites which are of great scientific, cultural and landscape value. It includes natural reserves for marine and coastal natural ecosystem, for marine species, for marine natural sites and non-living resources and special marine protected zones.

为保护珍稀、濒危海洋生物物种、经济生物物种及其栖息地以及有重大科学、文化和景观价值的海洋自然景观、自然生态系统和历史遗迹需要划定的海域,包括海洋和海岸自然生态系统自然保护区、海洋生物物种自然保护区、海洋自然遗迹和非生物资源自然保护区、海洋特别保护区。

marine reclamation 填海造地

It refers to a method to reclaim land from the sea by enclosing a sea area through embankment, thus an effective coastline being formed.

筑堤围割海域填成土地,并形成有效岸线的用海方式。

marine recreation and sport ship construction 海洋娱乐和运动船舶建造

It is the construction and repair of marine yachts and vessels for marine recreation or sports, such as racing yachts, sailing boats, yachts, powerboats, fishing boats and other recreational ships.

海上游艇和用于海上娱乐或运动的船只的制造与修理,如赛艇、帆船、游艇、汽艇、钓鱼船和其他娱乐船等。

marine recreational facilities and landscape construction 海洋娱乐设施与景观工程建筑

It refers to the construction of a marine park, amusement park, water amusement park and supporting facilities, or an exploitation project for marine landscapes.

海上公园、游乐园、水上游乐场所及配套设施,以及海洋景观开发工程的施工活动。

marine regression 海退

Marine regression is a geological process occurring when areas of submerged seafloor are exposed above sea level due to the elevation of earth crust and the fall of sea level. The opposite event, marine transgression, occurs when flooding from the sea covers previously exposed land.

地壳上升或海面下降引起海水后退的地质现象。

marine remote sensing services 海洋遥感服务

This term refers to the services of using sensors placed on an aircraft platform for acquiring images or data of marine landscape and elements.

利用装载于飞行器平台上的传感器,摄取海洋景观和海洋要素的图像或数据资料的服务。

marine resources 海洋资源

Marine resources refer to substance, energy and space quite exploitable for human beings. They can be divided into marine biotic resources, seafloor mineral resources, seawater resources, marine energy and marine space resources by attribute; into marine living resources and marine non-living resources by life and non-life; into marine renewable resources and marine non-renewable resources by regeneration.

海洋中可以被人类利用的物质、能量和空间。按其属性分为海洋生物资源、海底矿产资源、海水资源、海洋能源和海洋空间资源;按其有无生命分为海洋生物资源和海洋非生物资源;按其能否再生分为海洋可再生资源和海洋不可再生资源。

marine reverberation 海洋混响

It refers to signals produced when a sound wave spreads on a fluctuated sea surface, uneven sea floor or heterogeneous bodies within a seawater medium and scatters inversely at a receiving point.

起伏海面、不平整海底及海水介质内部不均匀体上,声波传播过程中反向散射在接收点上产生的信号。

marine salt producing industry 海洋盐业

It refers to an industry that produces

salt products predominantly containing sodium chloride from sea water.

利用海水生产以氯化钠为主要成分的盐产品的活动。

marine salt resources 海盐资源

They refer to salt resources that the seawater contains which can be easily extracted, such as sodium chloride.

海水中含有的并易直接提取的盐类资源,如氯化钠等。

marine salvage and diving equipment manufacturing 海上救捞及潜水装置制造

It refers to the manufacturing of marine salvage and diving equipments, including salvage hawser apparatus, quick check cable device, diving equipment, diving suits, diving communication devices, diving gas supply equipment, saturation diving systems, underwater work tools, etc.

海上救捞装备和潜水装置的制造。包括:救助抛缆器、快速止索器、潜水装备、潜水服、潜水通信装置、潜水供气设备、饱和潜水系统和水下作业工具等。

marine science and technology services subject 海洋科技服务课题

This is a subject concerning scientific research and experimental development, which is helpful to the production, transmission and application of scientific technology. It includes a demonstrative extension for the application scope of scientific achievements; a systematic service of providing scientific information and document for users; technology consulting service of providing feasibility reports, technical plans and suggestions as well as technology certifications for users; daily observation, monitoring of natural and biological phenomena as well as investigation and exploration of resources; a collection of general data concerning social, humane and economic phenomena, such as statistics and market investigations, as well as routine analysis and organization of these data; providing testing, standardization, measuring, calculating, quality control and patent services for society and the public, exclusive of the above-mentioned activities carried out by industrial and commercial enterprises for normal operation.

与科学研究与实验发展有关,并有助于科学技术知识的产生、传播和应用的活动。包括为扩大科技成果的使用范围而进行的示范性推广工作;为用户提供科技信息和文献服务的系统性工作;为用户提供可行性报告、技术方案、建议及进行技术论证等技术咨询工作;自然、生物现象的日常观测、监测,资源的考查和勘探;有关社会、人文、经济现象的通用资料的收集,如统计、市场调查等,以及这些资料的常规分析与整理;为社会和公众提供的测试、标准化、计量、计算、质量控制和专利服务,不包括工商企业为进行正常生产而开展的上述活动。计量单位:项。

marine science investigation services 海洋科学考察服务

It refers to scientific investigation services for seas and polar regions.

海洋和极地的科学考察活动。

marine science 海洋科学

Marine science, an important branch of earth science, is a body of knowledge concerning the study of marine waters, sea floors, coasts, as well as natural phenomenon, organisms, evolution laws of the atmosphere adjacent to the sea and marine exploitation and management.

研究海洋的水体、海底、海岸、海洋邻接大气的自然现象、机体、演化规律和海洋开发利用管理的知识体系,是地球科学的一个重要组成部分。

marine scientific and technological exchanges services 海洋科技交流服务

This term refers to social services and management for marine scientific and technological activities, including the exchange of marine scientific and technological information, consultation of marine technology, incubation of marine technology, assessment and verification of marine technology, etc.

为海洋科技活动提供的社会化服务与管理活动,包括海洋科技信息交流、海洋技术咨询、海洋技术孵化、海洋科技评估、海洋科技鉴定等。

marine scientific research 海洋科学研究

It refers to scientific research targeting the sea about its basic science, engineering technology, etc.

以海洋为对象,就其基础科学和工程技术等进行的科学研究活动。

marine section 海洋断面

It refers to vertical observational sections laid out within a sea area for surveys.

在调查海域中布设的垂直观测剖面。

marine sediments 海洋沉积物

It is a generic term for all marine sediments formed through marine sedimentation, which can be divided into pelagic sediments and terrigenous sediments.

各种海洋沉积作用所形成的海底沉积物的总称。可分为远洋沉积物和陆源沉积物。

marine seismic services 海洋地震服务

Marine seismic services include activities such as earthquake prevention and disaster mitigation for marine seismic detection, earthquakes and emergency rescues.

为海洋地震监测、震灾和紧急求援等进行的防震减灾活动。

marine ship repair and shipbreaking 海洋船舶修理及拆船

It refers to the repair of marine metal and non-metal ships and the breaking of used marine ships.

海洋金属及非金属船舶的修理与对废旧海洋船舶的拆卸活动。

marine skills training 海洋技能培训

It refers to marine skills training activities of schools with marine professions.

各种设有海洋专业学校的技能培训活动。

marine social groups and international organizations 海洋社会团体和国际组织

It is a generic term for marine-related

groups or organizations registered with a management authority in charge of social groups and organizations.

依法在社会团体登记管理机关登记的、与海洋相关的团体或组织。

marine social science research 海洋社会科学研究

It refers to a researching activity conducted for social sciences including population, society, economy, law, culture and history in a coastal area.

对沿海地区人口、社会、经济、法律、文化、历史等社会科学进行的研究活动。

marine social security 海洋社会保障

It refers to various social security activities for marine employees, including basic endowment insurance, unemployment insurance, medical insurance, etc.

为海洋从业人员提供的各种社会保障活动,包括基本养老保险、失业保险、医疗保险等。

marine soil 海积土

It refers to the soil deposited from a shallow sea to deep sea areas near the coast in the ocean.

海洋中靠近海岸的浅海至深海地带堆积形成的土。

marine sound channel 海洋声道

It refers to the channel in which a sound wave spreads at its lowest speed and to the farthest distance. When a sound wave travels in this channel, it can spread for thousands of miles without much loss of sound energy.

声波在海洋中传播速度最小而传播距离最远的通道。声波在海洋声道中传播时,声能损失小,传播距离可达数千英里。

marine sound velocity 海中声速

It refers to the travelling speed of sound in sea water. It is the function of salinity, temperature and pressure.

声音在海水中的传播速度。海中声速是盐度、温度和压力的函数。

marine space resources 海洋空间资源

It is a generic term for geological areas of the coast, on the sea, in the sea and under the sea concerning marine exploitation and development.

与海洋开发利用有关的海岸、海上、海中和海底的地理区域的总称。

marine special reserves 海洋特别保护区

Marine special reserves are areas with special geological conditions, ecosystems, biological and non-biological resources as well as value for marine exploitations, which require management with effective protective measures and scientific exploitations.

对具有特殊地理条件、生态系统、生物与非生物资源及海洋开发利用特殊需要的区域采取有效的保护措施和科学的开发方式进行特殊管理的区域。

marine special reserves management 海洋特别保护区管理

It refers to the management of areas that are of significance to a nation and its regions, who are in great need of management, protection and helpful to the realisation of the sustainable development of resources in aspects of natural resources, marine exploitation and marine ecology within a marine environment.

对海洋环境中在自然资源、海洋开发和海洋生态方面对国家和地方有特殊重要意义,需要特别管理和保护,实现资源可持续利用的区域的管理。

marine stratigraphy 海洋地层学

Marine stratigraphy is a scientific discipline about marine stratigraphic interdependence and regularity of distribution in space and time.

研究海底地层相互关系及其时空分布规律的学科。

marine technical services industry 海洋技术服务业

It provides professional technology and engineering skills for marine production and management. It also provides services for the promotion and exchange of science and technology.

为海洋生产与管理提供专业技术和工程技术,以及相应的科技推广与交流的服务活动。

marine technology detection 海洋技术检测

It includes activities such as monitoring, detection, testing, quality certification, achievement verification, and standard formulation of marine technology.

海洋技术监测、检验、测试、质量认证、成果鉴定、标准制定等活动。

marine technology 海洋技术

This is a generic term for methods, skills and equipment applied to studying marine natural phenomena and changes, the exploitation of marine resources and the protection of a marine environment.

研究海洋自然现象及其变化规律、开发利用海洋资源和保护海洋环境所使用的各种方法、技能和设备的总称。

marine terminological group 海洋专业团体

It includes social groups organized by members and experts in marine-related areas, such as Society of Oceanography and China Ocean Mineral Resources Research and Development Association (COMRA).

由海洋相关领域的成员、专家组成的社会团体,如海洋学会、大洋协会等。

marine territory 海洋国土

It covers a specific sea area as well as airspace above it, sea bed and subsoil within the jurisdiction of national sovereignty. It not only covers a nation's inner water and the water within a nation's territory under the jurisdiction of national sovereignty, but also covers the Exclusive Economic Zone and the continental shelf beyond the national sovereignty but under national jurisdiction, according to the *United Nations Convention on the Law of the Sea*. Marine territory is a collective and conceptual term for sea areas under jurisdiction, including inner waters, territory seas, contiguous seas, EEZ, continental shelves, etc.

在国家主权管辖下的一个特定的海域及其上空、海床和底土。它既包括一个国家的内海、领域中属于国家领土、归其主权管辖的海域,同时,按照《联合国海洋法公约》的规定,还包括该国管辖的不属于主权范围的专属经济区和大陆架。海洋国土是一国内海、邻海、毗连区、专属经济区、大陆架等所有管辖海域的形象总称,是一个集合概念。

marine thermodynamics 海洋热力学

It refers to a discipline which studies thermodynamic process of seawater movement and its change law.

研究海水运动的热力过程及其变化规律的学科。

marine tidal current 海潮流

Marine tidal current refers to a tidal current and ocean current of seawater. Tidal current is generated by a tide caused by astronomical tide-producing forces, while ocean current is generated by the factors of a monsoon, seawater density, earth's rotation, etc.

海水的潮流和海流。潮流是因天体引力引起的潮汐产生的,海流是由季节风、海水密度和地球自转等因素产生的。

marine tourism 海洋旅游

It is an industry providing services for human beings to make a trip or two along the seaside or on the sea.

为人类提供海滨或海上旅行、游览服务的产业。

marine tourism resources 海洋旅游资源

Marine tourism resources refer to natural and humanistic landscapes with a capacity of developing sightseeing, tourism, leisure, entertainment, vacations, sports, etc. on the seashore, island or ocean.

在海滨、海岛和海洋中,具有开展观光、游览、休闲、娱乐、度假和体育运动等活动的海洋自然景观和人文景观。

marine transect 海洋断面

It refers to vertical observational sections laid out within a sea area for surveys.

在调查海域中布设的垂直观测剖面。

marine trades group 海洋行业团体

It is a term for a social group established by marine trades, such as the Marine Fisheries Society, Fisheries Associations, Marine Shipping Associations, Marine Pharmaceutical Society, etc.

海洋行业组成的社会团体,如海洋渔业学会、渔业协会、海洋船舶协会、海洋药物学会等。

marine transportation 海洋交通运输业

It includes activities of marine transportation as well as related services, with ships as major tools.

以船舶为主要工具从事海洋运输以及为海洋运输提供服务的活动。

marine trap 海上圈闭

It refers to a geometric arrangement of any rock which can prevent the continuous migration of fluids in a reservoir stratum and gather them in seabed strata.

海底地层中,能够阻止流体在储集层中继续运移并将其聚集起来的任何岩石的几何排列。

marine value 海洋价值观

It is the general cognition of active roles the sea plays in human survival and development, which is acquired by means of correct and scientific generalization or abstraction based on knowledge accumulation through marine observation and exploitation. It is also the generalization of status and role of the sea in economic and social development.

人类对海洋观察和开发利用实践积累的全部知识,经过正确和科学地概括(或抽象)而形成的海洋对人类生存和发展,并产生所起积极作用的总体认识,是对海洋在经济与社会发展中的地位、作用的归纳。

marine vocational high school education 海洋职业中学教育

It refers to education activities by vocational high schools with marine professions.

各种设有海洋专业的职业中学教育活动。

marine vocational training 海洋职业培训

It refers to vocational training activities organized by institutions with marine professions.

各种设有海洋专业的职业培训活动。

marine waste disposal 海洋倾废

It refers to the disposal of waste and other substances into the ocean by means of ship, aircraft, platform and other carrying tools; dumpling abandoned ships, aircraft, platforms and other marine artificial structures into the ocean; disposing waste and other substances produced during the exploration and exploitation of marine mineral resources as well as waste and other substances from maritime production. In other words, it is about the disposal of waste through the self-purification ability of the ocean within a suitable marine space.

利用船舶、航空器、平台及其他载运工具,向海洋处置废弃物和其他物质;向海洋弃置船舶、航空器、平台和其他海上人工构造物,以及向海洋处置由于海底矿物资源的勘探开发及其相关的海上加工所产生的废弃物和其他物质。也即选择适宜的海洋空间,利用海洋的自净能力处理废弃物。

marine weather forecasting 海洋天气预报

It refers to qualitative and quantitative forecasting of weather and closely-related marine phenomenon over a specific sea area or seashore area within a period in the future.

对特定海区或沿海地区未来某一时期内的天气及其密切相关的海洋现象做出定性或定量的预测。

marine wholesale and retail industry 海洋批发和零售业

It refers to the wholesale and retail conducted during the circulation of marine commodities.

海洋商品在流通过程中的批发活动和零售活动。

maritime air mass 海洋气团

It refers to the air mass formed above seas and oceans.

广大的海面上及洋面上形成的气团。

maritime distress 海难

It refers to ship collisions, stranding, other navigation accidents, or other accidents happening in or outside a ship, which can cause serious damage or grave threat to ships or freights.

船只碰撞、搁浅或其他航行事故,或船上或船外所发生对船只或船货造成重大损害或极大威胁的其他事故。

maritime satellite services 海洋卫星服务

It refers to information services provided by man-made satellites for marine observation, marine research, marine environmental surveys, marine resources exploitation, etc.

人造卫星为海洋观测、海洋研究、海洋环境调查、海洋资源开发利用等提供的信息服务。

Maritime Traffic Safety Law of the People's Republic of China《中华人民共和国海上交通安全法》

It is a law for the Chinese government to manage the navigation of vessels at sea, berthing and operations, safety protection, rescues from disasters at sea and maritime traffic accidents, etc. Adopted at the *Second Meeting of the Standing Committee of the Sixth National People's Congress* on September 2, 1983, it came into effect as of January 1, 1984.

中国政府管理海上船舶航行、停泊和作业、安全保障、海难救助及海上交通事故等的法律。1983年9月2日第六届全国人民代表大会常务委员会第二次会议通过,1984年1月1日起施行。

market exposure 市场风险

It is the potential risk of losses arising from fluctuations in stock prices, interest rates, exchange rates, etc. Therefore it includes equity risk, currency risk, interest rate risk and commodity risk.

因股市价格、利率、汇率等的变动而导致价值未预料到的潜在损失的风险。因此,市场风险包括权益风险、汇率风险、利率风险以及商品风险。

market location theory 市场区位论

The research of location theory on market factors marks the transformation of classical location theory to modern location theory. Born in a monopoly capitalist era, market location theory holds that an industrial layout must take full account of the market factors and choose locations that maximise the profits.

区位论对市场因素的研究,标志着古典区位论向现代区位论的转化。市场区位论产生于垄断资本主义时代,这一学派的主要观点是:产业布局必须充分考虑市场因素,尽量将企业布局在利润最大的区位。

market risk 市场风险

It is the potential risk of losses arising from fluctuations in stock prices, interest rates, exchange rates, etc. Therefore it includes equity risk, currency risk, interest rate risk and commodity risk.

因股市价格、利率、汇率等的变动而导致价值未预料到的潜在损失的风险。因此,市场风险包括权益风险、汇率风险、利率风险以及商品风险。

Markov process 马尔可夫过程

It refers to a stochastic process which assumes that under a current (present) state, the probability of an occurrence of each event does not depend on the preceding outcome.

一类典型的随机过程,指在已知目前状态(现在)的条件下,事物未来的演变(将来)不依赖于它以往的演变(过去)。

marriage service at sea 海上婚姻服务

It refers to a marriage rites service at sea and under water.

在海上及水下所进行的婚姻礼仪服务。

material flow analysis(MFA) 物质流分析

It is an analytical method of quantifying flows and stocks of materials or substances in a system (a product system, economic system, social system, etc.). MFA is an important tool to assess the physical consequences of human activities and needs in the field of Industrial Ecology, where it is used on different spatial and temporal scales.

针对一个系统(产品系统、经济系统、社会系统等)的物质和能量的输入、迁移、转化、输出进行定量化的分析和评价的方法。

matter cycle 物质循环

[*ecology*] It refers to the to-and-fro flow in which the substances on the earth's surface are stored, transformed and migrate in or between ecological systems under the action of natural forces and biological activities.

[*resources technology*] A matter cycle is a process in which the basic elements, such as carbon, nitrogen, oxygen, phosphorus and sulphur, which constitute living organisms in a resource ecological system, circulate between organisms and between organisms and the environment.

[生态学] 地球表面物质在自然力和生物活动作用下,在生态系统内部或其间进行储存、转化、迁移的往返流动。

[资源科技] 资源生态系统中组成生物有机体的基本元素如碳、氮、氧、磷、硫等在生物与生物之间,生物与环境之间循环的过程。

maximum sustainable yield 最大持续渔获量

It is the highest annual catch sustainably obtained without prejudice to the population production capacity.

在不损害种群生产能力的条件下可以持续获得的最高年渔获量。

meadow saline soil 草甸盐土

It gradually develops from all kinds of meadow soil (or moisture soil). Influenced by the up and down movement of groundwater throughout the year, the salt-accumulation process and meadow process are concomitant. It can be developed and utilized as long as its surface has certain amount of organic accumulation, its subsoil has obvious rust streak, and it possesses salt elimination conditions, in addition to salt-accumulation features of the saline soil.

由各种类型的草甸土(或称为潮土)逐步演变而成。受地下水常年上下活动的影响,积盐过程和草甸过程相伴进行,除具有盐土的积盐特征外,表层有一定数量的有机积累,底土有明显的锈纹,具备降渍排盐条件即可开发利用。

mean sea level 平均海平面

It refers to the average height of tide level during some observation periods, which can be divided into daily average sea level, monthly average sea level, annual average sea level, years of average sea level, etc.

某观测时段的潮位平均值。可分为日平均、月平均、年平均和多年平均海平面等。

median rift 中央裂谷

It is a great rift valley extending along the mid-oceanic ridge axis, with an average depth of 3,000–6,000 metres, top and bottom elevation difference of 1,000–2,000 metres, and with a width of 15,000–50,000 metres.

沿大洋中脊轴延伸的巨大裂谷，一般水深3 000~6 000米，顶底高差1 000~2 000米，宽15 000~50 000米。

Mediterranean Sea 地中海

It is the world's largest inland sea surrounded by Europe, Asia, and Africa. It connects the Atlantic Ocean through the Strait of Gibraltar.

位于欧、亚、非三大洲之间，世界上最大的陆间海，大西洋的附属海。

medium of exchange 交换媒介

It refers to anything that a buyer pays to a seller for the purchase of goods and services.

买者在购买物品与劳务时给予卖者的东西。

menu costs 菜单成本

It stems from the cost of restaurants literally printing new menus. But economists use it to refer to the cost incurred by firms to change their prices, including hiring consultants to develop new pricing strategies, printing new menus, updating price lists or informing points of sale to change price tag, etc.

厂商对价格调整时所产生的成本负担，包括：研究和确定新价格的成本、重新编印价目表的成本、通知销售点更换价格标签的成本等。

microecosystem 微生态系统

In the range of specific space and time, it is an entity composed of biological groups of different species of individual 20–200 μm and their environment.

在特定的空间和时间范围内，由个体20~200微米不同种类组成的生物群与其环境组成的整体。

micro landform 微地貌

It is a secondary landform feature developed on a small landform unit, such as gullies, meanders, butterfly-shaped depressions, etc. Through the observation of a micro landform, the formation process of macro landforms can be further analyzed.

发育于小型地貌单元上的次一级地貌形态，如冲沟、河曲、蝶形洼地等。通过对微地貌的观测，可以进一步分析宏观地貌的形成过程。

micro-oceanography 微海洋学

It is a discipline which studies seawater, the micro-structure of the seabed sediments, as well as the evolution law of seabed sediment formation.

研究海水及海底沉积物的微细结构及其形成过程的演变规律的学科。

microstratification 微层化

It is the layered structure of ocean factors from a molecular dissipation scale to one metre on a vertical scale.

垂向尺度在分子耗散尺度至 1 米之间的海洋要素分层结构。

middle layer 中层

Lying below the upper ocean, with thickness between 1,000 to 1,500 metres thick, it is a layer of water having one or more thermoclines, haloclines and pycnoclines.

大洋上层以下，厚度约为 1 000 ~ 1 500 米，其温度、盐度、密度具有一个或多个跃层的水层。

middle mineralized groundwater 地下咸水

It refers to groundwater whose total mineralization is between 3.0 to 10.0 grams per litre.

总矿化度在 3.0 ~ 10.0 克/升之间的地下水。

mid-oceanic ridge 大洋中脊

Also called *mid-ocean ridge* or *mid-ocean rise*, it is a median mountain range on the floor of the ocean, extending through the North and South Atlantic Oceans, the Indian Ocean, and the South Pacific Ocean.

又称"中央海岭"、"洋中脊"，位于大洋中央，绵延全球海底的中央山脉。

mineralizing degree 矿化度

It refers to the total amount of ions, molecules and chemical compounds contained per unit volume, which is represented as "M".

单位体积中所含离子、分子与化合物的总量称为矿化度。以符号"M"表示。

mineral resources 矿产资源

They refer to natural minerals in solid, liquid and gaseous state formed under geological influence, which possess significant exploitation values under current and predictable future technological conditions.

由地质作用形成的，在当前和可预见将来的技术条件下，具有开发利用价值的，呈固态、液态和气态的自然矿物。

Mineral Resources Law of the People's Republic of China《中华人民共和国矿产资源法》

It was enacted in accordance with the *Constitution of the People's Republic of China*, with a view to developing the mining industry, promoting the exploration, development, utilization and protection of mineral resources and ensuring the present and long-term needs of the socialist modernization programme. Adopted at the 15*th Meeting of the Standing Committee of the Sixth National People's Congress* on March 19, 1986, it was revised at the 21*st Meeting of the Standing Committee of the Eighth National People's Congress* on August 29, 1996.

为了发展矿业，加强矿产资源的勘查、开发利用和保护工作，保障社会主义现代化建设的当前和长远的需要，根据中华人民共和国宪法，而制定的。1986 年 3 月 19 日第六届全国人民代表大会常务委员会第十五次会议通过，1996 年 8 月 29 日第八届全国人民代表大会常务委员会第二十一次会议修正。

mining of coastal coal 海滨煤矿开采

It refers to the exploration activity of coastal bitumite, anthracite, lignite and other coal.

对海滨烟煤和无烟煤、褐煤及其他煤炭的开采活动。

mining of seafloor hydrothermal deposit 海底热液矿床开采

It refers to the mining activities of a submarine multimetal sulphide deposit and multimetal mud, etc.

对海底多金属硫化物和多金属软泥等的开采活动。

mining of submarine chemical mineral 海底化学矿采选

It refers to the mining of a chemical mineral from abyssal continental shelf and a seabed of deep waters, such as sulphur deposits, barite, phosphorite, etc.

从深海大陆架及深海海域海底进行的化学矿的开采，如硫磺矿、重晶石、磷钙石等。

mining of submarine gas hydrate 海底可燃冰开采

It refers to the mining activities of a solid natural gas hydrate from the rock stratum in the deep sea.

从海洋深处岩层中提取固态形式的天然气水合物的开采活动。

mixed tide 混合潮

It is a tidal phenomenon in which the presence of a diurnal wave is conspicuous by a large inequality in the heights of either the two high tides or the two low tides usually occurring each tidal day.

全日潮占优势的不正规全日潮或半日潮占优势的不正规半日潮的潮汐现象。

moderate utilisation zone 适度利用区

It refers to waters or island areas available for human's moderate use according to the natural attributes and the status quo of the development. The precondition of moderate utilization is that development projects cannot damage geology and geomorphology, ecological environment, and resource characteristics of waters or islands.

根据自然属性和开发现状，可供人类适度利用的海域或海岛区域。适度利用是指开发项目不以破坏海域或海岛的地质地貌、生态环境和资源特征为前提。

moderate-energy coast 中能海岸

A medium-wave energy coast with an average wave-breaking height of 10 – 50 centimetres is called a moderate-energy coast. Characterized by coastal landforms such as sand beaches, barrier islands, spits, sand bars, shoals, etc., such a coast has a respective longshore sediment transport system.

平均破波高度为10~50厘米的中等波浪能量海岸称为中能海岸。此类海岸以沙滩、堡岛、沙嘴、沙坝和浅滩等滨岸地貌为特征，并具有各自的沿岸输沙系统。

monetary transmission mechanism 货币传递机制

It is about how currency changes via interest rates influence economy. To be more precise, it is the totality of organic relations between various intermediate links and causal relationships during the process in which the monetary authority starts operating by selecting certain policy tools after having determined monetary policy, until the realization of expected goal.

即货币的变动如何通过利率来影响整个经济。是指货币管理当局在确定货币政策以后,从选用一定的政策工具进行操作开始,到实现预期目标之间所经过的各种中间环节相互之间的有机联系及其因果关系的总和。

monitoring and forecasting services for marine environment 海洋环境监测预报服务

It includes all services such as observation, monitoring, research and forecasting conducted for marine environmental elements.

对海洋环境要素进行观测、监测、调查、预报等的服务活动。

monitoring and surveying of marine disasters 海洋灾害监测

It refers to the monitoring and surveying of disaster occurrence, development process and dynamic changes of related elements, including monitoring and surveying of disaster-causing factors such as ocean, atmosphere and crustal activity as well as monitoring and surveying of development process of all marine disasters.

监测、监视灾害发生、发展过程以及各种相关因素动态变化,包括对海洋、大气、地壳活动等致灾因子和各种海洋灾害发展过程的监测。

mouths of bays 海口

They refer to the entrances from oceans to bays, located between the two sides of natural entrances of the bays.

又称"海湾",自海洋到海湾的入口,位于海湾天然入口的两端之间。

Mozambique Channel 莫桑比克海峡

It is the longest channel in the world which is located between Africa and Madagascar Island.

位于非洲大陆与马达加斯加岛之间,世界上最长的海峡。

muddy coast 淤泥质海岸

It is a flat coast composed of fine particulate substances such as silt and clay, primarily under the influence of tides. It is generally distributed in a coastal environment with an abundant supply in fine sediments, concealed landforms and larger tidal ranges.

以潮汐作用为主的,由粉沙和黏土等细颗粒物质构成的低平海岸称为淤泥质海岸。一般分布在细颗粒泥沙供应丰富、地形比较隐蔽和潮差较大的海岸环境。

multiplier effect 乘数效应

It refers to an extra change to the total demand caused by the increase in consumption expenditure when the revenue is increased due to the expansive fiscal policy.

当扩张性财政政策增加了收入,从而增加了消费支出时引起的总需求的额外变动。

mutation 突变

It describes a phenomenon where a minor change in a control parameter occurring near a critical point can radically change the system state.

在临界点附近,控制参数发生微小改变可以从根本上改变系统状态的现象,叫做突变。

mutual fund 共同基金

It refers to an investment vehicle that is made up of a pool of funds collected from many investors for the purpose of investing in securities such as stocks, bonds, money market instruments and similar assets. Mutual funds are operated by money managers, who invest the fund's capital and attempt to produce capital gains and income for the fund's investors.

向公众出售股份,并用收入来购买股票与债券资产组合的机构。

N

Nansha Islands 南沙群岛

They refer to islands which are located in the southernmost part of China containing the most island reefs and the largest area among the four islands of the South China Sea, mainly including Taiping Island, Pagasa Island, Spratly Island, Swallow Reef, Tizard Banks, Vanguard Bank, etc. Zengmu Reef is the most southern point in Chinese territory. The People's Republic of China exercises territorial sovereignty over the Nansha Islands, whose administrative jurisdiction belongs to Sansha city of Hainan Province of China. Currently, with the exception of a few islands controlled by the Chinese Mainland and Taiwan, most islands are occupied by Vietnam, Philippines, Malaysia, etc.

南沙群岛是中国南海诸岛四大群岛中位置最南、岛礁最多、散布最广的群岛。主要岛屿有太平岛、中业岛、南威岛、弹丸礁、郑和群礁、万安滩等。曾母暗沙是中国领土最南点。南沙群岛领土主权属于中华人民共和国,行政管辖属中国海南省三沙市。目前除中国大陆和台湾控制少数岛屿外,主要岛屿均被越南、菲律宾、马来西亚等国侵占。

Nash Efficiency 纳什效率

It refers to a state in which one system subject can maximize its own benefit by accepting the arrangement of the system and ignoring the benefit level of other system subjects.

某一制度主体接受一种制度安排使得自身的得益最大化,而不考虑其他制度主体的得益水平的状态。

national marine functional zoning 全国海洋功能区划

With the joint efforts of the maritime administrative department of the State Council, relevant departments of the State Council and People's Government of coastal provinces, autonomous regions and municipalities under the direct control of the central government, it is defined and developed to delimit internal waters, territorial seas, islands, continental shelves and exclusive economic zones in the People's Republic of China in view of a geographic area (including necessary dependence on land area).

国务院海洋行政主管部门会同国务院有关部门和沿海省、自治区、直辖市人民政府开展的,以中华人民共和国内水、领海、海岛、大陆架、专属经济区位划分为对象,以地理区域(包括必要的依托陆域)为划分单元的海洋功能区划。

national plan on island protection 全国海岛保护规划

It is a plan formulated based on *the Island Protection Law of the People's Republic of China*, and related plans such as a national economic and social development plan, national marine functional zoning, an outline of the overall planning of land use (2006—2020) and a national ocean development plan. It was implemented to protect the ecological system of islands and adjacent sea areas, develop island resources properly, maintain national marine rights and interests and promote the sustainable development of economy and society. The *Plan* was approved by the State Council of the People's Republic of China and was announced formally by the State Oceanic Administration on April 19th, 2012, which covers six sections: current situation, guiding ideology, fundamental principle and planning target, island classification protection, key projects and safeguard measures on the implementation of the plan.

为保护海岛及其周边海域生态系统,合理开发利用海岛资源,维护国家海洋权益,促进经济社会可持续发展,依据《中华人民共和国海岛保护法》等法律法规、国民经济和社会发展规划、全国海洋功能区划,结合全国土地利用总体规划纲要(2006—2020 年)、国家海洋事业发展规划等相关规划,制定《全国海岛保护规划》。该《规划》经中华人民共和国国务院批准,2012 年 4 月 19 日由国家海洋局正式公布。《规划》分现状与形势,指导思想、基本原则和规划目标,海岛分类保护,重点工程,规划实施保障措施 6 部分。

national saving 国民储蓄

It is the balance when the final consumption expenditure is deducted from the gross national disposable income. It measures the maximum amount of capital of a debt-free country. National saving can fall into three categories: government savings, business savings and household savings.

是国民可支配总收入减去最终消费支出后的差额,它衡量一个国家在不举借外债的情况下,可用于投资的最大资金量。国民储蓄包括政府储蓄、企业储蓄和居民储蓄三个组成部分。

marine natural attribute 海洋自然属性

It refers to inherent natural attribute of the ocean.

海洋本身固有的自然属性。

natural gas hydrate 天然气水合物

It is also called *flammable ice*. Distributed in deep sea sediments, in which a large amount of methane is trapped within a crystal structure of water, it is a clathrate compound formed by water and gas molecules under high pressures and low temperatures, thus forming a solid similar to ice.

又称"可燃冰",分布于深海沉积物中,由天然气与水在高压低温条件下形成的类冰状的结晶物质。

natural rate of output 自然产量率

It refers to the production of goods and services achieved by an economy when the unemployment rate is at a normal level in the long run.

一个经济在长期中当失业处于其正常率时达到的物品与劳务的生产。

natural rate of unemployment 自然失业率

It refers to the unemployment rate of economic society under normal conditions when the labour market is in a steady state of supply and demand. The steady state here is considered to cause neither inflation nor deflation.

经济社会在正常情况下的失业率，它是劳动市场处于供求稳定状态时的失业率，这里的稳定状态被认为是既不会造成通货膨胀也不会导致通货紧缩的状态。

natural resource stock 资源储备

To address the possible supply delay and appreciation of natural resources, individuals, enterprises and government store resources on a certain scale in order to ensure the normal operation of future life, production and society.

个人、企业和政府为应付自然资源可能的供给时滞和升值，对资源采取一定规模的储存行为，以保证未来生活、生产和社会正常运转。

natural resource structure 自然资源结构

It refers to the composition and spatial combination of natural resources within a particular region.

在某一特定的地域范围内自然资源的组成及空间组合状况。

natural resources 自然资源

They are useful natural substances existing in nature. They are usable and naturally generated matter and energy, the material basis for human survival, mainly including five major resources such as climate, biology, water, land and mineral resources.

自然界存在的有用自然物。人类可以利用的、自然生成的物质与能量，是人类生存的物质基础。主要包括气候、生物、水、土地和矿产等5大资源。

natural resources evaluation 自然资源评价

It is quantitative or qualitative evaluation and appraisal on the amount, quality, regional combination, time and space distribution, development and utilization, management and protection of the natural resources in a country or region, according to a certain evaluation principle or basis.

按照一定的评价原则或依据，对一个国家或区域的自然资源的数量、质量、地域组合、时空分布、开发利用、治理保护等方面进行定量或定性的评定和估价。

natural-rate hypothesis 自然率假说

It is a concept of economic activity. According to this concept, the unemployment rate will eventually return to its normal or natural rate regardless of inflation rate.

一种认为无论通货膨胀率如何，失业最终要回到其正常率或自然率的观点。

nautical chart 航海图

It is a graphic representation of a maritime area and adjacent coastal regions to protect navigation safety. As a chart which was first developed with the greatest number, it is applied to making navigation plan, selecting anchorage, navigation location and plotting route.

保证航行安全的海图。一种发展最早、数量最多的海图，用于制定航行计划、选择锚地、航行定位与标绘航线。

nautical meteorology 航海气象学

It refers to a discipline which studies correlation between meteorological conditions and navigation. It is an important aspect which reflects how meteorology is applied to navigation in combination with oceanography.

研究气象条件与航海的相互关系的学科,为气象学结合海洋学应用于航海的一个重要方面。

nautical mile 海里

Also called *an international nautical mile*, it refers to a unit of length used in sea navigation. The nautical mile (the symbol is n miles) is a unit of distance that is about one minute of arc of latitude (where 1 degree is equal to 60 minutes and 1 circle is equal to 360 degrees) measured along any meridian. Since the terrestrial meridian is an oval, it has different curvature at different latitude. Therefore, the corresponding arc of the one minute of latitude is not the same. 1 nautical mile is equal to 1,852 metres.

又称"国际海里",航海上计量距离的单位,符号为 n mile。它等于地球椭圆子午线上纬度1分(1度等于60分,1圆围为360度)所对应的弧长。由于地球子午圈是一个椭圆,它的不同纬度的曲率是不同的,因此,纬度1分所对应的弧长也是不相等的。1 海里=1 852 米。

navigational markers 航标

They are assistant facilities of navigation for ship location, navigation and other purposes, including visual navigational marker, wireless navigation facility and acoustic navigational marker.

供船舶定位、导航或者用于其他专用目的的助航设施,包括视觉航标、无线电导航设施和音响航标。

nekton 游泳生物

It refers to the aggregate of actively swimming aquatic organisms in a body of water (usually oceans or lakes) able to move independently of water currents. It includes four categories, namely eunekton, planktonic nekton, benthonekton and xeronekton.

有发达的运动器官,在水层中能克服水流阻力自由游动的动物。包括真游泳生物、浮游游泳生物、底栖游泳生物和陆缘游泳生物四类。

net capital outflow 资本净流出

It is the net flow of funds being invested abroad by a country during a certain period of time (usually a year). A positive NCO means that the country invests outside more than the world invests in it and vice versa. NCO is one of two major ways of characterizing the nature of a country's financial and economic interaction with the rest of the world.

本国居民购买的外国资产减去外国人购买的国内资产。

net exports 净出口

They refer to the amount by which foreign spending on a home country's goods and services exceeds the home country's spending on foreign goods and services. Factors affecting net exports include prosperity abroad, tariffs and exchange rates.

又称"贸易余额",一国的出口值减去进口值。

neutral coast 中性海岸

It refers to a coast which is not vulnerable to the rise and fall of sea levels, such as estuaries, plains and fault coasts.

不受海面上升和下降影响的海岸。如河口岸、平原岸、断层岸等。

new elements theory 新要素学说

It is the enrichment and development of the Ohlin factor endowment theory. It greatly expands the category of production factors by introducing the laborers' intellectual investment, technological progress and innovation, and information acquisition convenience, thus deepening the connotation of production factors.

对俄林生产要素禀赋理论的丰富与发展。它大大地扩展了生产要素范围,把劳动者的智力投资、科技进步与创新、获取信息的便利程度都列入生产要素范畴,深化了生产要素的内涵。

new energy resources 新能源

They are renewable energies systematically developed and utilized on the basis of new technologies, such as nuclear energy, solar energy, wind energy, biomass energy, geothermal energy, ocean energy, hydrogen energy, etc.

在新技术基础上,系统地开发利用的可再生能源,如核能、太阳能、风能、生物质能、地热能、海洋能、氢能等。

new ice 初生冰

It is a general term used for recently-frozen sea water that does not yet make up solid ice. It may consist of frazil ice (plates or spicules of ice suspended in water), slush (water saturated snow), or shuga (spongy white ice lumps a few centimetres across). Other terms, such as grease ice and pancake ice, are used for ice crystal accumulations under the action of wind and waves.

最初形成的冰的总称。包括冰针、油脂状冰、黏冰和海绵状冰等。

new industrial district 新产业区

It is a network formed on the basis of a reasonable regional division of labour. Such a network is closely connected with the local labour market, practicing a specialized division of labour.

基于合理劳动地域分工基础上结成的网络,这些网络与本地的劳动力市场密切相连接,实行专业化分工。

new orders of marine ship 海洋船舶新承接订单量

It refers to contract orders of marine ships, which are officially valid within a reporting period (exclusive of option contract orders and unofficially valid contract orders which are not officially ratified). The unit of measurement is Dead Weight Tonnage (DWT).

报告期内正式生效的海洋船舶合同订单(不含选择权和未正式生效的合同订单)。计量单位:载重吨。

new road to industrialization 新型工业化道路

It was affirmed at the *16th National Congress of the Communist Party of China* in order to be distinguished from traditional industrialization road. That is, a new road to industrialization should be taken with high technology content, good economic returns, low resource consumption, little environmental pollution, and with human resources advantages being given full plan by sticking to the principle that informationization drives industrialization, and in-turn, industrialization promotes informationization.

中国共产党第十六次全国代表大会确定的区别于传统工业化道路的新的工业化道路。即坚持以信息化带动工业化,以工业化促进信息化,走出一条科技含量高、经济效益好、资源消耗低、环境污染少、人力资源优势得到充分发挥的新型工业化道路。

newly emerging marine industry 新兴海洋产业

It is a marine production and service industry developed since the 1960s, including offshore oil and gas industries, mariculture, marine tourism, coastal mining industries, seawater desalination industries, seawater chemical element extraction industries, etc.

20世纪60年代以来发展起来的海洋生产和服务行业。有海洋油气业、海水养殖业、海洋旅游业、海滨采矿业、海水淡化业及海水化学元素提取业等。

niche 生态位

It refers to the comprehensive situation of an organism in a biome including the living state, activity characteristics and its relationship with food and enemies, which is a specific part or minimum unit in its habitat.

一种生物在生物群落中的生活地位、活动特性以及它与食物、敌害的关系等的综合境况,是一种生物在其栖息环境中所占据的特定部分或最小的单位。

nilas 尼罗冰

It is an elastic ice sheet formed by the continuous growth of new ice, frozen with the thickness of about 10 centimetres. Under exogenic action, nilas is easy to bend and break into rectangular ice cakes.

初生冰继续增长,冻结成厚度10厘米左右有弹性的薄冰层,在外力的作用下,易弯曲,易被折碎成长方形冰块。

Nine Segment Lines 九段线

They are discrete boundary lines plotted on the periphery of the South China Sea islands with border symbols. In 1947, the Ministry of the Interior of the Chinese government of that day drew a line which is composed of 11 segment lines as an undefined border line on *The South China Sea Islands Map*. After the founding of new China, another line was marked on the same position of the map with the approval of the departments concerned. The eleven segment lines are modified as nine segment lines, which are usually called traditional border lines.

在地图上,用国界符号在中国南海诸岛外围标绘的断续界线。1947年,当时的中国政府内政部方域司在其编绘出版的《南海诸岛位置图》中,以未定国界线标绘了一条由11段断续线组成的线。新中国成立后,经政府有关部门审定出版的地图在同一位置上也标绘了这样一条线,而只是将11段断续线改为9段断续线。这一条线通常被称为传统疆界线。

nitrogenous fertilizer manufacturing 氮肥制造

It is the manufacturing of nitrogen fertilizer, such as synthetic ammonia, with ocean petrochemicals (for example, naphtha) as the raw materials.

以海洋石油化工产品(如石脑油)为原料,制造氮肥的生产活动,如合成氨。

nominal exchange rate 名义汇率

It refers to the price of one country's currency in relation to another's.

一个人可以用一国通货交换另一国通货的比率。

nominal Gross Domestic Product (nominal GDP) 名义 GDP

It equals the market value of total final products, which is calculated by the price of production goods and labour services for that year.

用生产物品和劳务的当年价格计算的全部最终产品的市场价值。

Non-inhabitant island 无居民海岛

It is an island, not serving as the domicile of a citizen, within the territorial sea and other sea areas under the jurisdiction of China.

在我国领海及管辖的其他海域内不作为公民户籍所在地的海岛。

non-equilibrium growth 非均衡增长

It is based on the non-equilibrium phenomenon which exists in the economy. The growth results can strengthen or undermine the different types of non-equilibrium phenomena, yet it is unlikely to make the total social supply and demand to achieve perfect equilibrium in economic growth. This means the actual growth and actually increased production capacity are in an unbalanced state.

以经济中存在的非均衡现象作为前提,增长的结果可以强化或削弱各种非均衡现象,但不可能使社会总供给和社会总需求在经济增长中达到完全均衡,意味着实际的增长量与实际增加的生产能力处于不均衡状态。

non-functional aggregation 非功能聚集

It refers to the aggregation of service functions like industrial infrastructure, culture, education and health, and other aspects like talent, information, etc.

工业基础设施、文教卫生等服务功能以及人才、信息等方面的聚集。

non-linearity 非线性

It refers to the relation between variables that is not proportional or linear. It represents irregular movement and mutation.

量与量之间不按比例、不成直线的关系,代表不规则的运动和突变。

non-ownership rights 他项权利

They are the rights of rent and mortgage formed when the right of sea area use is leased and pledged.

出租、抵押海域使用权形成的承租权和抵押权。

non-renewable energy resources 不可再生能源

In contrast to *renewable energy resources*, they generally refer to gasoline, oil, coal, natural gas, as well as other minerals without a sustainable yield after being exploited and utilized by human beings. They were formed inside the earth over millions of years from dead organisms. With large-scale exploitation and utilization, reserves are gradually reduced and can't be replaced.

"可再生能源"的对称。泛指人类开发利用后,在现阶段不能重复再生的能源资源,如煤炭、石油、天然气等矿藏能源。这类能源是经过漫长的地质年代逐步形成的,随着大规模的开采利用,蕴藏量逐渐减少,不能再生。

non-renewable marine resources 海洋不可再生资源

It refers to marine natural resources, the stock of which has gradually declined and consequently depleted after human exploitation.

人类开发利用后,其存量逐渐减少、衰退以致枯竭的海洋自然资源。

non-sensitive area in marine ecological environment 海洋生态环境非敏感区

It refers to the area that has a low target for marine eco-environmental functions and can recover its function after being damaged, including general industrial water area, port waters, etc.

海洋生态环境功能目标较低,且遭受损害后可以恢复其功能的海域,包括一般工业用水区、港口水域等。

non-vessel-operating services 无船承运业务

They are international marine transport operating activities in which a non-vessel-operating service operator, as a carrier, accepts consignors' goods, issues its own bills of lading or other transport documents, charges the consignors for transport, completes international marine goods transport via international shipping operators, and assumes the responsibilities of a carrier.

无船承运业务经营者以承运人身份接受托运人的货载,签发自己的提单或者其他运输单证,向托运人收取运费,通过国际船舶运输经营者完成国际海上货物运输,承担承运人责任的国际海上运输经营活动。

normal baseline 正常基线

It is a low water line along the coast marked in a large scale chart admitted by coastal state officials.

沿海国官方承认的大比例尺海图所标明的沿岸低潮线。

North Sea 北海

It is a marginal sea in the shape of a square located in the northeast of the Atlantic Ocean between the Great Britain and European Continent. It is 676 kilometres at the widest point from east to west and 1,126 kilometres at the longest point from south to north, with an area of around 575,000 square kilometres, the maximum water depth of 0.725 kilometres and average depth of 0.094 kilometres.

大西洋东北部的边缘海。呈方形,位于大不列颠岛与欧洲大陆之间,东西最宽676千米,南北最长1 126千米,面积约57.5万平方千米,最大水深0.725

千米,平均水深为 0.094 千米。

North-East Asian Regional-Global Ocean Observation System (NEAR – GOOS) 东北亚海洋观测系统

Since the IOC (*International Oceanographic Commission*) officially launched GOOS at the 17th conference convened in 1993, China, Japan, Korea, Russia, etc., initiated the *North-East Asian Regional-Global Ocean Observation System* (*NEAR-GOOS*) as a part of the international GOOS. So far, five specialists and regional conferences of NEAR-GOOS have been convened and Japan has set up a real-time data transmission centre and a delayed data centre. *The National Marine Data and Information Service of China* has also established the delayed data centre and relevant data can be exchanged via the internet. NEAR-GOOS has become one of the active regional GOOS projects of the IOC-GOOS.

继海委会 1993 年召开的第十七次大会决定正式发起 GOOS 之后,中国、日本、韩国、俄罗斯等国于 1994 年率先发起东北亚海洋观测系统(NEAR-GOOS),作为国际 GOOS 的一部分。迄今为止,召开过五次 NEAR-GOOS 专家、区域会议,其中日本已建立了 NEAR-GOOS 实时资料传输中心和延时资料中心。中国国家海洋信息中心也已建立了延时资料中心,有关资料可通过互联网交换。NEAR-GOOS 已成为海委会 GOOS 计划中较活跃的区域 GOOS 计划之一。

notice to mariners 航海通告

It refers to the documents reporting the changes in navigational markers, obstacles in sea area and publishing news of navigation books.

报道海区航标、障碍物等变化情况及航海图书出版消息的文件。

number of berths 泊位数

It refers to the actual number of berths within a reporting period. Berths are divided into quay berths and buoy berths. All berths for production and non-production within the scope of a port should be counted in, including the quay berths owned by port authorities, shipping departments, and exclusive berths belonging to enterprises.

报告期末泊位的实际数量。泊位分码头泊位和浮筒泊位,凡港区范围内所有的生产用和非生产用码头泊位均应进行统计,既包括产权属港务局的码头泊位,也包括产权属航运部门、企业的专用码头泊位。计量单位:个。

number of dumping zones 倾倒区个数

It refers to the quantity of marine dumping zones for toppling waste into the ocean within a reporting period.

报告期内海域倾倒废弃物所使用的海洋倾倒区数量。计量单位:个。

number of marine nature reserves 海洋类型自然保护区数量

It refers to the number of marine nature reserves established with approval in accordance with the *Nature Reserves Regulations* and *Marine Nature Reserves Management Procedures*, which is classified statistically by national level and regional level.

依照《自然保护区条例》和《海洋自然保护区管理办法》,批准建立的海洋自然保护区的数量。按照国家级和地方级分类统计。计量单位:个。

number of marine professions 海洋专业数量

It refers to the number of marine professions in higher education institutions and middle vocational schools. It is classified statistically by doctor degree or master degree in oceanography, undergraduate education, specialty education, adult higher education, and middle vocational education.

高等教育和中等职业教育所设的与海洋有关的专业数量。按照海洋专业博士、硕士、普通高等教育本科、专科,成人高等教育,中等职业教育分类统计。计量单位:个。

number of marine special reserves 海洋特别保护区数量

It refers to the number of marine special reserves established with approval in accordance with *Marine Special Reserves Management Procedures*, which is classified statistically by national level and regional level.

依照《海洋特别保护区管理办法》,批准建立的海洋特别保护区的数量。按照国家级和地方级分类统计。计量单位:个。

number of people visiting marine museum 海洋博物馆参观人次

It refers to the accumulative total of current-year visitors received by marine cultural institutions open to the public at the end of a reporting period. The unit of measurement is person-time.

报告期末向社会开放的海洋文化机构当年接待的所有参观人次的累计数。计量单位:次。

number of tourist agencies 旅行社数量

It refers to the number of business entities who are engaged in attracting, organizing and receiving tourists, tourists, providing tourism service, conducting domestic tourism, inbound tourism and outbound tourism business.

滨海地区从事招徕、组织、接待旅游者等活动,为旅游者提供相关旅游服务,开展国内旅游业务、入境旅游业务或者出境旅游业务的企业法人数量。计量单位:个。

O

occupied coastline length 利用岸线长度

It refers to a length of coastline occupied for production and service activities in a coastal area, which is aggregated according to national industries classification.

沿海地区进行生产与服务活动所占用海岸线的长度。按照国民经济行业分类汇总。

ocean 洋

The main part of vast, connected bodies of salt water on the earth's surface, it is far from the mainland and very deep, accounting for 89% of the total area of the world seas and oceans. The characteristics of oceans are as follows: hydrological features less affected by mainlands; relatively stable temperature and salinity; water colour closer to blue; great transparency; independent tidal wave system and current system.

地球表面上相连接的广大咸水水体的主体部分,远离大陆,深度较大,占全球海洋总面积的89%。水文特征受大陆影响较小,温度和盐度较稳定,水色高、透明度大,有独立的潮波系统和洋流系统。

ocean basin 洋盆

Located between mid-oceanic ridge and continental margin, it is a basin with oceanic crust. The water depth is generally 4,000–6,000 metres.

位于大洋中脊与大陆边缘之间,水深一般在4 000~6 000米,具有大洋型地壳的盆地。

ocean bottom survey 海底调查

It is a general term for observing and measuring a submarine landform, submarine bottom quality, submarine shallow structure, surface sediments, submarine heat flow and other aspects.

海底地形地貌、海洋底质、海底浅层结构和表层沉积物、海底热流等方面的观测与测量的统称。

ocean circulation 海洋环流

It refers to a circulation system or current rotation, which is head-tail connecting, relatively independent and formed by ocean currents in sea areas.

海域中的海流形成首尾相接且相对独立的环流系统或流旋。

ocean current 洋流

It refers to a large-scale aperiodic flow of ocean water at a relatively steady speed along a certain direction in the ocean.

又称"海流",海洋中,海水沿一定方向、以相对稳定的速度作大规模的非周期性流动。

ocean current energy 海流能

It refers to the energy stored in the ocean current in the ocean.

海洋中海流蕴藏的动能。

ocean drilling 深海钻井

It refers to a well drilling in the sea

area with a water depth of more than 300 metres for oil and gas exploration and development.

油气勘探或开发在水深大于 300 米的海域所钻的钻井。

ocean economy 海洋经济

It is sum total of various industry activities concerning marine exploitation and protection, as well as related activities.

开发、利用和保护海洋的各类产业活动，以及与之相关联的活动的总和。

ocean energy 海洋能

It refers to the renewable energy which is rich in the ocean, including tidal energy, wave energy, ocean thermal energy, ocean wave energy, tidal current energy and seawater salinity gradient energy, as well as ocean energy farm in a broad sense.

蕴藏在海洋中的可再生能源，包括潮汐能、波浪能、海洋温差能、海浪能、潮流能和海水盐差能，广义还包括海洋能农场。

ocean engineering 海洋工程

It refers to the engineering for new projects, reconstruction projects and extension projects, the engineering subjects or main construction activities of which are located oceanward along the coast. The engineering function is realized by a dint of changing marine environmental conditions, or that its environmental influence mainly works on marine environments.

工程主体或者工程主要作业活动位于海岸线向海一侧，或者需要借助、改变海洋环境条件实现工程功能，或其产生的环境影响主要作用于海洋环境的新建、改建、扩建工程。

ocean engineering management 海洋工程管理

It refers to the organization and management of marine equipment, submarine engineering and other engineering activities, including auditing and supervising submarine cables, pipe laying, etc.

对海洋设施、海底工程和其他工程开发活动的组织管理，包括审核和监督海底电缆、管道的铺设等。

ocean engineering construction industry 海洋工程建筑业

It refers to an industry for marine and coastal engineering construction such as harbours, navigation channels, coast power plants, coasts, dikes, etc.

从事海港、航道、滨海电站、海岸、堤坝等海洋和海岸工程建筑活动的行业。

ocean exploitation 海洋开发

It is about exploiting various marine resources and converting the potential value of ocean into an actual economic value, social benefit and ecological benefit by means of technology and facilities.

应用各种技术手段和设施，开发利用各种海洋，使海洋的潜在价值转化为实际经济价值、社会效益和生态效益的一切活动。

ocean hydrological chart 海洋水文图

It refers to a special chart, which mainly marks the physical property and dynamic property of sea water.

以表示海水的物理性质和动力性质为主要内容的专题海图。

ocean industry 海洋产业

It refers to production and service activities of ocean development, utilization and protection in the following five aspects: production and service activities for acquiring products directly from the sea; production processing and service activities for acquiring products directly from the sea; production and service activities for products directly applied to ocean and ocean exploitation; production and service activities for using seawater or ocean space as basic elements for processing; activities for marine science study, education, management and service.

开发、利用和保护海洋所进行的生产和服务活动。主要表现在以下五个方面:直接从海洋中获取产品的生产和服务活动;直接从海洋中获取的产品的一次加工生产和服务活动;直接应用于海洋和海洋开发活动的产品生产和服务活动;利用海水或海洋空间作为生产过程的基本要素所进行的生产和服务活动;海洋科学研究、教育、管理和服务活动。

ocean internal waves 海洋内波

They are waves that oscillate within the ocean.

在海洋内部发生的波动现象。

ocean observation technology 海洋观测技术

It is the technology necessary for the observation and survey of marine elements.

观察和测量海洋各种要素所用的技术。

ocean park 海洋公园

It refers to public places such as marine scenic spots, historic sites, and historical relics including marine geological relics, marine paleontological relics and marine natural landscapes for sightseeing and recreation.

海洋公园和海洋旅游景点、古迹、遗址(海洋地质遗迹、海洋古生物遗迹、海洋自然景观)的服务活动。

ocean remote sensing 海洋遥感

It refers to the use of sensing instruments placed on a platform far away from the target to measure marine elements without physical contact.

用装载在远离目标的平台上的遥感仪器对海洋的要素进行非接触测量的技术。

ocean stratification 海洋层化

It explains stratification where thermodynamic state parameters such as seawater temperature, salinity and density, are distributed along with depth.

海水的温度、盐度和密度等热力学状态参数随深度分布的层次结构。

ocean thermal energy 海水温差能

It is generated by the temperature difference between warmer shallow and cooler deep ocean waters.

又称"海洋热能",指由海洋表层温水与深层冷水之间的温差所蕴藏的能量。

ocean thermal energy power generation 海洋温差能发电

It refers to production activities of converting ocean thermal energy into electricity.

将海洋温差能转化成电能的生产活

动。

ocean wave 海浪

It refers to a wave phenomenon generated by wind at the ocean surface, mainly including wind wave and swell.

海面由风引起的波动现象。主要包括风浪和涌浪。

ocean waves angular spreading 海浪的角散

It refers to a propagation phenomenon of ocean waves spreading in different directions in the propagation process when the propagation directions of wave components are inconsistent.

海浪组成波的传播方向不一致时,在传播过程中向不同方向分散的传播现象。

ocean waves dispersion 海浪的弥散

It refers to the dispersion phenomenon of waves caused by the different speed of their wave components when ocean waves are superimposed during their propagation processes.

海浪在传播过程中,叠加在一起的波动,因其组成波速度不同而分散传播的现象。

ocean-atmosphere heat exchange 海-气热交换

It refers to heat exchange between ocean and atmosphere.

海洋与大气间的热量交换。

ocean-going cargo transportation 远洋货物运输

It refers to freight-oriented ocean transportation.

包括专门从事远洋货物运输的活动,以货运为主的远洋运输活动。

ocean-going fishing vessels 海洋远洋机动渔船

It refers to the vessels used for perennial or seasonal production within sea area beyond Chinese jurisdiction (such as foreign economic exclusive zone or open sea) by ocean fishery industries and production units following management methods of Chinese oceangoing fishery projects.

由各远洋渔业企业和各生产单位按我国远洋渔业项目管理办法,在非我国管辖水域(外国专属经济区水域或公海)进行常年或季节性生产的渔船。

oceangoing transportation of passengers 远洋旅客运输

It includes passenger transportation activities of ocean going passenger vessels, and passenger-oriented ocean going transport activities.

包括远洋客轮的旅客运输活动和以客运为主的远洋运输活动。

oceanic crust 洋壳

A special type of deep sea or ocean bottom, it is the crust which lies under the ocean basin. It is composed of a sedimentary layer and a silicon-magnesium layer from top to bottom, lacking a silicon-aluminum layer (granitic layer). Oceanic crust is thinner, (generally 5 - 8 kilometers), with an average thickness of 7.3 kilometers, and a mean density of 3.0 grams per cubic centimetre.

又称"大洋型地壳",界于大洋盆地之下的地壳,深海或大洋底部特有的类型。自上而下由沉积层和硅镁层组成,缺失硅铝层(花岗岩层),厚度较薄(5~8千米),平均厚 7.3 千米,平均密度为 3.0 克/厘米^3。

oceanic turbulence 海洋湍流

It refers to a phenomenon when unsteady vortices appear on many scales and interact with each other in ocean. In statistics, it refers to a pulsatory, irregular, instantaneous motion caused when seawater fluid blocks of different scales superimposed on each other based on the average movement within a period.

海洋中尺度不等的海水流体块,随时发生外观极不规律运动的现象。在统计学上认为是在一定时间的平均运动上,叠加瞬时不规则随机脉动的一种运动。

oceanic polymetallic nodule mining 大洋多金属结核开采

It refers to the mining or the exploitation of the ferrous metal auxiliary raw material mines, such as the oceanic polymetallic nodule.

大洋多金属结核等黑色金属辅助原料矿的采选活动。

oceanic province, oceanic region, oceanic zone 大洋区

[*marine science*] It refers to the deep and wide regions which are far away from the mainland.

[*ecology*] It is the region of open sea beyond the edge of the continental shelf where the water measures approximately 200 metres deep and it has relatively stable physical and chemical parameters.

[海洋科学]远离大陆,深度较大,面积广阔的区域。

[生态学]大陆架(水深大约200米)以外的远海大洋水层区。该区理化条件比较稳定。

oceanic ridge 海岭

It refers to an underwater mountain which stands in deep sea basins or on a continental slope.

耸立在深海盆地和大陆坡上的海底山脉。

oceanographic prediction 海洋预报

It is the prediction and information release carried out for marine conditions and marine phenomena such as tide, salinity, seawater temperature, sea wave, sea current, sea ice, tsunami, storm tsunami, sea-level change, coastal erosion and seawater intrusion.

对潮汐、盐度、海温、海浪、海流、海冰、海啸、风暴潮、海平面变化、海岸侵蚀、咸潮入侵等海洋状况和海洋现象开展的预测和信息发布的活动。

oceanographic survey 海洋调查

It refers to a survey or on investigation conducted in marine hydrology, marine meteorology, marine chemistry, marine biology, marine ecology, marine geology, etc. by research vessels, volunteer observing ships, buoys, etc.

通过调查船、志愿船、浮标等对海洋水文、气象、化学、生物、生态、地质等进行的测量或调查活动。

oceanographic observation 海洋观测

It refers to observations and surveys as well as related data acquisitions, transmissions, analyses and evaluations, targeting at tides, salinity, sea temperatures, sea waves, sea currents, seawater, tsunami waves, etc., in order to master and describe oceanographic conditions.

以掌握、描述海洋状况为目的,对潮汐、盐度、海温、海浪、海流、海冰、海啸波等进行的观察测量活动,以及对相关数据采集、传输、分析和评价的活动。

oceanographic environment observation 海洋观测环境

It refers to the smallest tridimensional space indispensable for the acquisition of continuous, accurate and representative observation data, with oceanographic observation stations/points as its centre, to ensure the normal operation of oceanographic observations.

为保证海洋观测活动正常进行,以海洋观测站(点)为中心,以获取连续、准确和具有代表性的海洋观测数据为目标所必需的最小立体空间。

oceanographic observation facilities 海洋观测设施

They include observation station houses, radar station houses, observation platforms, observation wells, observation ships, buoys, submerged buoys, seabed bases, observation marks, instruments and equipment, communication links, etc. as well as accessories for oceanographic observation stations/points.

海洋观测站(点)所使用的观测站房、雷达站房、观测平台、观测井、观测船、浮标、潜标、海床基、观测标志、仪器设备、通信线路等及附属设施。

oceanographic observation stations 海洋观测站

Also called *oceanographic observation points*, it is the oceanographic observation location established in the ocean, on an island or along the coast, for acquiring oceanographic observation data.

又称"海洋观测点",为获取海洋观测资料,在海洋、海岛和海岸设立的海洋观测场所。

oceanography 海洋学

It is a branch of earth science that studies marine natural phenomena, organisms, changing rules and its interaction with the atmosphere, coast and sea floor.

研究海洋的自然现象、机体、变化规律及其与大气、海岸、海底相互作用的一门学科。

oceanology 海洋学

Oceanography is a branch of earth science that studies marine natural phenomena, organisms, changing rules and its interaction with the atmosphere, coast and sea floor.

研究海洋的自然现象、机体、变化规律及其与大气、海岸、海底相互作用的一门学科。

ocean-related industry 海洋相关产业

It refers to an industry which has technological and economic connection with marine industry, with both the input and output as a bond.

以各种投入产出为联系纽带,与海洋产业构成技术经济联系的产业。

offshore 外滨

It refers to the relatively flat, irregularly shaped zone that extends outward from a breaker zone to the edge of a continental shelf.

破浪带的外界延伸到大陆架边缘的地带。

offshore area 近岸海域

It refers to a sea area which is close to a continental coast.

距大陆海岸较近的海域。

offshore evaluation well 海上评价井

It refers to a well drilled offshore where industrial oil and gas flow area has been ascertained after exploration. Before that, the amount of reserve ascertained should be submitted and materials for development plan are needed, including the evaluation on its hydrocarbon reservoirs, exploration of its characteristics, boundary containing oil and gas and reserve variation.

为评价油气藏,并探明其特征及含油气边界和储量变化,提交探明储量,获取油气田开发方案所需资料,而在海上勘探已获工业油气流面积上钻的井。

offshore hydrocarbon resource 海洋油气资源

It refers to the mineral assemblage of hydrocarbon with economic value, formed through geological processes.

由地质作用形成的具有经济意义的海底烃类矿物聚集体。

offshore oil and gas industry 海洋油气业

It refers to the production and services for exploring, extracting, delivering and processing of oil and natural gas in the ocean.

在海洋中勘探、开采、输送、加工石油和天然气的生产和服务活动。

offshore oil gas field 海上油气田

It refers to the aggregation of a number of oil and gas reservoirs in the same secondary tectonic belt at sea.

海上同一个二级构造带内若干油气藏的集合体。

offshore oil and gas production 海上采油

It refers to the whole processing technology of submarine oil and gas reservoir exploitation and the whole process of oil and gas processing and transmission.

海底油气藏开采的整套工艺技术和油气处理以及输送的全过程。

offshore oil and gas production system services 海上油气生产系统服务

It refers to the service activities provided for an offshore oil and gas production system, such as the service activities provided for a fixed production system, floating production system, underwater production system, etc.

为海上油气生产系统提供的服务活动,如为固定式生产系统、浮式生产系统、水下生产系统等提供的服务活动。

oil spill accident 溢油事故

It refers to the leakage of crude oil and its refined products in abnormal operation conditions. Oil spill accidents fall into three classes in terms of the oil spill volume, namely large, medium and small-sized ones. An oil spill volume of less than 10 tons is rated a small oil spill; an oil spill volume of 10-100 tons is rated a medium oil spill; an oil spill volume of more than 100 tons is rated a large oil spill.

非正常作业情况下原油及其炼制品的泄漏。溢油事故按其溢油量的多少分大、中、小三类,溢油量小于 10 吨的为小型溢油事故;溢油量在 10~100 吨的为中型溢油事故;溢油量大于 100 吨的为大型溢油事故。

oil spill disaster 溢油灾害

It refers to environmental disasters and ecological disasters at sea and along the shore, caused by large spills of oil or other oils due to offshore production activities or accidents.

由于海上生产活动或事故导致石油或其他油类等大量泄漏,造成的海上和岸边的环境灾害和生态灾害。

Okun's law 奥肯定律

In economics, Okun's law (named after Arthur Melvin Okun) is an empirically observed relationship relating unemployment to losses in a country's production. The "gap version" states that for every 1% increase in the unemployment rate, a country's GDP will be roughly an additional 2% lower than its potential GDP.

描述 GDP 变化和失业率变化之间存在的一种相当稳定的关系。奥肯定律的内容是:失业率每高于自然失业率 1 个百分点,实际 GDP 将低于潜在 GDP 的 2 个百分点。

old ice 老年冰

It refers to floating ice that is more than two years old. The surface of old ice is smoother than one-year-old ice.

至少经过一个夏天而未融化的两年冰,其表面比一年冰更平滑。

open complex giant systems 开放的复杂巨系统理论

They are a new scientific concept which was put forward by the famous Chinese scientist, Qian Xuesen, for the first time in 1990. According to this system complexity is the dynamic characteristic of these systems. Namely, elements of these systems are of a large number and variety with strong interrelationships and interactions. They are integrated as a whole in a hierarchy with vague boundaries, which proves the complexity of the systems.

1990 年,中国著名科学家钱学森首次向世人公布"开放的复杂巨系统"这一新的科学概念,认为复杂性实际上是开放的复杂巨系统的动力学特性。即构成系统的元素不仅数量巨大,而且种类极多,彼此之间的联系与作用很强,它们按照等级层次方式整合起来,不同层次之间界限模糊,甚至包含几个层次也不清楚,这种系统的动力学特性就是复杂性。

open economy 开放经济

Contrast to a closed economy, it refers to a state in which the economic activities of a country or area are closely related to an international or foreign market. A certain trade flow of import and export or flow of fund is involved in open economy as well.

"封闭经济"的对称。指一个国家或地区的经济活动与世界市场或外地市场有着密切联系的状况。在开放经济中,存在着一定规模的进出口贸易流或资本流。

open-end use of sea area 开放式用海

It refers to the direct utilisation of a sea area without sea reclamation, sea-closure or structure installation.

不进行填海造地、围海或设置构筑物,直接利用海域进行开发活动的用海方式。

opposite coasts 相向海岸

They are coasts with geographical relationships opposite to those of coastal states' waters. The waters of states with opposite coasts should have their boundaries delimited to avoid overlap.

沿海国海域的地理关系相互面对的海岸。海岸相向国家的海域,需划定其边界以避免重复。

optimal blowdown level 最优排污水平

It refers to the pollution level when economic activities produce maximum net social benefits, namely the pollution level when marginal private net income equals marginal external cost. An optimal blowdown level depends on the equilibrium of management and environmental costs.

经济活动产生最大社会净效益时的污染水平。即边际私人净收益与边际外部成本相等时的污染水平。最优排污水平取决于治理成本与环境成本的均衡。

ordered chaos 混沌有序

It refers to a complex state under which the condition or change of a matter is relatively obscure but its future development or change rule is relatively clear.

事物的状态或变化相对模糊,但未来发展或变化规律相对明晰的复杂状态。

ore-forming process in groundwater 地下水成矿作用

It refers to the process of the ore-forming ingredients which deposit or accumulate to form ore in local districts under the appropriate hydrogeochemical circumstances.

地下水中的成矿组分在适宜的水文地球化学环境中,在局部地段沉淀,富集形成矿床的过程。

organic agriculture 有机农业

It refers to crop farming that uses organic fertilizers in production to meet the nutritional needs of crops. Little or no synthetic fertilizers, pesticides, growth regulators, and livestock and poultry feed additives are applied. It also refers to livestock and poultry farming that uses organic feed to meet the nutritional needs of domestic animals.

在生产中完全或基本不用人工合成的肥料、农药、生长调节剂和畜禽饲料添加剂,而采用有机肥满足作物营养需求的种植业,或采用有机饲料满足畜禽营养需求的养殖业。

original value of fixed assets 固定资产原价

It refers to the total expenditure on the purchasing, construction, installation, reconstruction, extension and technical reform of a company. The unit of measurement is ten thousand yuan.

企业在购置、自行建造、安装、改建、

扩建、技术改造某项固定资产时所支出的全部支出总额。计量单位:万元。

orogeny 造山运动

It is the movement of materials in a crust or lithosphere roughly along the tangent to the earth's surface. Such movement often exhibits as the horizontal squeezing and stretching of rocks, in other words, horizontal displacement is generated, and folds and faults are produced to structurally form huge fold mountain systems, as well as grabens (or rift valleys), etc.

地壳或岩石圈物质大致沿地球表面切线方向进行的运动。这种运动常表现为岩石水平方向的挤压和拉伸,也就是产生水平方向的位移以及形成褶皱和断裂,在构造上形成巨大的褶皱山系和地堑、裂谷等。

osmoregulation 渗透压调节

It refers to a mechanism which can balance the salinity and body fluids inside organisms so as to adapt to various salinity environments.

在一定范围内,生物维持体内盐分和体液平衡而使之能适应不同盐度环境的机制。

Our Common Future《我们共同的未来》

It is a report on the future of mankind by the *World Commission on Environment and Development* (*WCED*). In February, 1987, this report was passed by the eighth WCED held in Tokyo, Japan. Later it was adopted by the 42nd General Assembly debate and was officially published in April, 1987. With "sustainable development" as the basic program, the report used abundant materials to discuss the problems in today's global environment and development, and it offered concrete and realistic action proposals to address these problems.

世界环境与发展委员会关于人类未来的报告。1987年2月,在日本东京召开的第八次世界环境与发展委员会上通过,后又经第42届联大辩论通过,于1987年4月正式出版。报告以"持续发展"为基本纲领,以丰富的资料论述了当今世界环境与发展方面存在的问题,并提出了处理这些问题的具体的和现实的行动建议。

outer harbor 外港

It is the part of a harbour out towards the sea, through which a vessel enters an inner harbour.

靠近外海一侧的港口部分。

outer limit of the territorial sea 领海外部界限

It refers to a boundary line between state sovereignty waters and state jurisdiction waters. In fact, it is a line, on which the distance from any point to the nearest point in baseline equals to the breadth of the territorial sea.

国家主权水域与国家管辖水域的分界线。实际上,为一条其每一点距领海基线上最近点的距离等于领海宽度的线。

overwintering ground 越冬场

It refers to waters where fish cluster and inhabit in winter. 鱼类冬季集群栖息的水域。

owners equity 所有者权益

It refers to the economic interest that owners enjoy in enterprise assets, which is equal to the balance of assets after deducting all liabilities. It includes paid-in capital (or equity), capital reserves, surplus reserves, undistributed profits, etc. The unit of measurement is ten thousand yuan.

所有者在企业资产中享有的经济利益,它等于企业资产减去负债后的余额。包括实收资本(或股本)、资本公积、盈余公积和未分配利润等。计量单位:万元。

ownership of marine resources 海洋资源所有权

It refers to ownership, right of use, usufruct and right of disposition on waters under jurisdiction of sovereignty by a nation.

国家对其主权所管辖海域的资源空间的所有权、使用权、收益权和处分权。

ownership of sea area 海域所有权

It refers to ownership of rights such as regulation, utilization, income and resources exploitation over marine area. According to Chinese law, the sea areas within Chinese territory are owned by the State, and the State Council exercises the right of ownership in the sea areas on behalf of the State.

对海洋区域的管制、使用、收益及资源开发等权利的归属。我国法律规定,中华人民共和国海域属于国家所有,国务院代表国家行使海域所有权。

Oyashio 亲潮

It refers to a cold current in the northwest of the Pacific Ocean, which flows from Kamchatka peninsula to the south along Kuril Islands and east of Hokkaido and joins Kuroshio in the northeast of Japan, reaching nearly 40°N and moves into the North Pacific current.

又称"千岛寒流"。太平洋西北部的一支寒流。自堪察加半岛沿千岛群岛和北海道东侧海域南下,在日本东北部海域约北纬40°附近与黑潮相汇合,并流入北太平洋暖流。

oyster reef 牡蛎礁

It is a kind of calcareous sediment which is mainly composed of deposit of organic remains, such as oysters.

一种钙质堆积体。主要由牡蛎等生物遗骸堆积而成。

P

Pacific Ocean 太平洋

Located among Asia, Oceania, America and Antarctica, it is the world's largest and deepest ocean with the most marginal seas and islands.

位于亚洲、大洋洲、美洲和南极洲之间的世界上最大、最深且边缘海和岛屿最多的大洋。

paleoceanography 古海洋学

It is a discipline which studies marine environments, incidents and evolution in geological history.

研究地质历史时期海洋环境和事件及其演变的学科。

panda standard 熊猫标准

It is a voluntary emission reduction standard designed specifically for the Chinese market. In a narrow sense, it defines the emission reduction test standards and principles; in a broad sense, it specifies the procedures, assessment bodies, regulation limits, etc. So as to perfect the market mechanisms. The standard was established to meet the demands of enterprises and individuals in China for taking action on climate issues. After emission reduction of a project is achieved and verified by a qualified third party, the enterprise will be awarded with corresponding quantity of panda standard credit quotas, which can be traded. It marks that China has begun to make its voice heard on global carbon trading.

专为中国市场设立的自愿减排标准,从狭义上确立减排量检测标准和原则,广义上规定流程、评定机构、规则限定等,以完善市场机制。它的设立是为了满足中国国内企业和个人就气候问题采取行动的需求,一些项目实现了减排或者清除之后,即遵循熊猫标准的原则,被合格的第三方机构核证,并通过注册,可以获得相应数量的熊猫标准信用额,信用额可以买卖。它标志着中国开始在全球碳交易中发出自己的声音。

Pangaea 泛大陆

It is a hypothetical supercontinent including all the landmass of the earth, surrounded by panthalassa, during the Late Paleozoic and early Mesozoic, when it split into Laurasia and Gondwanaland.

又称"联合古陆"。假定于古生代晚期和中生代早期,曾存在一个集地球上所有大陆为一体并被原始大洋围绕着的超级大陆。

panthalassa 泛大洋

It was the vast global ancestral ocean that surrounded the supercontinent Pangaea, during the late Paleozoic and early Mesozoic eras.

古生代晚期至中生代早期,围绕泛大陆的原始大洋。

Pareto Efficiency 帕累托效率

It is a state of allocation of resources in which it is impossible to make any one individual better off without making at least one individual worse off. The term is named after Vilfredo Pareto (1848—1923), an Italian engineer and economist who used the concept in his studies of economic efficiency and income distribution.

某一制度主体接受一种制度安排时,既是为了使自身的得益最大化又使得其他制度主体的得益最大化的状态。

partially mixed estuaries 部分混合河口

They refer to estuaries where there is no obvious interface between salt water and fresh water, the mixing degree of which is moderate. But there exists a density gradient in both horizontal and vertical directions. The action of both tidal current and runoff in these estuaries is strong. A relatively strong vertical circulation is formed due to the mixing of a large amount of salt water from a lower layer into an upper layer.

混合程度中等,咸淡水之间不存在明显的交界面,但在水平和垂直两个方向都存在密度梯度的河口。这种河口的潮流和径流作用都比较强,由于下层盐水大量掺入上层,形成较强的垂向环流。

passage 隘口

It refers to a long, narrow and steep zone growing on an ocean ridge or rise.

海岭或海隆上发育狭长、陡峭的地带。

passenger carrying capacity 客运量

It refers to the total number of passengers practically transported by ship. The unit of measurement is ten thousand people.

港口船舶实际运送的旅客人数。计量单位:万人。

passenger throughput 旅客吞吐量

It refers to the number of passengers who enter and depart ports by boat, including those who buy half-price tickets and take excursion vessels. Children eligible for free admission, ship crew and passengers who take short trips within a port are not counted in. The unit of measurement is ten thousand people.

经由水路乘船进、出沿海港区范围的旅客数量,包括购买半票的旅客人数和乘旅游船进、出沿海港区的旅客人数,不包括免票儿童、船舶船员人数、轮渡和港区内短途客运的旅客人数。计量单位:万人次。

passive continental margin 被动大陆边缘

Also called *Atlantic-type continental margin* or *stable continental margin*, the earth crust of passive continental margin is the transition from the oceanic crust to the continental crust. The continent and ocean are located in the belt of transition of the same rigid lithosphere plate, where the tensile and splitting effect is obvious and the fault basin is developed with no oceanic trench subduction zone, strong earthquake, volcano and orogeny.

又称"大西洋型大陆边缘"、"稳定大陆边缘",其地壳是洋壳到陆壳的过渡,大陆和海洋位于同一刚性岩石圈板块内的过渡带,是拉张裂离作用显著,断陷盆地发育,缺乏海沟俯冲带,无强烈的地震、火山和造山运动的大陆边缘。

patch reef 点礁

It is a small, irregular organic reef with a mound shape or a flat top forming a part of a reef complex, the span of which is less than 1 km.

一种呈土墩状或平顶状的珊瑚礁，跨度小于1千米。

path dependence 路径依赖

A system with positive feedback mechanism tends to evolve along certain path without being replaced by other potential or better systems once it is influenced by an occasional event outside, namely, the evolution of the system is "locked" on its current evolution way, structure and functions.

具有正反馈机制的体系一旦在外部偶然性事件的影响下被系统所采纳，便会沿着一定的路径发展演进，而很难为其他潜在的甚至更优的体系所替代，即制度的演化"锁定"在现有制度演化的方式及制度的结构和功能上。

patterns of sea area use 用海方式

It is the way of sea area use defined according to the features of sea area use and the degree of influence on the natural attribute of waters.

根据海域使用特征及对海域自然属性的影响程度划分的海域使用方式。

Pearl River 珠江

It is China's third longest river, and second largest by volume. With a length of 2,210 kilometres, a basin area of 442,585 square kilometres, the average annual runoff of 349.2 billion cubic metres, and an average annual runoff depth of 772 milimetres, the River flows through provinces such as Yunnan, Guizhou, Guangxi, Guangdong, etc. The main tributaries are Xi Jiang ("the West River"), Dong Jiang ("the East River"), and Bei Jiang ("the North River").

中国境内第三长河流，按年流量是中国第二大河流。长度2 210千米，流域面积442 585平方千米，平均年径流量3 492.00亿立方米，平均年径流深度772毫米，流经滇、黔、桂、粤等省（自治区），主要支流有西江、东江、北江。

Pearl River Delta 珠江三角洲

It is a piece of plain formed by thousands of years' erosion from the Pearl River. Starting north of Guangzhou, it spreads fanwise to the southeast and southwest. To its east are the Shenzhen special economic zone and the adjacent Dongguan, and to the southwest from north to south are Foshan, Jiangmen, Zhongshan and the Zhuhai special economic zone bordering Macao. Covering a land area of 41,700 square kilometres, the Delta includes 14 cities and counties: Guangzhou, Shenzhen, Zhuhai, Foshan, Jiangmen, Dongguan and Zhongshan, as well as Huizhou City and its three counties Huiyang, Huidong, Boluo, Zhaoqing City and its two county-level cities Gaoyao, Sihui.

珠江数千年来冲刷出来的一块平原，北起广州，呈扇形向东南和西南放射，东面有经济特区城市深圳和与之相邻的东莞，西南由北至南有：佛山、江门、中山以及与澳门接壤的经济特区城市珠海市。包括14个市县；广州、深圳、珠海、佛山、江门、东莞、中山等 7 市,以及惠州市的市区和高要、四会两市,其土地面积为 4.17 万平方千米。

pearl river delta economic zone 珠江三角洲经济区

It refers to the economic zone within the coastal areas of the Pearl River Delta, mainly including sea areas and land areas of Guangzhou, Shenzhen, Zhuhai and other cities under the jurisdiction of Guangdong Province.

珠江三角洲沿岸地区所组成的经济区域,主要包括广东省所辖的广州、深圳和珠海等城市的海域与陆域。

Penghu Islands 澎湖列岛

They are located in the south of the Taiwan Strait, consisting of 64 islands and covering an area of 126 square kilometres. Numerous islands, interlocked harbours and rugged terrains make the Penghu Islands a natural boundary between the East China Sea and the South China Sea, which have always been of vital military importance since ancient times and a springboard of introducing mainland culture to Taiwan.

位于台湾海峡的南部,由 64 个岛屿组成,面积约 126 平方千米,域内岛屿罗列,港湾交错,地势险要,是中国东海和南海的天然分界线,自古以来就为兵家必争之地,也是大陆文化传入台湾的跳板。

peninsula 半岛

It is a long narrow piece of land that sticks out from a larger piece of land into sea or lake and is bordered by water on three sides.

伸入海洋或湖泊,三面被水域包围的陆地。

perigean tide 近地点潮

It refers to a tide formed when the moon is approaching its perigee.

月球在近地点附近日期的潮汐。

periglacial landform 冰缘地貌

It refers to a landform in cold areas formed mainly by the action of ice melting.

主要由冻融作用塑造成的寒区地形。

peripheral area 外围区

It refers to an area with relatively slow economic development and relatively low development level, which is in a low economic and technical gradient and develops by receiving economic and technical radiation from a core area.

经济发展相对较缓慢、发展水平比较低的地区,处于经济技术低梯度上,接受核心区的经济技术辐射而得到发展。

Persian Gulf 波斯湾

It refers to an arm of the Arabian Sea between the Arabian Peninsula and southwest Iran. It has been an important trade route since ancient times and gained added strategic significance after the discovery of oil in the Gulf States in the 1930s.

印度洋阿拉伯海西北海湾,位于阿拉伯半岛与伊朗高原之间。

persistent organic pollutant (POP) 持久性有机污染物

It refers to organic pollutant that has a high level of toxicity and can persist in the ocean environment. They can be accumulated gathered in organisms through the food chain and do harm to human health.

毒性极高,在海洋环境中持久存在,能通过食物链在生物体内富集并危害人体健康的有机污染物。

Petty-Clark's Law 配第-克拉克定律

It indicates that with the development of economy and increase of per capita income, labour force moves from a primary industry to a secondary industry and a tertiary industry in order. In view of the distribution of labour force in these three industries, the proportion of labour force in the primary industry declines, while the labour force in the secondary, and especially in the tertiary industry, grows in number.

随着经济的发展和人均国民收入水平的提高,劳动力首先由第一产业向第二产业转移,进而再向第三产业转移;从劳动力在三次产业之间的分布状况看,第一产业的劳动力比重逐渐下降,第二产业特别是第三产业劳动力的比重则呈现出增加的趋势。

Phillips curve 菲利普斯曲线

It demonstrated the relationship between rates of unemployment and corresponding rates of inflation that result in an economy. Stated simply, decreased unemployment, (i.e., increased levels of employment) in an economy will correlate with higher rates of inflation. The curve indicates when the rate of unemployment is lower, there will be higher increase in money salary. Conversely, when the rate of unemployment is higher, there will be lower increase in money salary, even negative increase.

以横轴表示失业率、纵轴表示货币工资增长率的坐标系中,画出一条向右下方倾斜的曲线,这就是最初的菲利普斯曲线。曲线表明,当失业率较低时,货币工资增长率较高;反之,当失业率较高时,货币工资增长率低,甚至为负数。

physical capital 物质资本

It is the stock of equipment and building used in the production of goods and services.

用于生产物品与劳务的设备和建筑物存量。

physical oceanography 物理海洋学

In a narrow sense, it is a discipline that studies the force field, the thermohaline structure as well as the associated temporal and spatial changes of various motions, material exchange, energy exchange and conversion in the ocean by applying the viewpoint and method of physics.

In a broad sense, it is a discipline that studies a physical phenomenon in the ocean and its change rules, as well as the interaction between ocean water and the atmosphere, lithosphere and biosphere with the theory, method and technology of physics.

狭义上指运用物理学的观点和方法研究海洋中的力场、热盐结构以及因之而产生的各种运动的时空变化,海洋中的物质交换、能量交换和转换的学科。

广义上指以物理学的理论、方法和技术,研究海洋中的物理现象及其变化规律,并研究海洋水体与大气圈、岩石圈和生物圈的相互作用的学科。

Pigouvian Taxes 庇古税

It is an environmental tax levied on waste polluters according to the damage caused by pollution. The tax is used to bridge the gap between private cost and social cost so as to make them equal. This tax is named after economist Arthur Pigou.

根据污染所造成的危害对排污者征税,用税收来弥补私人成本和社会成本之间的差距,使两者相等,这种税便被称为"庇古税"。

pipeline transportation for marine oil and gas 海洋油气管道运输

It refers to the transportation of marine oil and natural gas through pipes.

通过管道对海洋石油和天然气的运输活动。

planktology 浮游生物学

It is a science which studies the biological phenomena and activities of plankton. Planktology studies the biological content of plankton such as the morphology structure, phyletic classification, chemical composition, and biological activities. In addition, it involves the species composition and quantitive variation of plankton in time and space and their mutual relations with marine environments, such as marine hydrology, marine geology, marine physics, chemistry, etc.

研究浮游生物的生命现象和活动规律的科学。浮游生物学一方面研究浮游生物的形态构造、系统分类、化学组成和生命活动等生物学内容;另一方面也涉及浮游生物在时间空间上的种类组成和数量变动与海洋水文、地质、物理、化学等海洋环境的相互关系。

plankton 浮游生物

It is a diverse group of organisms that lack advanced locomotive organs, having no or little weak motion ability. They provide a crucial source of food to many large aquatic organisms, such as fish and whales. They float in the water column or drift with the water current, including phytoplankton and zooplankton.

缺乏发达的运动器官,没有或仅有微弱的运动能力,悬浮在水层中,常随水流移动的生物。包括浮游植物和浮游动物两大类。

planned expenditure 计划支出

It is the amount of money spent willingly on products and services by families, enterprises and government.

家庭、企业和政府愿意花在产品和服务上的数额。

planning area 规划区

It refers to the well-built areas in a city, town or village, and regions where planning control should be implemented for the needs of urban and rural construction and development.

城市、镇和村庄的建成区以及因城乡建设和发展需要,必须实行规划控制的区域。

plateau 坪

It refers to the extensive area of a flat or plain in a mountain or hilly area.

泛指山区或丘陵区局部的平地或平原。

polar science 极地科学

It is a discipline that studies ice and

snow, geology, geophysics, marine hydrography, meteorology, chemistry, organisms, environment and others in polar area.

研究南北极地区的冰雪、地质、地球物理、海洋水文、气象、化学、生物、环境等的学科。

polarization effect 极化效应

It refers to a phenomenon when areas and regions whose economic activities are expanding will attract net population inflow, capital inflow and trade activities to accelerate development, leading to reduced pace in development of surrounding areas.

又称"回波效应",指经济活动正在扩张的地点和地区将会从其他地区吸引净人口流入、资本流入和贸易活动,从而加快自身发展,并使其周边地区发展速度降低的现象。

pollution stress 污染压力

It refers to the stress put by sea pollutants on the structure and function of a marine ecosystem.

入海污染物质对海洋生态系统结构和功能的胁迫。

polymetallic nodule 多金属结核

Also called *manganese nodule*, it refers to the nodule-like manganese, iron oxide and oxide deposits formed in the sediments on ocean floors.

又称"锰结核",自生于海底表层沉积物中的呈结核状的铁锰氢氧化物和氧化物矿床。

polymorphism 多态现象

It refers to the occurrence of more than one form of individual in a single species within an interbreeding population.

同种生物的个体对某些形态、性质等所表现的多样性状态。

population supporting capacity of land 土地人口承载力

It is the number of a population at a certain consumption level supported by the food produced in a certain area of land resources.

一定面积土地资源生产的食物所供养的一定消费水平的人口数量。

port and harbour planning 港口规划

It refers to the prediction of the passenger and cargo throughput in different periods and the corresponding proposed planning of the construction scale, facility layout and stage construction of the port and harbour.

对港口不同时期客货吞吐量的预测及相应拟定的建设规模、设施布置和分期建设安排等。

port and harbour works 港口工程

It refers to the engineering activities and corresponding works of building, extending and rebuilding constructions of the port and harbours.

兴建、扩建或改建港口的建筑物的工程活动及相关设施。

port general plans 港口总体规划

They refer to the specific plans for a port at a given period of time, including the water area and land area of the port, the division of the port area, the throughput of the port, the property and function of the port, the use of the water area and land area, the construction of the port facilities, the use of the coastline, the configuration of the construction land, the stage construction sequence, etc. The general plans for the port should conform to the layout plans.

一个港口在一定时期的具体规划,包括港口的水域和陆域范围、港区划分、吞吐量和到港船型、港口的性质和功能、水域和陆域使用、港口设施建设岸线使用、建设用地配置以及分期建设序列等内容。港口总体规划应当符合港口布局规划。

Port Law of the People's Republic of China《中华人民共和国港口法》

It was enacted with a view to strengthening port administration, maintaining port safety and operational order, protecting the legitimate rights and interests of the parties and promoting the construction and development of ports. Adopted at the *Third Meeting of the Standing Committee of the Tenth National People's Congress of the PRC* on June 28, 2003, it came into effect as of January 1, 2004.

为了加强港口管理,维护港口的安全与经营秩序,保护当事人的合法权益,促进港口的建设与发展,制定本法。2003年6月28日中华人民共和国第十届全国人民代表大会常务委员会第三次会议通过,2004年1月1日起施行。

port layout plans 港口布局规划

They encompass the port layout plans of a municipality directly under the central government, the autonomous region, the province and the nation.

港口的分布规划,包括全国港口布局规划和省、自治区、直辖市港口布局规划。

port resources 港口资源

They refer to the natural resources which can be used for construction and development, such as ports and their peripheral coasts, bays, river banks, islands, etc.

港口及其周边的海岸、海湾、河岸、岛屿等可供建设与发展的天然资源。

port shipping area 港口航运区

It refers to the sea area delimited for the ships' safe navigation, mooring, loading and unloading operation or wind shelter, including ports, sea-routes and anchorages.

为满足船舶安全航行、停靠,进行装卸作业或避风所划定的海域,包括港口、航道和锚地。

possibility space 可能性空间

It is a collection of all possibilities which occur in the process of a matter's development and changes.

事物发展变化中各种可能性集合称为这个事物的可能性空间。

potassic fertilizer 钾肥制造

It is the manufacturing of potassic fertilizer by extracting potassium salt such as potassium chloride and potassium sulfate from sea water and bittern.

从海水、苦卤中提取氯化钾、硫酸钾等钾盐,用于制造钾肥的生产活动。

potassium chloride manufacturing 氯化钾制造

It refers to production activities of extracting potassium chloride from bittern and subsurface brine.

以苦卤、地下卤水为原料提取氯化钾的生产活动。

potential density 位密

It is the density that seawater micelles at an isobaric surface (depth) would acquire when they adiabatically rise to the sea surface.

某一等压面(深度)处的海水微团绝热上升到海面时所具有的密度。

potential temperature 位温

It is the temperature that seawater micelles at an isobaric surface (depth) would acquire when they adiabatically rise to the sea surface.

海洋中某一等压面(深度)处的海水微团绝热上升到海面时所具有的温度。

power generation from ocean tidal energy 海洋潮汐能发电

It refers to production activities of converting ocean tidal energy power into electricity.

将海洋潮汐能转化成电能的生产活动。

power generation from ocean wave energy 海洋波浪能发电

It is a production activity of converting ocean wave power into electricity.

将海洋波浪能转化成电能的生产活动。

precipitation recharge 降水补给

It is the process during which precipitation infiltrates and provides supplies to ground water.

降水入渗补给地下水的过程。

preferential attachment 择优连接

New institutional nodes are constantly added into an institutional network, but probabilities of new nodes attaching to existing nodes in the network are different. Not all nodes can successfully get connected, and the addition rule priority connection, that is so called competitiveness in selection of attachment nodes.

制度网络中不停地有新的制度节点加入,但是新节点连接到网络中现有节点的概率是有差异的,并不是所有的节点都能同样成功地获取连接,加入的规则是择优连接,即所谓连接节点选择上的竞争性。

present value 现值

Also known as *present discounted value*, it is a future amount of money that has been discounted to reflect its current value, as if it existed today.

用现行利率产生一定量未来货币所需要的现在货币量。

prevention and treatment of marine land-based pollutions 海洋陆源排污治理

It refers to the comprehensive treatment of pollution in coastal waters and sea-going rivers.

沿海水域、入海河流的污染综合治理活动。

prevention of pollution 污染预防

It is the use of various processes, practices, materials or products to avoid, reduce or control pollution. It may include recycling, treatment, process change, control mechanism, efficient use of resources and material substitution, etc.

旨在避免、减少或控制污染而对各种过程、惯例、材料或产品的采用,可包括再循环、处理、过程更改、控制机制、资源的有效利用和材料替代等。

primary production 初级生产量

It is the production of organic compounds from atmospheric or aquatic carbon dioxide. It may occur through the process of photosynthesis, using light as a source of energy, or chemosynthesis, using the oxidation or reduction of chemical compounds as a source of energy. Almost all life on earth is directly or indirectly reliant on primary production. The organisms responsible for primary production are known as primary producers or autotrophs, and form the base of the food chain. In terrestrial ecoregions, these are mainly plants, while in aquatic ecoregions algae are primarily responsible. Primary production is distinguished as either net or gross, the former accounts for losses to processes such as cellular respiration, the latter does not.

自养生物通过光合作用或化能合成作用所固定的太阳能量或所制造的有机物的量,初级生产量分总初级生产量和净初级生产量,后者是总初级生产量减去自养生物在光合作用或化能合成作用的同时因呼吸作用所消耗的量。

primary productivity 初级生产力

It refers to an autotroph's ability to convert inorganic carbon into organic carbon through photosynthesis using solar energy or through chemosynthesis using chemical energy.

自养生物利用太阳能进行光合作用,或利用化学能进行化能合成作用,同化无机碳为有机碳的能力。

principle of recycling and regeneration 循环再生原理

All products in the world will eventually become waste, and any "waste" in the world must be the "raw material" or buffering agent useful for a particular component or ecological process in the biosphere. All human behaviour will eventually feed back to the actor itself in the form of a message, either beneficial or harmful. Recycling and regeneration of materials, and feedback and regulation of information are the fundamental motives for the sustainable development of a complex ecosystem.

世间一切产品最终都要变成废物,世间任一"废物"必然是对生物圈中某一组分或生态过程有用的"原料"或缓冲剂;人类一切行为最终都会以某种信息的形式反馈到作用者本身,或者有利、或者有害。物资的循环再生和信息的反馈调节是复合生态系统持续发展的根本动因。

private saving 私人储蓄

It is the remaining income of a family after the payment of taxes and spending.

家庭在支付了税收和消费之后剩下来的收入。

processing and manufacturing of marine synthetic crude oil 海洋人造原油加工制造

It refers to the production activities of extracting crude oil from coastal oil forming shale.

从海滨油母页岩中提炼原油的生产活动。

processing of marine flavorings 海洋调味品加工

It refers to the processing of flavourings with marine aquatic products as accessories, such as seafood sauce, oyster sauce and shrimp sauce.

以海洋水产品为辅料生产各种调味品的活动,如海鲜酱油、耗油、虾酱等。

processing of marine instant food 海洋方便食品加工

It refers to the processing of quick-frozen food and instant food using marine aquatic products as the raw material, such as quick-frozen fish balls, fish fillets and seaweed.

以海洋水产品为原料制成的各种速冻食品、方便食品的制造。如速冻鱼丸、鱼片、海苔等。

producer price index(PPI) 生产者物价指数

It refers to a comprehensive index of wholesale price changes, often viewed as an indicator of future retail price changes. The PPI tracks prices of foods, metals, lumber, oil and gas, and many other commodities, but does not measure the price of services. Economists look at trends in the PPI as an accurate precursor to changes in the CPI, since an upward or downward pressure on wholesale prices is usually passed through to consumer prices over time.

企业购买的一篮子物品与劳务的费用的衡量指标。

product ecology 产品生态学

It is a discipline that, through identification and diagnosis, can specify ecological environment parameters which influence the product competitive power and formulate ecological standards for product for sales, thus incorporating the ecological environment value into the commercial value of the whole product, such as the energy-saving and fluoride-free refrigerator.

通过辨识和诊断,确定影响产品竞争能力的生态环境参数,制定产品进入市场的产品生态规范,使整个产品商业价值中包含生态环境价值的学科,如低能耗、无氟的冰箱等。

product life cycle 产品生命周期

It is a life chain of a whole process composed of raw material collection and preparation, product manufacturing and processing, packaging, transportation, retail sale, consumer use, recycling, maintenance, final recycle or treatment as waste.

一种产品从原料采集、原料制备、产品制造和加工、包装、运输、分销,消费者使用、回用和维修,最终再循环或作为废物处理等环节组成的整个过程的生命链。

product life cycle design 产品生命周期设计

It comprehensively takes all the environment factors in a product's whole life into consideration and incorporates them into the design in a product development period, aiming at minimizing the impact of the whole life cycle of product on the environment which finally leads to and produces a manufacturing and consuming system with more sustainability.

在产品开发阶段,综合考虑产品整个生命周期过程中的环境因子,并将其纳入设计之中,以求产品整个生命周期过程中的环境影响最小化,最终引导产生更具有可持续性的生产和消费系统。

production factor endowments 生产要素禀赋

It refers to the relative abundance of various production factors within an area.

区域内各种生产要素的相对丰裕程度。

production of aquatic products 水产品产量

It is the number of aquatic products from businesses of state-owned enterprises, public institutions, cooperative economic organizations of urban and rural collective ownership, individual fishing and sideline businesses of farming families, as well as joint operations of a variety of joint ventures or cooperations. Namely, it is the number of aquatic products of the whole society. The unit of measurement is ten thousand tons.

由国有企业、事业单位、城乡集体所有制合作经济组织、个体渔业和农民家庭副业经营以及各种合资、合作联合经营的水产品数量,即全社会的水产品数量。计量单位:万吨。

production of marine natural gas 海洋天然气产量

It refers to the sales volume of natural gas through a gathering network and the overall production of natural gas for local use. The production is calculated by volume of export sales and volume for enterprise's self-use. The unit of measurement is ten thousand cubic metres.

包括进入集输管网的销售量和就地利用的全部气量。天然气产量=外输(销)量+企业自用量。计量单位:万立方米。

production of marine petrochemical products 海洋石油化工产品产量

It refers to the amount of products processed through distillation, catalytic cracking, hydrocracking, alkylation, hydrofining, electro-chemical refining, lubricant processing, etc., using crude oil from the beach or the ocean as a raw material. The products include organic intermediate, hydrocarbon and its halogenated derivatives, etc. The unit of measurement is tons.

以取自滩海或海上的原油为原材料,通过蒸馏、催化裂化、加氢裂化、烷基化、加氢精制、电化学精制及润滑油加工等生产的产品数量。海洋石油化工产品包括有机中间体、烃类及其卤化衍生物等。计量单位:吨。

production of metal placer 金属砂矿产量

It refers to the production quantity of metal ores, including ferrous (mainly iron ore) and non-ferrous metals (copper ore, lead-zinc ore, nickel-cobalt ore, tin ore and magnesium ore etc.), precious metals (gold ore and silver ore), rare metal ores (lanthanide metal ore, and metal ores which are similar to lanthanide metal ore), oceanic polymetallic nodules, submarine polymetallic sulphides, polymetallic soft muds etc. The unit of measurement is tons.

包括黑色(主要指铁矿)和有色金属(铜矿、铅锌矿、镍钴矿、锡矿、镁矿等)、贵金属(金矿、银矿等)、稀土金属矿(镧系金属矿、镧系金属性质相近的金属矿)、大洋多金属结核、海底多金属硫化物和多金属软泥等金属矿生产数量。计量单位:吨。

production of natural gas 天然气产量

It is the total volume of the natural gas entering the gathering pipeline network for sales and the gas utilized in situ. The unit of measurement is ten thousand cubic metres.

进入集输管网的销售量和就地利用的全部气量。计量单位:万立方米。

production of oceangoing fishing 远洋捕捞产量

It refers to the output of aquatic products caught in a sea area beyond the jurisdiction of a country (foreign exclusive economic zone, continental shelf, or high seas). The unit of measurement is ten thousand tons.

在非本国管辖海域(外国专属经济区、大陆架或公海)捕捞的水产品产量。计量单位:万吨。

production of salt from seawater 海水制盐

It refers to activities of producing salt after solarization, concentration and crystallization with seawater (including underground brine in coastal shallow water) as a raw material.

以海水(含沿海浅层地下卤水)为原料,经日晒、浓缩、结晶生产海盐的活动。

production of sea salt chemical products 海盐化工产品产量

It refers to the number of products which are processed using chemicals directly extracted from the sea, such as sea salt, bromine, potassium and magnesium as raw materials. It includes the processing of caustic soda (sodium hydroxide), soda ash (sodium bicarbonate) and other alkalis and chemicals; the processing of potassium chloride and potassium sulphate with salt-manufacturing by-products as raw materials; or the processing of bromine products and other elements such as iodine. The unit of measurement is tons.

以海盐、溴素、钾、镁等直接从海水中提取的化学物质作为原料进行的一次加工产品数量。包括烧碱(氢氧化钠)、纯碱(碳酸氢钠)和其他碱类和化学原料的生产;及以制盐副产物为原料进行的氯化钾和硫酸钾的生产;或溴素加工产品以及碘等其他元素的加工生产。计量单位:吨。

production of seawater desalination 海水淡化产量

It refers to the production of freshwater from seawater desalination by using various desalination technologies in the reporting period. The unit of measurement is ten thousand tons.

报告期内利用各种淡化技术将海水处理为淡化水的产量。计量单位:万吨。

production of seaweed chemical products 海藻化工产品产量

It refers to the amount of products produced with chemical matters such as iodine, potassium iodate, seaweed gel, mannitol and carrageenan extracted from marine algae as raw materials.

利用化学方法从海藻中提取碘、碘酸钾、海藻胶、甘露醇、卡拉胶等化学物质作为原料进行的一次加工产品数量。计量单位:吨。

productions of marine crude oil 海洋原油产量

It refers to marine crude oil output directly for sale and for self use, calculated by crude oil output. Production volume of marine crude oil within marine petroleum system is currently calculated by retrograde method. The unit of measurement is ten thousand tons.

按净原油量来计算的,能直接用于销售和生产自用的海洋原油量。目前海洋石油系统原油产量计算方法采用倒算法。计量单位:万吨。

productions of marine drugs and biological products 海洋药物和生物制品产量

It refers to the amount of products when objects of medical value are extracted from marine organisms to produce medicines and biological products.

从海洋生物提取药用价值物体,加工制作医药和生物制品的产品数量。

productive area of salt pan 盐田生产面积

It is the area provided directly for sea salt production, including a crystallization area, evaporation area, brine keeping area, area of ditches, trenches, pools and ridges within a beach, as well as an area of beach lumps. The unit of measurement is hectares.

直接提供给海盐生产的面积,包括结晶面积、蒸发面积、保卤面积、滩内的沟、壕、池、埝面积及滩坨面积。计量单位:公顷。

productivity 生产率

It refers to the amount of goods and services produced by the labour input per unit.

每单位劳动投入所生产的物品和劳务的数量。

professional technical personnel 专业技术人员

They refer to people who are engaged in professional technical work and professional technical managerial work, namely personnel appointed to professional technical posts and engaged in professional technical work and professional technical managerial work, as well as those not appointed to professional technical posts but currently working in professional technical positions in enterprises or institutions. Professional technical posts involve 17 categories, namely engineering and technical personnel, agricultural technical personnel, scientific research personnel, health technical personnel, teaching staff, economic personnel, accountants, statisticians, translators and interpreters, personnel working in the field of library materials, archives, cultural relics and museums, press and publication staff, lawyers and notaries, radio and television broadcasters, arts and crafts personnel, sports personnel, artistic personnel, art personnel, as well as personnel of enterprises' political and ideological work.

从事专业技术工作和专业技术管理工作的人员,即企事业单位中已经聘任专业技术职务从事专业技术工作和专业技术管理工作的人员,以及未聘任专业技术职务,现在专业技术岗位上工作的人员数。包括工程技术人员、农业技术人员、科学研究人员、卫生技术人员、教学人员、经济人员、会计人员、统计人员、翻译人员、图书资料、档案、文博人员、新闻出版人员、律师、公证人员、广播电视播音人员、工艺美术人员、体育人员、艺术人员及企业政治思想工作人员,共17个专业技术职务类别。

prograding coast 淤涨岸

It is a coast with a shoreline advancing to the sea, caused by the fact that the longshore sediment supply is greater than the number of sediments marine power can move.

因沿岸泥沙供应数量大于海洋动力所能搬移的数量导致滨线向海推进的海岸,称为淤涨岸。

Program on the Global Ecology and Oceanography of Harmful Algal Blooms(GEOHAB) 全球有害藻华的生态学与海洋学研究计划

It is an international coordination project on harmful algal blooms observation and prediction cooperation study organized by the *Scientific Committee on Oceanic Research* in 2001.

2001年由海洋研究科学委员会领导和组织的有害藻华观测和预报合作研究的国际协调计划。

proluvium fan 洪积扇

It is a fan-shaped accumulation landform formed by ephemeral torrent in the valley outlet, which is widely distributed in dry or semi-dry regions. It has gentle slope with coarse gravels accumulating on the top of proluvium fan and turning thinner towards marginal matters.

暂时性流水在沟谷出口处形成的扇状堆积地貌,多分布于干旱、半干旱地区。坡度比较平缓,粗大砾石堆积于洪积扇顶部,向边缘物质逐渐变细。

property rights 产权

In economics, they usually refer to the ownership and stewardship of land, capital, and other goods, broadly including the right to ownership, possession, control, use, usufruct and disposition.

是经济所有制关系的法律表现形式。它包括财产的所有权、占有权、支配权、使用权、收益权和处置权。

protection of marine forest tree 海洋林木养护

It refers to measures for accelerating the growth and development of forests in a tidal mud flat.

促进海涂林业生长发育的措施活动。

provincial marine functional zoning 省级海洋功能区划

It is approved by the maritime administration department of the provincial people's government and the relevant departments of the people's government at the corresponding level. Based on the geographic area and national marine functional zoning, it divides waters and islands under the jurisdiction of the people's government at the corresponding level. It ranges from the coastline (MHWST) to the outer limit of the territorial sea, and may extend landward according to the actual situation.

省级人民政府海洋行政主管部门会同本级人民政府有关部门,依据全国海洋功能区划开展的,以本级人民政府所辖海域及海岛为划分对象,以地理区域和海洋功能区为划分单元的海洋功能区划,其范围为自海岸线(平均大潮高潮线)至领海的外部界限,可根据实际情况向陆地适当延伸。

proxy services of marine transportation 海洋运输代理服务

It refers to proxy services related to marine transportation.

与海洋运输有关的代理及服务活动。

public investment policy 公共投资政策

It is an important composing part of the regional economic development policy. Public investment policy is formulated to offer special support to regions lagging behind in transportation, communications, power supply, water supply, education and culture, in order to enhance the balanced development of the regional economy and avoid disparity in these fields.

区域经济发展政策的重要组成部分,是为促进区域间经济的均衡发展而制定的对交通、通信、供电、供水、教育、文化落后的区域给予特别支持,避免在这些领域差距悬殊的政策。

public saving 公共储蓄

It refers to the tax revenue remaining after the government pays its expenditure.

政府在支付其支出后剩下的税收收入。

publication of marine audio and video products 海洋音像制品出版

It refers to the publication of marine audio and video products.

海洋活动的音像制品出版活动。

publication of marine propaganda materials 海洋宣传品出版

It refers to the publication of marine publicity materials, such as posters and publicity pictures.

海洋活动的宣传品出版活动,如宣传画、宣传图片等。

purchasing-power parity 购买力平价

Also called *law of one price*, it is an economic concept which posits that "a good must sell for the same price in all locations". It constitutes the basis of the theory of purchasing power parity, an assumption that in some circumstances (for example, as a long-run tendency) it would cost exactly the same number of, for example, US dollars to buy euros and then to use the proceeds to buy a market basket of goods as it would cost to use those dollars directly in purchasing the market basket of goods.

The purchasing power parity (PPP)

is an equivalent coefficient between currencies calculated based on different price levels in different countries, which allows us to make a reasonable comparison among gross domestic products of different countries in economics. There is a large deviation between theoretical exchange rate and real exchange rate.

又称"一价定律",经济学的一个著名假说,该定律说明同样的产品在同一时间在不同地点不能以不同的价格出售。购买力平价是一种根据各国不同的价格水平计算出来的货币之间的等值系数,使我们能够在经济学上对各国的国内生产总值进行合理比较,这种理论汇率与实际汇率可能有很大的差距。

pyramid distribution urban system 金字塔型分布的城镇体系

Also called *smooth-scale distribution urban system*, it refers to the fact that the number of cities and towns decreases with the increase in size of an urban system.

又称"顺序-规模分布型城镇体系",指城镇的数量随着城镇规模的增加而减少。

pyramid of production rate 生产率金字塔

It refers to a pyramid-shaped nutritive layer composed of autotrophic plants, herbivorous animals and carnivorous animals in the order of productivity from high to low in a stable ecology.

在一个稳定的生态系统中,由最低层生产率最大的自养植物、上一层的植食性动物和最上层生产率最小的肉食性动物形成的金字塔状的营养层。

Q

Qiongzhou Strait 琼州海峡

Located between the southern Chinese islands, Hainan and the Leizhou peninsula, the Qiongzhou Strait is a channel of great significance which connects the Beibu gulf and the South China Sea.

位于中国海南岛与雷州半岛之间,沟通北部湾与南海的重要通道。

quasi-real right 准物权

It refers to a real right similar to civil property right due to its dominance, absoluteness and exclusiveness in that it takes as object other properties distinct from things like personal or movable possessions. It mainly includes the right to use sea areas, the mineral prospecting right, the mining right, the water intake right and the right to use water areas or tidal flats for breeding or fishery, all of which are part of the usufructuary right.

以物之外的其他财产为客体的具有支配性、绝对性和排他性因而类似于物权的民事财产权。准物权具体包括海域使用权、探矿权、采矿权、取水权和使用水域、滩涂从事养殖、捕捞的权利,是用益物权的一部分。

quotient of location 区位熵

It is a quantitative tool used for the analysis of industry efficiency and profit. As a general way of group recognition, it is employed to measure the relative concentration degree of some aspect of an industry in a specific area.

产业的效率与效益分析的定量工具,是一种较为普遍的集群识别方法,是用来衡量某一产业的某一方面,在一特定区域的相对集中程度。

R

random walk 随机行走

It means the path of changes in a variable is unpredictable.

一种变量变动的路径是不可预期的。

randomness 随机性

The uncertain causal relationship among things makes the result of an incident uncertain. Events or data series are said to be random when they lack specific direction, and cannot be explained or predicted except by chance or in a probabilistic or statistical manner.

事物本身是确定的,但因为事物的因果关系不确定,从而导致事件发生的结果不确定。

rare element resources 稀有元素资源

They are a variety of rare element resources contained in sea water, such as bromine, uranium, and heavy water. Rare elements refer to the elements of very low abundance on the earth, especially within the earth's crust.

海水中所含有的各种稀有元素资源,如溴、铀、重水。稀有元素指地球上,特别是地壳内丰度极低的元素。

rate of coastal erosion 海岸侵蚀速率

It refers to the magnitude of the recession of a coastline or the magnitude of the downcutting of the intertidal zone and offshore slope per unit time. In some regions, the coastline can be substituted by a zero water depth contour of the nautical map or other depth contour, but the selection of the depth contour cannot surpass the 20-metre-deep range of the nautical map.

单位时间内海岸线位置后退的幅度或潮间带和水下岸坡下蚀的幅度。在某些地区,海岸线可以海图零米等深线或其他等深线代替,但等深线的选取不能超过海图 20 米水深的范围。

rate of groundwater level decline 地下水水位下降速率

It refers to the decline value of groundwater level (or water head) in unit time.

单位时间内地下水水位(水头)下降值。

ratio of material flow 物质流通率

It is the movement amount of material in an ecological system in unit time, unit area or unit volume.

生态系统中的物质在单位时间、单位面积或体积的移动量。

rational expectation 理性预期

It is a hypothesis assuming that agents can optimize all information, including information theories about government policies to make future predictions.

当人们在预测未来时可以充分运用他们所拥有的全部信息,包括有关政府政策的信息的理论。

real exchange rate 实际汇率

It is the purchasing power of a currency relative to another at current exchange rates and prices. It is the ratio of the number of units of a given country's currency necessary for buying a market basket of goods in the other country, after acquiring the other country's currency in the foreign exchange market, to the number of units of the given country's currency that would be necessary to buy that market basket directly in the given country. Thus the real exchange rate is the exchange rate times the relative price of a market basket of goods in the two countries.

又称"真实汇率",是名义汇率用两国价格水平调整后的汇率,即外国商品与本国商品的相对价格,它反映了本国商品的国际竞争力。实际汇率有时称为贸易条件,它告诉我们能按什么比率用一国的产品交换另一国的产品。

recession 衰退

It refers to a downward trend in a business cycle characterised by a decline in production and rise in unemployment.

经济经历产出下降和失业上升的时期时,经济被称为处于衰退中。

recession coast 后退海岸

It is caused by erosion due to sea level rise and coast settlement, leading to reduction of land area and increase of sea area, which is widespread in seismic active areas and delta bay.

海面上升或海岸沉降引起的海岸侵蚀后退,使陆地面积减少,海域面积增大,多分布于地震活动频繁区和三角湾地区。

recharge area 补给区

It refers to an area in which water is absorbed so that it eventually reaches a zone of saturation in one or more aquifers or areas that are close to the earth's surface receiving atmospheric precipitation and surface water.

含水层出露或接近地表接受大气降水和地表水等入渗补给的地区。

reclaim land from the sea 围海造田

It is the reclamation of a gulf shoaly land frequently submersed in water. Feasibility study, demonstration, planning and design shall be conducted using the overall planning of land utilisation as a basis, and a complete irrigation and drainage system must be constructed in the reclamation area to facilitate the discharge of surface water, lower groundwater level, or to leach salt and irrigate crops.

对经常淹没在水下的海湾浅滩地的围垦。围海造田必须以土地利用总体规划为依据,进行可行性研究论证和规划设计,围垦区内须修建完整的排灌系统,以便排除地面水和降低地下水位,或借以淋洗盐分和灌溉作物。

reclamation area 填海面积

It is an accumulated reclamation area of coastal regions for production and service activities. It is aggregated according to the classification of national economic industries. The unit of measurement is square kilometres.

沿海地区进行生产与服务活动累计填海面积。按照国民经济行业分类汇总。计量单位:平方千米。

reclamation engineering constructions 围填海工程建筑

They are the construction activities of building a cofferdam with a certain height in a coastal zone to enclose waters with a certain range and filling the area with sand, soil or stone to form land.

在海岸带建造一定高度的围堰、圈围一定范围的海域,填以泥沙或土石形成陆地的施工活动。

recreational fisheries 休闲渔业

They are fisheries activities aiming at entertainment and sports, such as game fisheries, and ornamental fish culture.

以休闲娱乐和体育运动为目的的渔业活动。如游钓、观赏鱼养殖等。

recycling effect 再循环效应

It refers to cycles of resources between main bodies, thus maximizing the use of limited resources.

资源在主体间的循环往复,从而使有限资源得到最大限度的利用。

Red Sea 红海

It is an inland sea located in the northwest of Indian Ocean between Asia and Africa.

位于亚洲与非洲之间,印度洋西北的陆内海。

red tide 赤潮

It refers to an ecologically abnormal phenomenon caused by explosive proliferation of marine microalgae. These algae, known as phytoplankton, are single-celled protists, plant-like organisms that can form dense, visible patches near the water's surface and do harm to other organisms in the ocean. When the algae are present in high concentrations, the water appears to be discoloured or murky, varying in colour from pink, red, green or yellowish brown.

一定环境条件下,海洋微藻、原生动物或细菌暴发性增殖,聚积达到某一水平而引起的水体变色或对海洋其他生物产生危害的生态异常现象以及海洋近岸水域微型与小型浮游生物,高度密集而导致表层水变为粉红色、赤色、绿色或黄褐色的现象。

reductionism 还原论

Adopting the method of system decomposition, it is a philosophical position which holds that a complex system is nothing but the sum of its parts, and that an account of it can be reduced to an account of individual constituents. It is based on superposition principle in physics that the whole is the sum of its parts.

还原论采用的是系统分解的方法,即将一个复杂系统分解成若干个相对简单的子系统,只要研究清楚各个子系统的性质就可以获得整个系统的性质,其理论基础是物理上的叠加原理:整体等于部分之和。

reef flat 礁滩

It refers to a sea beach which is composed of pulverised corals, gravels and coral reefs in the intertidal zone of a coral coast.

在珊瑚海岸潮间带,由珊瑚碎屑、沙砾与珊瑚礁体胶结而成的海滩。

regeneration 再生

It is a process in which part of an organism regenerates a complete body.

生物体的一部分重新生成完整机体的过程。

region 区域

It refers to a large space which holds multiple kinds of resources and where a variety of productive and non-productive social and economic activities can be operated.

拥有多种类型的资源、可以进行多种生产性和非生产性社会经济活动的一片相对较大的空间范围。

region reasonable scale 地域合理规模

The region reasonable scale of industrial distribution is based on the reasonable scale of the enterprises. The optimal combination of the production facilities and equipment of the enterprises will increase the production capacity and yield, and decrease the product costs, thus bringing large-scale saving benefits. The regional collaboration of these enterprises can bring greater economies of scale. The regional aggregation scale of the enterprises which brings large-scale saving benefits is called the region reasonable scale.

产业布局的地域合理规模,是以企业的合理规模为基础的。企业生产装置和设备的最佳组合,会使生产能力和产量大幅度增加,产品成本下降,从而带来大规模的节约效益,具有这种节约效益的企业在地域上的相互协作,又可以带来更大的规模效益,这种能够带来大规模节约效益的企业在某个地方聚集规模,我们称之为地域合理规模。

region sustainable development 区域可持续发展

It refers to the maintenance of harmonious, efficient, orderly and long-term development abilities by coordinating the relationship between population, resources, environment and economic and social development.

协调好区域内人口、资源、环境与经济、社会发展之间的关系与行为,使区域保持和谐、高效、有序、长期的发展能力。

regional development 区域开发

It refers to the process of achieving regional economic growth and development by utilizing resources, which indicates the comprehensive use of various resources, including natural, economic, technological, cultural and social resources in a specific region by a development subject for realizing maximum economic development and social progress while maintaining the harmony and unity of regional resources, environment, economy and society.

人类开发利用各种资源、谋取区域经济增长和区域经济发展的过程,指一定的开发主体对特定区域的自然、经济、技术、文化、社会等各种资源进行综合利用,在保持区域资源、环境、经济、社会和谐统一的前提下,求得最大的经济发展与社会进步。

regional economics 区域经济学

In a narrow sense, it is a field of science concerned with regional economic development and interregional relationship. Broadly speaking, it studies the general rule of regional economic development, which targets a specific geographic space and explores the development and changing rules of various economic phenomena in the geographic space.

狭义的区域经济学是研究区域经济发展和区际关系的科学,广义的区域经济学是研究区域经济发展一般规律的科学,即它是以特定的地理空间为研究对象,探究各种经济现象在地理空间上发展变化规律的科学。

regional economic cooperation 区域经济合作

It refers to the redistribution and combination of production factors between different regions so as to maximize the economic and social benefits which are performed by economic subjects of various regions in accordance with certain agreements or contracts. In nature, it is the intangible merchandize trade between different regions.

不同地区的经济主体,依据一定的协议章程或合同,将生产要素在地区之间重新配置、组合,以便获取最大的经济效益和社会效益的活动。区域经济合作实质上是区域之间的非物质商品贸易。

regional economic core area 区域经济核心区

It refers to an area which plays a leading role in national or regional economic development with rapid economic growth and high-quality development. It is a core area for realizing national or regional economic development, where the per capita GDP is much higher than the national average level.

在全国经济或者区域经济发展中居于主导地位、经济增长快、发展质量较高的地区,是一个国家或地区实现经济发展的主要地区,通常其人均GDP要大大高于全国人均水平。

regional economic development policy 区域经济发展政策

It is the combination of a series of regional economic development policies for the realization of specific objectives in a certain period and background

是国家在一定的时期和背景下,为实现特定的目标而制定的一系列关于区域经济发展的政策的总和。

regional economic development strategy 区域经济发展战略

It refers to the ideas and plans for the overall development of a regional economy. Based on the distribution of production factors in various regions and their status and functions in a national economic system, the development objectives, direction and general plans are set to guide the regional economic development and promote a regional economic boom.

对区域经济总体发展的设想、思路和谋划。它根据不同地区生产要素条件的分布情况和该地区在国家经济体系中的地位和作用,对地区未来发展的目标、方向和总体思路进行谋划,以达到指导地区经济发展、促进地区经济腾飞的作用。

regional economic development 区域经济发展

It refers to the improvement of regional economic development quality through technological innovation, upgrading of industrial structures and social progress, including the increase of per capita income, upgrading of industrial structures based on technological advances and improvement of urbanization levels.

通过技术创新、产业结构升级以及社会进步实现区域经济发展质量的提高,包括人均收入水平的提高、以技术进步为基础的产业结构升级和城市化水平的提高三层含义。

regional economic development 区域经济开发

It is a regional economic process in which human beings make full use of various economic development methods to promote the development of a certain region. It is a practical theory based on regional economic growth theory and has significant use value.

人类运用发展经济的各种手段作用于特定区域的区域经济过程,它是以区域经济增长理论为基础、具有更大使用价值的应用理论。

regional economic growth 区域经济增长

In a narrow sense, it refers to the increase of the entire social wealth in a region, which equals the increase of GDP and total production quantity in terms of currency and entity respectively. Broadly speaking, it includes the control of population, growth of per capita GNP, increase of product demand, etc.

狭义的区域经济增长是指一个区域内的社会总财富的增加,用货币形式表示,就是国内生产总值的增加,用实物表示,就是各种产品生产总量的增加。广义的区域经济增长则还包括对人口数量的控制、人均国民生产总值的提高,以及产品需求量的增加等。

regional economic offset policy 区域经济补偿政策

It refers to the redistribution of financial funds between governments, namely, financial transfers and payout policies. As an important aid, it plays a significant role in improving public infrastructure in underdeveloped areas, constructing favourable investment environments and narrowing the gap with developed areas.

财政资金在政府间的再分配,即财政转移支付政策。财政转移支付作为一种重要的援助手段,对于欠发达地区改善公共基础设施、创造良好的投资环境、缩小与发达区域的差距意义重大。

regional economic planning 区域经济规划

It refers to the detailed planning for regional and industrial development under the guidance of regional economic development strategies.

区域经济发展战略指导下的区域和产业发展的详细安排。

regional economic policy 区域经济政策

It refers to the combination of a series of policies on regional problems established by the government, which focuses on regional economic development. It aims at regulating market mechanisms in a spatial allocation of resources for realizing the optimal spatial allocation of resources and promoting the harmonious development of the regional economy.

政府针对区域问题而制定的一系列政策的总和,它的着重点是区域经济发展,它的必要性是纠正市场机制在资源的空间配置方面的不足,它的目标是实现资源在空间上的优化配置和促进区域经济的协调发展。

regional economic structure 区域经济结构

As a significant factor influencing regional economic growth, it refers to an internal relationship and proportion between resource allocation within a region and various industries. Regional economic growth and development largely relies on the advancement and reasonability of regional industry structure on the premise of a certain economic system and enterprise efficiency.

区域内资源配置和各类产业之间的内在联系和比例关系,是影响区域经济增长的重要因素之一。在一定经济体制和企业效率的前提下,区域经济增长与发展状况在很大程度上取决于区域产业结构的先进性和合理性。

regional environment 区域环境

It is a combination of both natural and social elements in a certain region and a complicated and multi-functional environment.

一定地域范围内的自然和社会因素的总和,是一种结构复杂、功能多样的环境。

regional image 地区形象

It is the comprehensive evaluation and overall impression from the public on a certain region.

公众对某一地区的综合评价和总体印象。

regional image shaping 地区形象塑造

It is the scientific summary and design of a regional image. It summarizes and generalizes the already existing regional characteristics, then designs a distinct way of expression for the convenience of the public inside and outside the region to know and understand it.

地区形象的科学的总结和设计,是将已经存在的地区特征归纳、总结出来,并设计出一个鲜明的表达方式,便于区内、区外公众对该区域的认识和了解。

regional industrial structure 区域产业结构

It refers to the proportion of all economic factors in a specific area.

特定区域内各经济要素之间的比例关系。

regional innovation 区域创新

It refers to the process of studying, developing, utilizing and transmitting new technology and knowledge in a region to promote its technological advances, upgrading of industrial structures and regional competitiveness.

一个地区研究、开发、运用和扩散新技术和新知识,并以此促进地区技术进步和产业结构升级,提升区域竞争力的过程。

regional investment environment 区域投资环境

It is a generic term for all factors outside an enterprise, which can influence the production and operating activities of an enterprise and their consequences within an invested region.

存在于受资区域内,能够影响企业生产经营活动过程及其结果的一切企业外部因素的总称。

regional oceanography 区域海洋学

It is a comprehensive discipline which studies various marine phenomena in a sea area.

综合研究一个海区的各种海洋现象的学科。

regional planning 区域规划

It is a detailed arrangement of regional and industry development guided by regional economic development policies, including regional development, industry development, land utilisation, urban systems, etc.

区域经济发展战略指导下的区域和产业发展的详细的安排。它包括区域发展、产业发展、土地利用、城镇体系等多方面的内容。

regional policy 区域政策

It refers to a total of a series of policies formulated by the government for regional problems. It focuses on regional economic development and stresses the correction of market mechanism deficiencies on spatial allocation of resources. As a typical governmental behavior, regional policy aims at realizing the optimal spatial allocation of resources and promoting the harmonious development of regional economy.

由政府针对区域问题而制定的一系列政策的总和,它的着重点是区域经济发展,它的必要性是纠正市场机制在资源的空间配置方面的不足,它的目标是实现资源在空间上的优化配置和促进区域经济的协调发展。区域经济政策是一种典型的政府行为。

regional spatial structure 区域空间结构

It refers to the interrelationship of economic development levels between various regions in the process of a nation's economic development.

一个国家经济发展中各个地区之间经济发展水平的相互关系。

regional specialization of production 地区生产专业化

It is the manifestation of highly centralized production in space. It makes use of the special advantages of a certain industry or production in a special region according to the law of area division of labour, concentrates on the development of a certain industry or certain products on a large scale, and then outputs to outside of the region for maximum economic benefits.

生产在空间上高度集中的表现形式,它按照劳动地域分工规律,利用特定区域某类产业或产品生产的特殊有利条件,大规模集中地发展某个行业或某类产品,然后向区外输出,以求最大经济效益。

regional leading industry 地区主导产业

With regional resource advantages as

the basis, it is the industry which can represent the development direction of the regional economy, and support or dominate the regional economic development to some degree.

以地区资源优势为基础,能够代表区域经济发展方向,并且在一定程度上能够支撑、主宰区域经济发展的产业。

regionalization of natural resources 自然资源区域性

It refers to the differences between all kinds of natural resources in the geographical spatial distribution.

各类自然资源在地理空间分布上的差异性。

regular lattices 规则网络

Every node in a lattice has the same degree and clustering coefficient. The most common regular lattices are the loop networks composed of N nodes and among which, every node of the lattice only connects to the nearest K nodes.

网络中每个节点具有相同的度和簇系数。最常见的规则网络是由 N 个节点组成的环状网络,网络中每个节点只与它最近的 K 个节点连接。

relationship of resource property right 资源权属关系

It is the reflection of resource property relations, ownership relations or power relations of economic subject to resource assets in the law.

资源的财产关系、所有制关系或经济主体对资源资产的权力关系在法律上的反映。

relic island 残岛

It refers to a part of an island which is temporarily left in the oceans, lakes or rivers. It will gradually shrink and finally disappear under the erosive action of sea, lake wave and flowing river water.

岛屿受海蚀、湖浪侵蚀或河流流水冲蚀,逐渐缩小,最后消失,暂时残留在海洋、湖泊或河流中的部分。

relic sea 残海

It refers to the sea areas where a geosynclines type of sea basin is at ageing and degradation stage, such as each branch of the Mediterranean Sea. Due to a very large area, the Mediterranean Sea, located among the Asian, European and African continents, retains some of the characteristics of the ocean; the Black Sea has become a semi-enclosed sea; the Caspian Sea and the Aral Sea have become landlocked seas, representing the final stage of development of the ocean.

又称"残留海",地槽型海盆处于衰老退化阶段的海域。如地中海地槽的各个支海。亚、欧、非三大洲之间的地中海,由于面积很大,保留着大洋的一些特征;黑海则成为半封闭的;里海和咸海则已隔离出来,代表着海域发展的最后阶段。

renewability of natural resources 自然资源可更新性

It is a characteristic of natural resources that can be constantly renewed for sustainable use through their own reproduction or recovery.

自然资源通过自身繁殖或复原,得

以不断推陈出新,从而能被持续利用的特性。

renewable energy independent power system 可再生能源独立电力系统

It is a renewable energy power system which can operate independently without connecting to a power grid.

不与电网连接的单独运行的可再生能源电力系统。

renewable energy resources 可再生能源

They refer to resources which can be regenerated and supplied or recycled regularly in nature, such as solar energy, wind energy, water energy, biomass energy, and tidal energy.

在自然界中可以不断再生并有规律地得到补充或重复利用的能源。例如太阳能、风能、水能、生物质能和潮汐能等。

renewable marine resources 海洋可再生资源

They are marine natural resources that are self-recovering and sustainable.

具有自我恢复原有特性,并可持续利用的一类海洋自然资源。

renewable resources 可再生资源

It is a generic term for all materials which are stored in various forms. After losing their original use value, they obtain a new use value through different processing ways including social production, communication and consumption.

又称"可更新资源",指在社会生产、流通、消费过程中的物质,不再具有原使用价值而以各种形式储存,但可通过不同加工途径而使其重新获得使用价值的各种物料的总称。

rent of sea area to the national government 海域租金

All entities and individuals who use sea area should pay fees to the State annually, in accordance with *The law of PRC on Administration of the Use of Sea Areas*. The rent actually is land rent.

使用国家海域的单位和个人,依照《海域使用管理法》的规定,按年度逐年向国家缴纳海域使用金,其实质是地租。

repair of fixed and floating marine equipment 海洋固定及浮动装置修理

It refers to the maintenance, major inspections and repairs for shipyards and marine floating structures.

对船厂或浮于海面结构的保养、大检和修理活动。

research for marine biological engineering technology 海洋生物工程技术研究

It refers to research and experimental development for marine biological engineering technology.

对海洋生物工程技术进行的研究与试验发展活动。

research for marine chemical engineering technology 海洋化学工程技术研究

It is about research and experimental development for marine chemical engineering technology.

对海洋化学工程技术进行的研究与试验发展活动。

research for marine environmental engineering technology 海洋环境工程技术研究

It is about research and experimental development for marine environmental engineering technology.

对海洋环境工程技术进行的研究与试验发展活动。

research for marine transportation engineering technology 海洋交通运输工程技术研究

It is about research and experimental development for marine transportation engineering technology.

对海洋交通运输工程技术进行的研究与试验发展活动。

reservation area 保留区

It refers to the sea areas which haven't been developed and utilized, and there is no plan for exploitation during the zoning period.

目前尚未开发利用,且在区划期限内也无计划开发利用的海域。

reserve ratio 准备金率

It is the proportion of deposits that banks hold as reserves.

银行作为准备金持有的存款比例。

reserve requirement 法定准备金

It is a central bank regulation that sets the minimum fraction of customer deposits and notes that each commercial bank must hold as reserves.

商业银行按照法律规定必须存在中央银行里的自身所吸收存款的一个最低限度的准备金。

reserves 准备金

They are vault cash physically held in commercial banks, plus deposits in accounts with the central bank according to a certain proportion. They are set aside against unexpected events such as unusually large net withdrawals by customers or even bank runs.

商业银行库存的现金和按比例存放在中央银行的存款。实行准备金的目的是为了确保商业银行在遇到突然大量提取银行存款时,能有相当充足的清偿能力。

residual sea surface height 剩余海面高

It refers to the instantaneous sea surface topography with various kinds of errors included, which plays an important role in the balancing of intersections.

包含有各种误差在内的瞬时海面地形,对交叉点进行平差有重要的作用。

resource allocation 资源配置

It is the process of reasonable allocation of resources to individual users according to certain principles.

根据一定原则合理分配各种资源到各个用户的过程。

resource asset management 资源资产管理

It is the management of resource assets as a means of production and wealth that can be used to obtain profits, which emphasizes both material management, and property management and value management of resources.

将资源资产作为能够获取收益的生

产资料和财富来进行管理。这种管理不仅强调资源的实物管理,而且强调产权管理和价值管理。

resource distribution 资源分布

It refers to the distribution of resources, including land, water, minerals, fuel and wealth in general among corresponding geographic entities (states, countries, etc.).

资源在空间所处的位置及其格局的特征。

resource ecology 资源生态学

It is a study of the formation, distribution, flow and consumption of natural resources as well as their processes and rules from the perspective of ecology, with emphasis on studying the impact of these processes on an eco-environment and the theories and methods of maintenance and reconstruction of natural resources.

从生态学的角度研究自然资源形成、分布、流动、消耗及其过程和规律的学科,并强调研究这些过程产生的生态环境影响及其自然资源维护与重建的理论与方法。

resource layout 资源布局

It refers to production arrangement of regional resources according to the allocation of resources.

根据资源配置而进行的区域资源生产安排。

resource planning 资源规划

It refers to the comparison, choice and arrangement made on resources exploitation, utilization and conservation based on the principle of sustainable development.

根据可持续发展的原则,对资源的开发利用与保育方案,做出比选与安排的活动过程。

resource policy 资源政策

It refers to the strategy made by a state for guiding activities such as development, utilization, management, protection of the resources in order to achieve strategic goal for socio-economic development within a certain period.

国家为实现一定时期内社会经济发展战略目标而制定的指导资源开发、利用、管理、保护等活动的策略。

resource strategy 资源战略

It refers to the strategies for major issues such as resource development, utilization and conservation from the global, long-term context and the aspect of internal links, external environment, etc.

从全局、长远、内部联系和外部环境等方面,对资源开发利用与保育等重大问题进行谋划而制定的方略。

resource substitute 资源替代

It is a behaviour where people gradually replace existing resources with similar or more effective resources through constant comparison, selection and re-understanding of various resources.

人类通过在各类资源间不断进行的比较选择和重新认识,逐步采用具有相似或更高效用的资源取代现有资源的行为。

resource zoning 资源分区

It refers to division of hierarchical systems in resource areas.

资源区域的等级系统划分。

resource-based city 资源型城市

It is a city developed on the basis of the exploitation of mineral resources and energy.

在开发矿产资源和能源的基础上发展起来的城市。

resource-intensive industry 资源密集型产业

Also called *land-intensive industry*, it is an industry that can be carried out only when more land and other natural resources are used in the input of production factors. Land resources as a production factor refer to all kinds of natural resources, including land, primeval forests, rivers, lakes and seas, and various mineral resources.

又称"土地密集型产业"。在生产要素的投入中需要使用较多的土地等自然资源才能进行生产的产业。土地资源作为一种生产要素泛指各种自然资源，包括土地、原始森林、江河湖海和各种矿产资源。

resources 资源

It is a generic term for various environment elements or things characterized by scarcity and social constraints. It can be developed and utilized under certain historical conditions by human beings to improve the welfare standards and survival ability. The fundamental nature of resources is social utility and relative scarcity to human beings.

在一定历史条件下，能被人类开发利用以提高自己福利水平或生存能力的、具有某种稀缺性的、受社会约束的各种环境要素或事物的总称。资源的根本性质是社会化的效用性和对于人类的相对稀缺性。

resources carrying capacity 资源承载力

It is the ability and capacity of resources to bear the population size and the corresponding total economic and social activities, under the condition that rational development and utilization of resources and virtuous cycle of eco-environment are ensured in a region within a certain period.

某区域一定时期内在确保资源合理开发利用和生态环境良性循环的条件下，资源能够承载的人口数量及相应的经济社会活动总量的能力和容量。

resources economy 资源经济

It is an economy featuring exploration, development, processing, utilization, circulation and reuse of various natural resources as the leading industry.

从事各类自然资源的勘查、开发、加工、利用、流通以及再利用为主导产业的经济。

resources of international seabed area 国际海底资源

It refers to the material, energy and space which can be utilized by human beings and distributed at the seabed, ocean floor, and its subsoil beyond the jurisdiction of the state.

分布在国家管辖范围以外海床和洋底及其底土上的、可以被人类利用的物质、能量和空间。

retail of marine biological medicines 海洋生物药品零售

It refers to the retail exclusively for

marine bio-chemical medicine, traditional Chinese medicinal materials, Chinese patent medicine, etc.

专门经营海洋生物化学药品、中药材及中成药等的零售活动。

retail of marine functional food 海洋保健品零售

It refers to the retailing of marine functional food.

海洋保健食品的零售活动。

retail of marine petroleum products 海洋石油产品零售

It refers to the retail exclusively for marine liquefied petroleum gas and other products.

专门经营海洋石油液化气等的零售活动。

retail of seawater desalination products 海水淡化产品零售

It refers to the retail activities of barreled water and bottled water produced through seawater desalination.

海水淡化后的桶装及瓶装水的零售活动。

reverse osmosis method 反渗透法

When external pressure higher than the osmotic pressure of solute is applied to the raw water of a membrane, only water passes where other substances are removed to stay at the surface of the membrane, as the raw water permeates the semipermeable membrane. The above process is called the reverse osmosis method or the reverse osmosis process.

在膜的原水一侧施加比溶液渗透压高的外界压力,原水透过半透膜时,只允许水透过,其他物质不能透过而被截留在膜表面的过程。

reverse osmosis process 反渗透法

When external pressure higher than the osmotic pressure of solute is applied to the raw water of a membrane, only water passes where other substances are removed to stay at the surface of the membrane, as the raw water permeates the semipermeable membrane. The above process is called the reverse osmosis method or the reverse osmosis process.

在膜的原水一侧施加比溶液渗透压高的外界压力,原水透过半透膜时,只允许水透过,其他物质不能透过而被截留在膜表面的过程。

revetment 护岸工程

It is the construction work to support bank slopes and slope bases with impact-proof material for preventing invasion by water flows, waves and others.

河、湖、海堤的岸坡和坡脚用耐冲材料保护,防止水流、波浪等侵袭破坏的工程。

ridge 山脊

It refers to the highest part of a mountain extending in a line.

海岭上呈线状延伸的最高部分。

right in rem 物权

It is the right of a holder to enjoy the direct, exclusive control of specific items, including proprietary right, usufructuary right and security interest.

权利人对特定的物享有直接支配和排他的权力,包括所有权、用益物权和担保物权。

right of innocent passage 无害通过权

It is the right of innocent passage through territorial seas that vessels of all states, whether coastal or landlocked, shall enjoy. Innocent passage is a concept in law of the sea which allows for a vessel to pass through the territorial waters of another state subject to certain restrictions. The UN Convention on the Law of the Sea defines innocent passage as: "Passage is innocent so long as it is not prejudicial to the peace, good order or security of the coastal state. Such passage shall take place in conformity with this *Convention* and with other rules of international law."

所有国家包括沿海国或内陆国,其船舶均享有无害通过领海的权利。无害通过是指不损害沿海国的和平、良好秩序和安全。

right of sea area use transfer 海域使用权转让

It refers to a behaviour in which owners of the right to sea area use obtain certain profits when transferring legally-obtained right to others by trade, reciprocity, donation, etc.

是指海域使用权人将依法取得的海域使用权,通过买卖、互易、赠与等途径让与他人,从而获取一定利益的行为。

right of sea area use 海域使用权

It refers to exclusive right of control on a state-owned sea area obtained by any entity or individual through legal procedures.

单位或个人以法定方式取得的对国家所有的特定海域的排他性支配权利。

rill mark 沙沟

It is a long and shallow sandy ridge which develops on the continental shelf.

大陆架上发育的长条状沙质浅沟。

Rio+20 summit 里约+20 峰会

The United Nations Conference on Sustainable Development was hosted by Brazil in Rio de Janeiro in June of 2012. Rio+20 summit was a 20-year follow-up to the 1992 *United Nations Conference on Environment and Development* (UNCED) held in the same city, which is commonly referred to as the "Rio+20 summit". In view of the UN schedule, the Rio conference has two main themes: to develop green economy to achieve sustainable development and lift people out of poverty; and to improve sustainable political management and institutional frameworks.

2012 年 6 月,"联合国可持续发展大会"将在巴西里约热内卢举行。此次会议与 1992 年在里约热内卢召开的"联合国环境和发展大会"正好时隔 20 年,因此也被称为"里约+20 峰会"。根据联合国的安排,里约大会的主题有两个:一是在可持续发展和消除贫困的背景下发展绿色经济;二是关于可持续政治治理与制度框架。

rip current 裂流

Also called offshore compensation current, it refers to a strong channel of surface current formed when the water that usually moves along the shore flows seaward because of breaking waves.

离岸补偿水流。破波从岸边分裂向岸的水流。

ripple marks 波痕

It refers to a micro geomorphology similar to the sections between a wave crest and tough, formed by non-viscous materials (mainly loose sandy sediments) under the action of water flow, wave or wind. It is widely distributed on a river bed, intertidal section, subtidal zone and the surface of sand dunes.

非黏结性物质(主要是松散的砂质沉积物)在水流、波浪或风的作用下形成的一种波脊与波谷相间的微地貌形态。波痕广泛出现在河床、潮间带、潮下带以及风成沙丘的表面。

rise 海隆

It refers to the gently protruded elevation at the ocean bottom generally shaped as long strips. The rise is found rising from the seafloor to several hundred metres in height and seamounts may develop on part of these rises.

大洋底部中平缓突起的高地,通常呈长条状,高出海底数百米,部分海隆上发育海山。

risk aversion 风险规避

It refers to an adverse attitude of investors towards risk. When faced with two investments with a similar expected return but different risks, a risk-averse investor will prefer the one with the lower risk.

又称"风险厌恶",指投资者对投资风险反感的态度,即在降低风险的成本与收益的权益过程中,厌恶风险的人们在相同的成本下更倾向于作出低风险的选择。

risk of sea area use 用海风险

It is probability and damage degree of incidents that possibly damage, destroy and even ruin sea area functions or adjacent exploitation and utilization activities due to the sea use of projects (caused by human or natural factors).

由于项目用海(人为或自然因素引起的)对海域功能或相邻开发利用活动可能造成损害、破坏乃至毁灭性事件的发生概率及其损害的程度。

river 河流

It is a natural watercourse, usually freshwater, flowing towards an ocean, a lake, a sea, or another river. It is also defined as frequent or periodic water flow accumulated in depressions formed along the surface line.

沿地表线形低凹部分集中的经常性或周期性水流。

river bed 河床

It is an area usually covered by water in the river valley.

河谷中经常性被水所淹没的部分。

river mouth 河口

It is an outlet where the river flows into waters (ocean, lake, reservoir and main stream).

河流注入受水体(海、湖、水库、干流)的出口。

river port 河港

It refers to a port along the banks of rivers, which serves as a base for berthing, formation, and refueling of ships and as a terminal for passengers transportation and cargos.

位于江河沿岸的港口。它是内河运输船舶停泊、编队、补给燃料的基地,也是旅客交通和货物的集散地。

river source; headwaters 河源

It is considered to be the source stream of rivers, such as stream, spring, glacier, marsh and lake.

河流的发源地。可为溪、泉、冰川、沼泽或湖泊等。

river tide 江潮

It is a tidal phenomenon at lower reaches of a river. It occurs when tidal wave from seas goes upstream along the water course. Through friction with river bed and water discharging, the duration of the rising tide is short while that of the falling tide is long. With the distance back to upstream increasing, the occurrence of high tide and low tide is postponed, and the range of the tide decreases.

江河下游的潮汐现象。因外海潮波沿河道上溯而发生,并受河床摩擦和河水下泄的影响,涨潮历时较短,落潮历时较长。随着上溯距离的增加,高潮和低潮出现时刻逐渐推迟,潮差逐渐减小。

river valley 河谷

It is a depression through which river flows, which is formed by flowing water, including valley slope and valley bottom. Valley bottom is divided into river bed and valley flat while valley slope is the slope along the banks of the valley, sometimes with distribution of river terrace. River valley usually becomes lower and wider from upstream downwards.

河流所流经的长条形凹地,主要由流水作用形成,包括谷坡、谷底两部分。谷底通常又分为河床和河漫滩,谷坡即河谷两侧的斜坡,有时分布有河流阶地。河谷一般自上游逐渐降低、变宽。

robustness 鲁棒性

Robustness is a key concept in modern control systems which describes system parameters, structural changes and whether system remains effective when perturbed by an external interference.

现代控制系统的一个重要概念。它是刻画系统参数或结构变化以及受外干扰作用时,系统性质是否保持的一个概念。

rock bench 岩滩

It is a beach composed of bedrock.

一种由基岩构成的海滩。

rock uncovered 明礁

It is an isolated rock which appears when mean springs are at a high-tide level.

平均大潮高潮面时露出的孤立岩石。

root definition 根定义

It means to give a clear and concise description of the basic nature of a system with a certain thought or idea.

运用某种思想观点来得到对系统根本性质的清楚简洁的描述。

roughness 粗糙性

It refers to the indistinguishability between events due to insufficient and inadequate knowledge (or information) describing the events. All the indistinguishable events are grouped in one boundary region by rough set. Therefore, the indeterminacy of the rough set is a conception based on boundary, and when the boundary region is a null set, the question will become deterministic.

因为描述事件的知识(或信息)不充分、不完全,导致事件间的不可分辨性。粗糙集把那些不可分辨的事件都归属一个边界域。因此,粗糙集中的不确定性是基于一种边界的概念,当边界域为一空集时,则问题就变为确定性的。

round coast 圆形海岸

It is a coast of rounded morphology formed when islands along the coast have been continually eroded, deposited and rounded under the action of marine erosion and deposition.

岛屿沿岸在海蚀、海积作用下,不断被冲蚀、堆积及圆化,形成呈圆形形态的海岸。

route 航路

It is the navigation route of ships from the point of departure to the intended destination.

又称"航线",船只由起航点至预定到达点的航行路线。

route 航线

It is the navigation route of ships from the point of departure to the intended destination.

又称"航路",船只由起航点至预定到达点的航行路线。

runoff area 径流区

It refers to a range where underground water in an aquifer flows from a recharge area to a discharge area.

含水层的地下水从补给区至排泄区的流经范围。

rural-urban integration 城乡一体化

It is a new rural-urban relationship where rural and urban areas support, benefit from and promote each other so as to achieve coordinated development and common prosperity with the city as the centre, the town as the bond, and the village as the basis.

以城市为中心、小城镇为纽带、乡村为基础,城乡依托、互利互惠、相互促进、协调发展、共同繁荣的新型城乡关系。

S

saddle 鞍部
　　It refers to broad low-lying areas similar to the shape of saddle between ocean ridge and adjacent highlands.
　　海底山脊或相邻高地间类似马鞍形状的宽阔低洼地带。

sales income 销售收入
　　It is the sales revenue achieved by an enterprise's sale of products in a reporting year. The unit of measurement is ten thousand yuan.
　　报告年度本企业销售产品实现的销售收入。计量单位:万元。

saline soil 盐碱土
　　It is the high-salinity soil characterised by corrosiveness, collapsibility and salt expansibility, whose particle is the coagulation of gypsum, mirabilite, halite, etc.
　　又称"盐渍土"。含盐量大于一定量的土。土粒为石膏、芒硝、岩盐等凝结,具有腐蚀、溶陷和盐胀等特性的土。

saline wasteland 盐碱化荒地
　　A type of land combination dominated by saline soil and moderately-strong salinized soil, it is a wasteland type with the largest quantity, which occurs at geomorphological positions with arid climate and impeded drainage, and in sections intruded by seawater.
　　以盐土和中强盐渍化土壤为主的一种土地组合,是数量最多的一种荒地类型,发生在气候干旱和排水不良的地貌部位和受海水侵袭的地段。

saline-alkali land 盐碱地
　　It is the land which is not conducive to growing crops because its soil contains more soluble salts.
　　土壤中含有较多的可溶性盐分,不利于作物生长的土地。

salinity 盐度
　　It is the saltiness or dissolved salt content of a body of water, represented by the symbol "S". It is a total mass (in grams) of solids contained in every kilogram of seawater after hydrochloric acid is converted into oxide, bromine and iodine are replaced by equivalent chlorine and all organic matters are oxidized.
　　海水中含盐量的标度。每千克海水中在盐酸转化为氧化物、溴和碘被等量的氯置换、有机物全部被氧化后,所含固体物质的总克数。以符号"S"表示。

salinity tongue 盐舌
　　It describes the phenomenon of tongue-shaped salinity distribution on a plane or vertical salinity distribution map.
　　在盐度平面或垂向分布图上,盐度呈舌状分布的现象。

salinization damage 盐碱害
　　When salinity is greater than 10 and alkalinity exists, soil salinization occurs rapidly and crops die due to the corrosion on their roots.

盐度大于 10,并有碱度存在时,土壤迅速盐碱化,造成农作物根部受腐蚀而死亡的危害。

salinization of soil 土地盐渍化

It is a process or phenomenon where the groundwater level rises enhancing the surface accumulation of salt, so that the original non-saline soil turns to saline soil or aggravates the salinization of the original soil. This is generally a result of improper human activities, such as improper hydraulic engineering technical measures.

由于人为活动不当,主要是采取的水利工程技术措施不当,导致地下水位升高,盐分表聚加强,使原来非盐渍土演变成盐渍土或使原土壤盐渍化加重的过程和现象。

salt damage 盐害

It mainly refers to the harm caused by sodium chloride and sodium sulphate to crops and soil.

主要指氯化钠和硫酸钠对农作物和土壤的危害。

salt finger 盐指

It is a phenomenon of salinity distribution downwards in a finger shape, occurring at an interface when a relatively warm and salty water layer overlies a relatively colder and fresher water layer.

热而高盐的水层位于冷而低盐的水层之上时,在界面处发生盐度向下呈指状分布的现象。

salt groundwater 地下盐水

It refers to groundwater whose total mineralization is between 10.0 and 50.0 grams per litre.

总矿化度在 10.0~50.0 克/升之间的地下水。

salt pan 盐田

It is a site for obtaining sea salt by evaporation.

蒸发法制取海盐的场地。

salt water cooling 海水冷却

It can be completed directly via heat-exchange equipment with ocean water as the cooling medium.

直接以海水为冷却介质,经换热设备完成冷却。

saltwater intrusion 海水入侵

Due to natural or man-made causes, the change of hydrodynamic conditions in coastal regions may disrupt the equilibrium state between freshwater and seawater in the aquifer, and in turn result in the intrusion of saltwater or highly mineralized saline groundwater which has hydraulic connections with seawater towards the land along the aquifer.

由于自然或人为原因,海滨地区水动力条件发生变化,使海滨地区含水层中的淡水与海水之间的平衡状态遭到破坏,导致海水或与海水有水力联系的高矿化地下咸水沿含水层向陆地方向扩侵的现象。

salvage at sea 海上救助打捞活动

It refers to the salvage activities of a ship or person in distress and cargoes fallen in seawater.

对海上遇险船只和人员的救助,以及落水货物的打捞活动。

sand ridge 沙脊

It refers to a strip-like sandy deposit

body on the continental shelf, which ranges from thousands of meters to dozens of kilometers long with the relative elevation of a few meters.

大陆架上长条状沙质堆积体,一般可长达数千米至数十千米,相对高差数米。

sand ridges 沙脊群

They refer to the geographic entities on the continental shelf which are developed with an interphase arrangement of numerous striped sand ridges and erosion gullies in the radial form.

陆架上多个条带状沙脊和冲刷沟谷相间排列而成的地理实体,其形态多呈辐射状。

sand spit 沙嘴

It is a deposition landform found off coasts. At one end, spits connect to a head, and extend into the nose. A spit is a type of bar or beach that develops where a re-entrant occurs, such as at cove's headlands, by the process of longshore drift. Longshore drift (also called littoral drift) occurs due to waves meeting the beach at an oblique angle, and backwashing perpendicular to the shore, moving sediment down the beach in a zigzag pattern. Longshore drift is complemented by longshore currents, which transport sediment through the water alongside the beach. These currents are set in motion by the same oblique angle of entering waves that causes littoral drift and transport sediment in a similar process.

向海突出的一种低平狭隘的海岸堆积地貌,略似镰刀形,基部依附陆地,前端突出海中。主要由泥沙经沿岸流搬运堆积而成,常见于海湾岬角处和河口附近。

satellite oceanography 卫星海洋学

It is a discipline the studies theories and methods of exploring the ocean by using a satellite; processing, transmitting and utilizing satellite data; and the application of a satellite in oceanographic research and marine forecasting.

研究利用卫星探测海洋的理论和方法,卫星资料的处理、传输和利用,卫星在海洋学研究和海洋预报中应用的学科。

sawtooth-like coast 锯齿形海岸

It is a coast where hard rocks and soft rocks interlock with each other in the shape of a saw tooth.

坚硬与松软的岩石交替相间组成的海岸,呈锯齿形,凹凸相间。

scale free network 无标度网络

It is a network whose node degree follows the power-law distribution.

节点度服从幂律函数分布的网络称为无标度网络。

sea 海

It refers to the margin parts where a continuous body of saline water covering most of the earth's surface is encircled or separated by lands, atolls and peninsulas. The sea is composed of three types: marginal sea, intercontinental sea and intracontinental sea.

地球表面上相连接的广大咸水水体被陆地、岛礁、半岛包围或分隔的边缘部分。可分为边缘海、陆间海和陆内海三种类型。

sea and coast natural ecosystem protection 海洋和海岸自然生态系统保护

It covers management of natural reserves such as estuarine ecosystems, intertidal zone ecosystem salt mash ecosystems (saline water, brackish water), mangrove ecosystems, bay ecosystems, seagrass bed ecosystems, coral-reef ecosystems, upwelling ecosystems, continental shelf ecosystems, and islands ecosystems.

对包括河口生态系统、潮间带生态系统、盐沼(咸水、半咸水)生态系统、红树林生态系统、海湾生态系统、海草床生态系统、珊瑚礁生态系统、上升流生态系统、大陆架生态系统、岛屿生态系统等类型的自然保护区进行的管理活动。

sea arch 海穹

It refers to an arch-shaped land feature formed by the meeting of two eroded sea caves from opposite sides when wave erosion acts upon the two sides of a headland.

又称"海蚀拱桥",岬角处因两侧受海水侵蚀作用,而形成的两个方向相反、被蚀穿的海蚀洞互相贯通的拱桥状地貌。

sea area 海域

It refers to specific water scope in the ocean.

海洋中特定的水体范围。

sea area bearing capacity 海域承载力

It refers to the maximum extent or threshold value of certain sea area to sustain coordinated development among population, environment and economy by self maintenance and self adjustment. It aims to maximize social welfare on the condition of ensuring reasonable utilization of marine resources and virtuous cycle of marine ecological environment within the foreseeable period.

海域承载力是指一定的海洋区域在可预见期内,在确保海洋资源合理利用和海洋生态环境良性循环的条件下,为实现社会福利最大化,通过自我维持与自我调节,特定海域能够支持人口、环境和经济协调发展的最大程度或阈值。

sea area used for agricultural by sea reclamation 农业填海造地用海

It refers to a sea area which is impoldered into land by building dikes to enclose waters, for agricultural, forestry and livestock production. The way of using the sea is marine reclamation land.

通过筑堤围割海域,填成土地后用于农、林、牧业生产的海域,用海方式为农业填海造地。

sea area used for amusement ground 游乐场用海

It is a sea area used for marine recreational activities such as yachting, windsurfing, surfing, diving, underwater sightseeing and fishing. The way of using the sea is amusement ground.

开展游艇、帆板、冲浪、潜水、水下观光及垂钓等海上娱乐活动所使用的海域,用海方式为游乐场。

sea area used for anchorage area 锚地用海

It refers to a sea area used for vessels' tide and moor awaiting, joint inspection, wind sheltering, lightering operations at sea, etc.

船舶候潮、待泊、联检、避风及进行水上过驳作业等所使用的海域。

sea area used for aquaculture reclamation 围海养殖用海

It is a sea area for closed or semi-closed aquaculture production by building dikes to enclose and cut waters. The way of using the sea is aquaculture reclamation.

筑堤围割海域进行封闭或半封闭式养殖生产的海域,用海方式为围海养殖。

sea area used for artificial fish reefs 人工鱼礁用海

It refers to a sea area which is used for aquaculture production by the construction of artificial fish reefs.

通过构筑人工鱼礁进行增养殖生产的海域。

sea area used for bathing beach 浴场用海

It refers to the waters specifically for visitors to swim and paddle in. The way of using the sea is bathing beach.

专供游人游泳、嬉水的海域,用海方式为浴场。

sea area used for cable conduit 电缆管道用海

It refers to the sea area used to bury or lay submarine optical (or electric) communications cables, power cables, deep sea sewage pipes, water pipes, and other tubular installations to transport other materials, excluding the sea area used for oil pipelines to extract oil and gas. The way of using the sea is submarine cable conduit.

埋(架)设海底通信光(电)缆、电力电缆、深海排污管道、输水管道及输送其他物质的管状设施等所使用的海域,不包括油气开采输油管道所使用的海域。用海方式为海底电缆管道。

sea area used for coast protection engineering 海岸防护工程用海

It refers to the sea area used to build coastal defenses for the precaution of the erosion of waves and alongshore currents, and for the prevention of an attack of natural disasters such as a typhoon, cyclone, and heavy wind from cold waves, etc.

防范海浪、沿岸流的侵蚀及台风、气旋和寒潮大风等自然灾害的侵袭,建造海岸防护工程所使用的海域。

sea area used for comprehensive utilization of seawater 海水综合利用用海

It is a sea area used for developing desalination, comprehensive utilization of seawater, etc., including seawater desalination factories, alkali plants, comprehensive utilisation of seawater factories, water intakes and outlets, impounding reservoirs, sedimentation tanks, etc.

开展海水淡化和海水化学资源综合利用等所使用的海域,包括海水淡化厂、制碱厂及其他海水综合利用工厂的厂区、取排水口、蓄水池及沉淀池等所使用的海域。

sea area used for dumping zones 倾倒区用海

It refers to a sea area occupied by dumping zones, whose way of using the sea is toppling.

倾倒区所占用的海域,用海方式为倾倒。

sea area used for electric power industry 电力工业用海

It refers to the sea area used for the electric power industry, encompassing the sea area used for plant areas of power plants, nuclear power plants, wind power plants, tidal and wave power plants, docks, approach bridges, platforms, harbour basins (including waters for ship berthing and swirling at open-type wharf frontiers), dykes, pier base of wind fans and towers, underwater power generating equipment, water intakes and outlets, impounding reservoirs, sedimentation tanks and thermal drainage areas.

电力生产所使用的海域,包括电厂、核电站、风电场、潮汐及波浪发电站等的厂区、码头、引桥、平台、港池(含开敞式码头前沿船舶靠泊和回旋水域)、堤坝、风机座墩和塔架、水下发电设施、取排水口、蓄水池、沉淀池及温排水区等所使用的海域。

sea area used for engineering construction 工程用海区

It refers to a designated sea area which can be used to meet the needs of engineering construction projects, including projects which utilize the water surface, water body, seabed, subsoil, etc.

为满足工程建设项目用海需要划定的海域,包括占用水面、水体、海床或底土的工程建设项目。

sea area used for fishery 渔业用海

It is the waters used for the development and utilization of fishery resources, and for the development of marine fishery production.

为开发利用渔业资源、开展海洋渔业生产所使用的海域。

sea area used for fishery infrastructures 渔业基础设施用海

It refers to the waters used for fishing boat docking, handling operations, taking shelter from winds, and breeding important offspring seeds. It includes waters used for fishery docks, approach bridges, dams, fishing harbours (including waters for ship berthing and swirling at open-type wharf frontiers), fishing port channels, attached warehouses, important offspring seed breeding sites, as well as water intakes and outlets of seawater farms on lands extending into the sea.

用于渔船停靠、进行装卸作业和避风,以及用以繁殖重要苗种的海域,包括渔业码头、引桥、堤坝、渔港港池(含开敞式码头前沿船舶靠泊和回旋水域)、渔港航道、附属的仓储地、重要苗种繁殖场所及陆上海水养殖场延伸入海的取排水口等所使用的海域。

sea area used for industry 工业用海

It refers to a sea area used for industrial production.

开展工业生产所使用的海域。

sea area used for land reclamation projects 造地工程用海

It refers to the waters with effective coastlines formed through marine reclamation land in order to meet the needs of urban construction, agricultural production and waste disposal.

为满足城镇建设、农业生产和废弃物处置需要,通过筑堤围割海域并最终填成土地,形成有效岸线的海域。

sea area used for marine protected areas 海洋保护区用海

It refers to a sea area used by various marine protected areas.

各类涉海保护区所使用的海域。

sea area used for military affairs 军事用海

It refers to a sea area used for establishing military facilities and conducting military activities.

建设军事设施和开展军事活动所使用的海域。

sea area used for non-permeable structures 非透水构筑物用海

It refers to a method of using sea in a non-permeable way to construct a dock, croy, approach embankment, breakwater or subgrade which do not form the sea reclamation or an effective coastline.

采用非透水方式构筑不形成围填海事实或有效岸线的码头、突堤、引堤、防波堤、路基等构筑物的用海方式。

sea area used for oil and gas exploration 油气开采用海

It is a sea area used for the exploitation of oil and gas resources, including oil platforms, trestles used in oil and gas exploitation, floating oil storage units, oil pipelines, artificial islands used in oil and gas exploitation and roads connecting those artificial islands with other lands or islands.

开采油气资源所使用的海域,包括石油平台、油气开采用栈桥、浮式储油装置、输油管道、油气开采用人工岛及其连陆或连岛道路等所使用的海域。

sea area used for open culturing 开放式养殖用海

It refers to a sea area used for open cultivation without building dikes to enclose waters, such as raft culturing, cage culturing, manual seeding without manmade facilities or natural proliferation production. The way of using the sea is open culturing.

无须筑堤围割海域,在开敞条件下进行养殖生产所使用的海域,包括筏式养殖、网箱养殖及无人工设施的人工投苗或自然增殖生产等所使用的海域。用海方式为开放式养殖。

sea area used for permeable structures 透水构筑物用海

It refers to a method of using sea areas to build structures such as piers, trestles, water chalets, artificial reefs, etc. through permeable methods.

采用透水方式构筑码头、海面栈桥、高脚屋、人工鱼礁等构筑物的用海方式。

sea area used for port 港口用海

It refers to a sea area used for ships' mooring, loading and unloading operations, wind sheltering, transferring, etc. It includes a sea area used for harbor wharfs (including open-type cargo or passenger terminals), approach bridges, platforms, harbor basins (including waters for ship berthing and swirling at open-type wharf frontiers), dikes, yards, etc.

供船舶停靠、进行装卸作业、避风和调动等所使用的海域,包括港口码头(含开敞式的货运和客运码头)、引桥、平台、港池(含开敞式码头前沿船舶靠泊和回旋水域)、堤坝及堆场等所使用的海域。

sea area used for road-bridges 路桥用海

It refers to a sea area used for road-bridge engineering of connecting land and island, etc., including areas used for sea-crossing bridges, sea-crossing roads, coastal roads and their ancillary facilities, with the exception of sea areas connecting land and island for oil-gas exploitation, piers, etc.

连陆、连岛等路桥工程所使用的海域,包括跨海桥梁、跨海和顺岸道路等及其附属设施所使用的海域,不包括油气开采用连陆、连岛道路和栈桥等所使用的海域。

sea area used for salt industry 盐业用海

It is a sea area used for salt production, including waters used for salt pans, water intakes and outlets of salt pans, reservoirs, salt wharfs, approach bridges and harbour basins (including waters for ship berthing and swirling at open-type wharf frontiers).

用于盐业生产的海域,包括盐田、盐田取排水口、蓄水池、盐业码头、引桥及港池(船舶靠泊和回旋水域)等所使用的海域。

sea area used for salvaged material by sea reclamation 废弃物处置填海造地用海

It refers to a sea area which is impoldered into land by building dikes to enclose waters, for the disposal of industrial waste residues, urban construction wastes, household garbage, dredged materials, etc.

通过筑堤围割海域,用于处置工业废渣、城市建筑垃圾、生活垃圾及疏浚物等废弃物,并最终形成土地的海域。

sea area used for scientific research and teaching 科研教学用海

It refers to a sea area which is particularly used for scientific study, test or teaching activities.

专门用于科学研究、试验及教学活动的海域。

sea area used for sewage dumping 排污倾倒用海

It refers to a sea area used for discharging sewage and dumping waste.

用来排放污水和倾倒废弃物的海域。

sea area used for shipbuilding industry 船舶工业用海

It refers to a sea area used for the manufacturing, repairing and disassembling of ships (including fishing boats), including waters for shipyard production areas, wharfs, approach bridges, platforms, docks, slipways, dikes, harbor basins (including waters for ship berthing and swirling at open-type wharf frontiers, the frontier waters for docks, slipways, etc.) and other facilities.

船舶(含渔船)制造、修理、拆解等所使用的海域,包括船厂的厂区、码头、引桥、平台、船坞、滑道、堤坝、港池(含开敞式码头前沿船舶靠泊和回旋水域,船坞、滑道等的前沿水域及)其他设施等所使用的海域。

sea area used for solid mineral mining 固体矿产开采用海

Sea area used for solid mineral mining refers to the sea area used for mining the sea sand and other solid mineral resources, including the offshore area and the area used for mining solid mineral resources when digging from land to the ocean floor.

开采海砂及其他固体矿产资源所使用的海域,包括海上以及通过陆地挖至海底进行固体矿产开采所使用的海域。

sea area used for special purpose 特殊用海

It is a sea area used for particular purposes such as scientific research and teaching, military affairs, natural reserves, coastal protection works, etc.

用于科研教学、军事、自然保护区及海岸防护工程等用途的海域。

sea area used for standard sewage emissions 污水达标排放用海

It is a sea area for receiving specified standard sewage. The way of using the sea is standard sewage discharge.

受纳指定达标污水的海域,用海方式为污水达标排放。

sea area used for submarine engineering 海底工程用海

It refers to a sea area used for the construction of submarine engineering facilities.

建设海底工程设施所使用的海域。

sea area used for submarine tunnel 海底隧道用海

It refers to a sea area used for the construction of submarine tunnels and ancillary facilities, including the main body of the tunnel, its submarine ancillary facilities, ventilating shaft and other impermeable facilities.

建设海底隧道及其附属设施所使用的海域,包括隧道主体及其海底附属设施,以及通风竖井等非透水设施所使用的海域。

sea area used for submarine venues 海底场馆用海

It refers to a sea area used to construct submarine aquariums, submarine warehouses, storage tanks and other ancillary facilities.

建设海底水族馆、海底仓库及储罐等及其附属设施所使用的海域。

sea area used for tourism and recreation 旅游娱乐用海

It refers to a sea area used to develop coastal and marine tourism resources, and conduct marine recreational activities.

开发利用滨海和海上旅游资源,开展海上娱乐活动所使用的海域。

sea area used for tourism infrastructures 旅游基础设施用海

It refers to a sea area used for the

construction of supporting facilities in a tourist area so as to satisfy the needs of marine travelling, sightseeing and recreational activities. It includes a sea area for building tourist docks, marinas, approach bridges, harbor basins (including waters for ship berthing and swirling at open-type wharf frontiers), dikes, recreational facilities, landscape architectures, travel platforms, water chalets, artificial islands for tourism, restaurants and hotels.

旅游区内为满足游人旅行、游览和开展娱乐活动需要而建设的配套工程设施所使用的海域,包括旅游码头、游艇码头、引桥、港池(含开敞式码头前沿船舶靠泊和回旋水域)、堤坝、游乐设施、景观建筑、旅游平台、高脚屋、旅游用人工岛及宾馆饭店等所使用的海域。

sea area used for town's construction by sea reclamation 城镇建设填海造地用海

It refers to a sea area which is impoldered into land by building dikes to enclose waters for construction of cities and towns (including the industrial park). The way of using the sea is marine reclamation land.

通过筑堤围割海域,填成土地后用于城镇(含工业园区)建设的海域,用海方式为建设填海造地。

sea area used for transportation 交通运输用海

It refers to a sea area used for transportation purposes, such as ports, shipping, road-bridge, etc.

为满足港口、航运、路桥等交通需要所使用的海域。

sea area used for waterway 航道用海

It refers to the navigable sea area delimited by traffic department for vessels (including the waters for marine navigation light, such as light beacons, top markers and floating pharos), exclusive of sea area for fishing port waterway. The way of sea use can be categorized into special waterway, anchorage and other open uses.

交通部门划定的供船只航行使用的海域(含灯桩、立标及浮式航标灯等海上航行标志所使用的海域),不包括渔港航道所使用的海域。用海方式为专用航道、锚地及其他开放式。

sea book 海籍

It refers to the books or maps used to record the location, boundary mark, ownership, area, type, function, ways of sea use, service life, the scale of sea area and collection standards of the sea area use fee and other basic information for sea use in different programs.

记载各项用海的位置、界址、权属、面积、类型、用途、用海方式、使用期限、海域等级、海域使用金征收标准等基本情况的簿册和图件。

sea breeze 海风

It is a general term for the wind blowing from sea to sea and the wind blowing from sea onto land.

从海面吹向海面与从海面吹向陆地的风的统称。

sea chart cartography 海图制图学

A branch of map cartography, it is a theoretical and technological discipline which studies chart developing, chart content and expression method, chart composing and drawing, duplicating, utilizing and updating.

地图制图学的一个分支。研究海图的发展过程,海图内容及其表示方法,海图的编绘、复制、使用和更新的一门理论和技术的学科。

sea cliff 海蚀崖

It refers to a steep cliff formed by wave erosion. Due to wave impact, grooves are formed at the edge of a coast and upper rocks hang in the air and slide to form a sea cliff, frequently seen at the sea coast with steeper rock slopes and stronger wave actions. Sea cliffs constantly back away as waves strike, and many rocks accumulate at the foot of the slope. When waves cannot scour the foot of the slope directly, sea cliffs will stop backing away.

波浪中冲蚀作用形成的陡崖。因波浪冲击,海岸边形成凹槽,上部岩石悬空坍落而成,多见于岩坡较陡、波浪作用较强的岸段。海蚀崖随波浪冲击而不断后退,在坡脚有大量坍落岩块聚积,波浪不能直接冲蚀坡脚时,后退停止。

sea current 海流

It refers to a large-scale aperiodic flow of ocean water at a relatively steady speed along a certain direction in the ocean.

又称"洋流"。海洋中,海水沿一定方向、以相对稳定的速度作大规模的非周期性流动。

sea damage 海损

It refers to direct and indirect damage when shipping vessels or cargoes suffer from natural disasters or other shipwrecks.

航运船舶或货物因自然灾害或其他海滩事故所遭受的直接和间接损失。

sea dike 海塘

It is a an embankment constructed along downstream banks in estuaries to protect farm lands along the banks from being eroded by ocean tides. Considering tidal invasion, the structure of the dike should be more solid than an ordinary one, and therefore the beach lands beyond the dike require protection as well.

又称"海堤"。在江、河下游入海地段两岸修筑的堤防。作用是保护两岸农田免受海潮浸渍。因有潮汐的侵袭,海塘的结构要比普通的堤防更加坚固,海塘外的滩地也需要加以保护。

sea fog 海雾

It refers to advection fog above the sea.

海面上空的平流雾。

sea ice 海冰

It refers to the salty ice formed by the freezing of seawater. In a broad sense, sea ice refers to all the ice on the ocean, including brine ice, river ice and icebergs.

海水冻结而成的咸水冰。广义指海洋上所有的冰,包括咸水冰、河冰、冰山等。

sea ice disaster 海冰灾害

It refers to a marine disaster caused by the freezing of seawater. Sea ice, especially drift ice, is astonishingly destructive, which can pose extreme threats to sailing vessels, marine resources development facilities and personnel security.

海水结冰而造成的海洋灾害。海冰破坏力尤其是流冰的破坏力相当惊人,对航行船舶和海洋资源开发设施以及人员安全都会构成极大的威胁。

sea ice remote sensing 海冰遥控

It refers to the technology of observing sea ice with remote sensors and extracting the parameters such as the type, thickness, temperature, density and distribution of the sea ice.

通过遥感器对海冰进行观测,提取海冰的类型、厚度、温度、密度及其分布等参数的技术。

sea island area 海岛面积

It refers to the land area of a single island. The island refers to all the islands with economic activities, including inhabited islands and uninhabited islands. The unit of measurement is square kilometres.

单个海岛的土地面积。海岛是指一切有经济活动的海岛,包括有居民海岛和无居民海岛。计量单位:平方千米。

sea islands used for national defense 国防用途海岛

They refer to the islands used for maintaining national defense security.

用于维护国防安全目的的岛屿。

sea level change 海平面变化

It refers to a periodic and aperiodic change of the sea level caused by climate, weather, ocean, geophysics, astronomy and other factors, and a change in sea surface height caused by climate change or crustal tectonic movement, etc.

气候、天气、海洋、地球物理和天文等因素引起海平面发生的周期性或非周期性的变化,气候变化与地壳的构造运动等原因引起的海面高度变化。

sea level rise disaster 海面上升灾害

It refers to a (delayed) disaster caused by a global warming induced sea level rise (absolute rise of sea level) or ground subsidence of a coastal area, coastal crustal movement and a relative change of sea level.

全球气候变暖导致的海平面上升(海平面绝对上升)或沿海地面沉降、沿海地壳运动及海平面相对变化造成的(缓发性)灾害。

sea lot 宗海

It refers to the same type of unit where the sea is closed by an ownership boundary line.

被权属界址线所封闭的同类型用海单元。

sea noise 海洋噪声

It refers to the noise generated by natural factors, such as waves, rain falling, turbulent currents, biological activities and molecular thermal motions.

由于自然原因,如波浪、降雨、湍流、生物活动、分子热运动等在海洋中产生的噪声。

Sea of Okhotsk 鄂霍茨克海

It is a marginal sea of the northwestern Pacific Ocean, lying between the Asian continent, Sakhalin Island, Kuril Islands, and Hokkaido.

位于亚洲大陆、萨哈林岛与千岛群岛、北海道岛之间,太平洋西北部的边缘海。

sea reclamation 围海

It is a method of using a sea area for marine development activities by enclosing and cutting waters in a fully or partially closed form through an embankment or other means.

通过筑堤或其他手段,以全部或部分闭合形式围割海域进行海洋开发活动的用海方式。

sea related financial and tax services 涉海财务及税务服务

They refer to the accounting, audit and tax services offered to marine institutions, such as marine asset evaluation.

为海洋企事业机构提供会计、审计及税务服务,如海洋资产评估等。

sea salt 海盐

It is a type of salt (sodium chloride) extracted from seawater.

食盐(氯化钠)的一种,由海水获得。

sea salt processing 海盐加工

It refers to the processing and production of salt products which are obtained from sea salt after a series of processes including halogenating, washing, grinding, drying and screening, or adding potassium iodate and flavouring during processing. Salt products include washed salt, refined salt, iodized salt, seasoning salt, nutrition fortified salt, feed salt, etc.

以海盐为原料,经过化卤、洗涤、粉碎、干燥、筛分等工序,或在其中添加碘酸钾及调味品等加工制成盐产品的生产活动,包括洗涤盐、精制盐、加碘盐、调味盐、强化营养盐、饲料盐等。

sea salt production 海盐产量

It refers to quality production in compliance with quality standards after acceptance inspection within a reporting period, when products with sodium chloride as the main component are obtained by solarization using seawater (including underground brine in coastal shallow waters) as raw material. The unit of measurement is ten thousand tons.

报告期内以海水(含沿海浅层地下卤水)为原料经晒制而成的以氯化钠为主要成分的产品,并经验收后符合质量标准的合格产量。计量单位:万吨。

sea salt wholesale 海盐批发

It refers to wholesale as well as import and export of sea salt.

海盐的批发和进出口。

sea spray 海洋飞沫

It refers to tiny wave drops and salt spray formed when ocean waves crash and water drops evaporate.

海浪破碎及水滴蒸发形成的细小浪滴和盐沫。

sea state 海况

It refers to a numerical or written description of ocean-surface conditions under wind action.

以数值与文字描述的风力作用下的海洋表面状况。

sea surface control network 海面控制点

It refers to a marine geodetic control network at the sea surface, and it is usually a buoy used to fix the position.

海面上的海洋大地测量控制点,一

般为固定位置的浮标。

sea surface height 海面高

It refers to the height between a sea surface and level ellipsoid, corresponding to the geodetic height of earth's surface.

海面与水准椭(圆)球之间的高度,与地面的大地高相应。

sea surface topography 海面地形

It refers to sea surface height relative to Earth's geoid and it gets its name from land topography.

海面与大地水液面之间的差距,是借用陆地地形的含义而得名。

sea water ice point 海水冰点

It refers to the temperature at which seawater freezes, and it decreases with the increase of seawater salinity.

指海水结冰时的温度。海水冰点随海水的盐度增加而降低。

sea water mixing 海水混合

It refers to the process in which the adjacent seawater gradually tends to be uniform in property when seawater with different characteristics permeates and converts in neighboring areas.

具有不同特征的海水,在其邻接区域会发生彼此渗透、转化,从而相邻海水的性质逐渐趋向均一,这一过程称为海水混合。

sea wave 海浪

It refers to a wave phenomenon generated by wind at the ocean surface, mainly including wind wave and swell.

海面由风引起的波动现象。主要包括风浪和涌浪。

seabed 海床

It refers to the sand, lithoherm and sludge on top of marine subsoil or the surface layer on top of other substances, which can be territorial waters, archipelagic waters, exclusive economic zones, continental shelf and high seas or the "zones".

海洋底土上面的沙、岩礁、淤泥或其他物质上部的表层,既可以是领海、群岛水域、专属经济区、大陆架、公海,也可以是"区域"。

sea-burial services 海葬服务

They cover all services related to sea burial.

与海葬有关的各类服务。

sea-crossing bridge management 跨海桥梁管理

It refers to the charging and management of sea-crossing bridges.

跨海桥梁的收费与管理活动。

sea-coast 海岸

It refers to a narrow strip of land of indefinite width landward of the shore between the mean lower water line and the upper bound of wave action.

自多年平均低潮线向陆到达波浪作用上界之间的狭长地带。

seafloor spreading 海底扩张

It refers to tectonic movement that new seafloor crust forms when mantle materials penetrate the lithosphere fissures or rifts along mid-ocean ridges, and constantly spreads out laterally away from them, rises up or fill up the rift.

地幔物质沿大洋中部脊,穿透岩石圈裂缝或裂谷,向两侧不断扩展,上升、充填裂谷,而产生新的海底地壳的构造运动。

seafloor template 海底基盘
It refers to a steel frame structure with orientation functions, on which the seafloor wellhead system, blowout prevention system or seafloor production system can be installed. More than two wellhead grooves can be placed on the seabed.

具有导向功能,其上能安装海底井口系统和防喷系统或海底生产系统钢质框架结构物,海床上安放两个以上的井口槽。

seaknolls 海底丘陵
They refer to widely distributed geographic entities which are composed of many sea knolls. The overall landform is undulated, with peaks and valleys alternating with each other.

由众多海丘组合而成、分布面积较广的地理实体,整体地形高低起伏,峰谷相间。

sea-land ecological assessment 海岛生态评估
It refers to ecological recognition and assessment to the current situation of an ecosystem of islands, the impact or damage which has already been caused or can possibly be caused to the ecology of an island by human activities and the ecological restoration of damaged islands.

对海岛生态系统现状、人类活动已经或可能对海岛生态造成的影响或损害以及受损害海岛的生态修复进行生态识别和评估。

sea-land ecosystem 海岛生态系统
It can be divided into a natural island ecosystem and a nature-economy-society complex island ecosystem. The former one refers to an island and an ecosystem mutually formed by the biocenosis of an island and the surrounding seas and abiotic environments; and the latter one refers to an ecosystem formed by human intervention.

可分为海岛自然生态系统和海岛自然-经济-社会复合生态系统。前者指海岛及其周围海域的生物群落和非生物环境共同构成的生态系统;后者指经人类干预而构成的生态系统。

sea-land integration 海陆一体化
It means to integrate a sea system and land system which are originally relatively isolated into a new unified whole with the application of system theory and synergetics theory, in accordance with the internal connection between the two geographic units of sea and land, in order to achieve a more effective allocation of the sea and land resources by integrated planning, linkage development, and the combination and integrated management of industry chains.

根据海、陆两个地理单元的内在联系,运用系统论和协同论的思想,通过统一规划、联动开发、产业链的组接和综合管理,把本来相对孤立的海陆系统,整合为一个新的统一整体,实现海陆资源的更有效配置。

sea-land interface 海-陆界面
It refers to an interface connecting seawater and the beach and tidal flat sedi-

ments distributed at the front edge of a continent, or the sediments distributed at a continental shelf.

海水和分布于大陆前缘的海滩、潮滩沉积物以及分布于大陆架上的沉积物之间的交界面，称为海-陆界面。

sea-land landscape 海岛景观

It refers to a natural landscape and artificial scenery which are of ornamental value on the islands.

在海岛上具有观赏价值的自然景色和人工景物。

seamount chain 海山链

It refers to seamounts with a chain-shape distribution, which are composed of a series of relatively isolated seamounts and seaknolls.

由一系列相对独立的海山、海丘组成，呈链状分布的海山群体。

seamounts 海山群

They refer to seamount groups consisting of large stretches of numerous underwater mountains.

由众多海山组成、成片分布的海山群体。

seaport material handling 海港装卸搬运

It refers to the self-governed (or relatively self-governed) non-ship handling services in a harbour.

海港内独立（或相对独立）的非船舶装卸服务。

sea-related ad services 涉海广告服务

They refer to the various paid publicity campaigns for marine enterprises or products.

为海洋企业或产品提供的各种有偿宣传活动。

sea-related building engineering installation 涉海建筑工程安装

It refers to the equipment installation inside a building after the completion of the main building of sea-related engineering, such as the installation of lighting and power equipment in ports and docks.

涉海工程建筑物主体工程竣工后，建筑物内的各种设备安装。如港口、码头的照明和电力设备等安装。

sea-related construction and installation industry 涉海建筑与安装业

It refers to the construction of sea-related housing and the installation of its equipment.

涉海单位房屋建筑的施工活动及其设备的安装。

sea-related enterprise investment services 涉海企业投资服务

They refer to the asset and coordination management of marine enterprises with legal personality, as well as the investment and asset management related to marine production.

具有法人资格的海洋企业的资产管理、协调管理及与海洋生产相关的投资与资产管理活动。

sea-related exhibition services 涉海会展服务

They refer to the exhibition and conference services provided for the circulation and demonstration of marine products, trade negotiation and communication, international exchanges, etc.

为海洋产品流通和展示、经贸洽谈和交流、国际往来等举办的展览和会议服务。

sea-related housing construction 涉海房屋建筑

It refers to the construction of main housing in coastal areas or sea-related units, including factories, office spaces, hotels, restaurants, etc.

沿海地区及涉海单位房屋主体工程的施工活动,包括厂房、办公用房、宾馆饭店等。

sea-related intellectual property services 涉海知识产权服务

They refer to the agency, assignment, registration, appraisal, assessment, authentication, consultation and retrieval of marine patents, trademarks, copyrights, software, etc.

对海洋专利、商标、版权、著作权、软件等的代理、转让、登记、鉴定、评估、认证、咨询、检索等服务。

sea-related legal and notary services 涉海法律及公证服务

They refer to the legal, notary and arbitration services offered to marine enterprises.

为海洋企业提供的法律、公证、仲裁等服务。

sea-related life insurance 涉海人寿保险

It refers to the life insurance which specifically targets marine personnel and coastal tourists.

专门为海洋从业人员和滨海游客提供的人寿保险活动。

sea-related market research services 涉海市场调查服务

They refer to the economic and social investigation activities provided for marine management, such as marine economic surveys, marine statistical surveys, etc.

为海洋管理提供的社会经济调查活动,如海洋经济调查、海洋统计调查等。

sea-related non-life insurance 涉海非人寿保险

It refers to an insurance service provided for the protection of marine production and management, such as ocean transportation insurance.

为保障海洋生产和管理提供的保险活动,如海洋交通运输保险。

sea-related service industry 涉海服务业

It covers various sea-related service activities such as catering services, coastal public transportation services, marine financial services, sea-related featured services, sea-related business services, etc.

包括海洋餐饮服务、滨海公共运输服务、海洋金融服务、涉海特色服务和涉海商务服务等涉海服务活动。

sea-river interface 海-河界面

It refers to a series of medium salinity brackish water areas formed when sea water at an estuary and freshwater from rivers mix.

河口地区海水与来自河流的淡水相混合,形成一系列具有中等盐度的半咸水水域,称为海-河界面。

seascarp 海崖

It is a steep linear escarpment on the

continental slope or ocean basin.

大陆坡或大洋盆地上陡峭的线状崖壁。

seaside exposure 海边地区腐蚀

It refers to the environmental corrosion of a land area by sea water. It can fall into three categories: (1) offshore slope; (2) tidal zone; (3) narrow offshore land area close to the coastal area and coastline, whose corrosion is more serious, since it is of abundant sunlight, high humidity and it contains sea salt particles.

海水对陆地作用地带的环境腐蚀。分为:(1)水下岸坡区;(2)潮差区;(3)海岸区——海岸线以上狭窄的近海的陆地地区,由于该地区的光线充足,湿度高而且含有海盐粒子,因此腐蚀较为严重。

seaside travel agency 滨海旅行社

It refers to business enterprises engaged in tourism for the purpose of profit in coastal areas.

沿海地区有营利目的,从事旅游业务的企业。

seawater 海水

Seawater refers to the water that makes up the oceans and seas. It is mixed liquid dissolving various mineral salts, organic substances, gases, and containing many other suspended materials.

构成海洋水体的水。溶解有多种无机盐、有机物质和气体以及含有许多悬浮物质的混合液体。

seawater chemical element extraction 海水化学元素提取

It refers to the activities of extracting chemical substances directly from seawater, such as bromine, potassium, uranium, heavy water, etc.

直接从海水中提取化学物质,如溴素、钾、铀、重水等的活动。

seawater chemical elements resources 海水化学元素资源

They refer to the vast amounts of chemical elements contained in seawater, among which halogen elements are the most abundant. Currently, seawater chemical elements resources which have been widely utilized are mainly bromine and iodine of the halogen elements; potassium and magnesium of the alkali metal elements; and uranium and heavy water of the radioactive elements.

海水中含有的大量化学元素,其中以卤族元素含量最为丰富。目前,已被广泛利用的海水化学资源主要有卤族元素溴、碘,碱金属元素钾、镁,放射性元素铀和重水。

seawater chemical resources 海水化学资源

They refer to the dissolved chemical resources in seawater which can be exploited and utilized.

海水中溶存的可供开发利用的化学物质。

seawater circulating and cooling technology 海水循环冷却技术

It refers to a technology of using seawater as a cooling medium, by which seawater can be reused through the cooling of heat-exchange equipment and then through the cooling of a cooling tower.

以海水为冷却介质,海水经换热设

备完成一次冷却后,经冷却塔冷却循环使用。

seawater comprehensive utilization project construction 海水综合利用工程建筑

It refers to the building activities of construction engineering about desalination and comprehensive utilization of seawater.

海水淡化及综合利用建筑工程的施工活动。

seawater density 海水密度

It refers to the mass of seawater per unit volume.

海水单位容积的质量。

seawater desalination 海水淡化

It refers to the process of removing soluble salts from seawater to produce fresh water. The four principal methods used for desalination include a thermal energy method (distillation and freezing), mechanical energy method (pressure dialysis and reverse osmosis), electric energy method (electrodialysis) and chemical energy method (solvent extraction and ion exchange).

海水脱除盐分变为淡水的过程。主要方法有4种,即:热能法(蒸馏法和冷冻法)、机械能法(压透析法和反渗透法)、电能法(电渗析法)和化学能法(溶媒抽出法和离子交换法等)。

seawater desalination resources 海水淡化水资源

Seawater desalination resources refer to the freshwater resources obtained from the seawater after desalination.

对海水进行淡化处理后,从海水中获得的淡水资源。

seawater desalting automatic control system equipment manufacturing 海水淡化自动控制系统装置制造

It refers to the manufacturing of an industrial control system, instruments, regulation, and control equipment, etc., during the process of the seawater desalting.

海水淡化生产过程中工业控制系统、仪表和调节控制等装置的制造。

seawater desalting special separation equipment manufacturing 海水淡化专用分离设备制造

It refers to the manufacturing of the separation, filtration, purification and distillation equipment used for seawater desalting.

用于海水淡化使用的分离、过滤、净化、蒸馏设备的制造。

seawater desalting supply instrument manufacturing 海水淡化供应用仪表制造

It refers to the manufacturing of a metering instrument, automatic adjustment and control instruments and devices used during the supply of desalted seawater.

淡化的海水供应过程中使用的计量仪表、自动调节及控制仪器及装置的制造。

seawater diffusion 海水扩散

It refers to the diffusion phenomenon of substance molecule or vortex motion occurring due to the irregular movement of molecules or fluids in seawater.

海水中由于分子或流体不规则地运动,所产生的物质分子或涡动的扩散现象。

seawater direct utilization 海水直接利用

It is a generic term about utilization of seawater as raw water, to directly replace freshwater as industrial or domestic water, such as seawater cooling, seawater desulphurisation, seawater toilet flushing seawater for domestic use, aquaculture, etc.

以海水为原水,直接代替淡水作为工业用水或生活用水等的总称。如海水冷却、海水脱硫、海水冲厕(大生活用海水)、海水养殖等。

seawater encroachment 海水倒灌

It refers to the backflow phenomenon of seawater towards the land due to the fact that the water level of a riverway in land is lower than the sea level in coastal regions.

沿海地区由于陆地内河道水位低于海平面,从而引起海水向陆地回流的现象。

seawater fish fry and fingerling services 海水鱼苗及鱼种服务

They refer to services of providing fish fries and fish seeds for mariculture.

为海水养殖提供鱼苗、鱼种的服务活动。

seawater intrusion 咸潮入侵

It is the phenomenon of a rise in salinity of the upper reaches of a river, which occurs in the course of high salinity seawater running upstream along a river mouth under the tidal effect. Saltwater intrusion at a river mouth can also cause the groundwater in the river mouth area to become salty as the groundwater on both sides of the river receives replenishment from river water infiltration.

在潮汐作用下,高盐度海水沿河口上溯过程中,造成上游河水盐度升高的现象。河口咸潮入侵,因河流两侧地下水接受河水渗入补给,也会造成河口地区的地下水变咸。

seawater irrigation 海水灌溉

It refers to an activity of irrigating the alkali-resisting crops with seawater.

利用海水灌溉耐碱农作物的活动。

seawater material resources 海水物质资源

They refer to all the useful marine resources which include seawater and chemicals dissolved in it, various mineral sediments stored on the seafloor and living organisms in the ocean.

海洋中一切有用的物质,包括海水本身及溶解于其中的化学物质、沉积蕴藏于海底的各种矿物质资源以及生活在海洋中的各种生物体。

seawater pearl processing 海水珍珠加工

It refers to the processing of seawater pearls.

海水珍珠的加工活动。

seawater resources 海水资源

They refer to all available substances existent in seawater, including chemicals produced during seawater desalination, sea ice utilization and direct seawater utilization and chemicals extracted from seawater.

海洋水体中存在的可供利用的物质。包括海水淡化、海冰利用、海水直接利用和从海水中提取的化学元素。

seawater resources utilization zone 海水资源利用区

It is a designated sea area needed for the development and utilization of seawater resources or direct utilization of underground brine, including salt pan, special industrial water area, general industrial water area, etc.

为开发利用海水资源或直接利用地下卤水需要划定的海域,包括盐田区、特殊工业用水区和一般工业用水区等。

seawater salinity gradient energy 海水盐差能

It refers to chemical potential difference between seawater and fresh water or between seawater with different salinity. It mainly lies in the junction where the river meets the sea.

海水和淡水之间或两种含盐浓度不同的海水之间的化学电位差能。主要存在于河海交接处。

seawater temperature 海水温度

It is a physical quantity for measuring the thermal condition of seawater which mainly absorbs heat from the short-wave radiation of the sun and the long-wave radiation of the atmosphere.

为度量海水热状况的一个物理量,其中海水吸热以来自太阳的短波辐射及大气的长波辐射为主。

seawater transparency 海水透明度

It is a measure of seawater visibility, that is, the ratio between light intensity and former light intensity after light travels in the water for a certain distance.

表示海水能见程度的一个量度,即光线在水中传播一定距离后,其光能强度与原来光能强度之比。

seawater treatment special equipment manufacturing 海水处理专用设备制造

It refers to the manufacturing of centrifugal machine, solid-liquid separator and other equipment when seawater is directly utilized.

海水直接利用所用离心机、固液分离机等设备的制造。

seawater viscosity 海水黏滞性

It refers to the molecule viscosity or turbulence viscosity occurring when momentum exchange causes internal frictional stress, due to the irregular movement of the molecule or fluid parcel in sea water.

海水中由于分子或流体块作不规则的运动,而发生动量交换引起内摩擦应力时所出现的分子黏滞性或湍流黏滞性。

seaway dredging 海运航道疏浚

It refers to the dredging, maintenance and management to ensure a clear seaway.

为保证海运航道畅通所进行的疏浚与护理活动。

seaweed chemical product manufacturing 海藻化工产品制造

It is the manufacturing of marine chemical medicine, with chemical substances such as iodine, potassium iodate, seaweed gel, mannitol and carrageenan extracted from seaweed by chemical methods.

利用化学方法从海藻中提取碘、碘酸钾、海藻胶、甘露醇、卡拉胶等化学物质,用于海洋化学药品原药的生产。

second law of thermodynamics 热力学第二定律

It claims that heat can never pass from a colder to a warmer body without some other change; it is impossible, by means of an inanimate material agency, to derive mechanical effects from any portion of matter by cooling it below the temperature of the coldest of the surrounding objects; the increment of entropy in an irreversible thermodynamics process is always bigger than zero.

不可能把热从低温物体传到高温物体而不产生其他影响,不可能从单一热源取热使之完全转换为有用的功而不产生其他影响,不可逆热力过程中熵的微增量总是大于零。

secondary environment 次生环境

It refers to an environment formed after the original natural environment has been changed by social production activities of human beings.

由于人类社会生产活动,导致原生自然环境改变后形成的环境。

secondary production 次级生产量

It refers to organic substances and fixed energy synthesized by animals and other heterotrophs through consuming the producer's primary production. In the ocean, secondary production refers to the production of the phytophagy zooplankton or the phytophagy and detritivore zoobenthos, which can be estimated through a number of different methods including on-site determination (egg laying ratio method), population dynamics model and P/B.

动物和其他异养生物靠消耗生产者的初级生产量制造的有机物质或固定的能量。在海洋中次级生产量指植食性浮游动物或植食及碎屑食性底栖动物的产量,可以用现场测定法(产卵率法)、种群动力学模型方法,以及 P/B 法测定和估算。

secondary productivity 次级生产力

It refers to the process where consumers transform chemical energy in food into chemical energy of their own tissues. In this process, the capacity of consumers transforming energy into organic substances is the secondary productivity.

消费者将食物中的化学能转化为自身组织中的化学能的过程称为次级生产过程。在此过程中,消费者转化能量合成有机物质的能力即为次级生产力。

self-organization 自组织

It is an evolutionary process in which a system spontaneously increases its activity organization and structure order degree through its own power. It progresses without the intervention or control of the external environment and other external systems. The ordered and relatively complex system formed accordingly is called a self-organizing system.

系统通过自身的力量自发地增加它的活动组织性和结构的有序度的进化过程,它是在不需要外界环境和其他外界系统的干预或控制下进行的。由此而形成的有序的较为复杂的系统称为自组织系统。

self-organization criticality (SOC) 自组织的临界性

An open, dynamic and complex system far away from an equilibrium state can evolve into a critical state through a lengthy self-organizing process; after the system reaches this state, its spatiotemporal dynamic behaviour no longer has characteristic spatiotemporal scales but exhibits a spatiotemporal correlation characteristic covering the whole system and satisfying the power law distribution.

一类开放的、动力学的、远离平衡态的复杂系统通过一个漫长的自组织过程能够演化到一个临界状态,达到这个状态以后,系统的时空动力学行为不再具有特征的时空尺度,而是表现出覆盖整个系统的满足幂率分布的时空关联特征。

semi-structured question 半结构化问题

They fall in between structured questions and non-structured questions. There are some rules to follow in solving processes and solving method; however, these rules are not completely determined. In other words, these questions have both quantitative factors and non-quantitative factors; therefore appropriate models can be built, though the optimal plan cannot be determined.

半结构化问题介于结构化问题和非结构化问题之间,其求解过程和求解方法有一定规律可以遵循,但又不能完全确定。也就是说这种问题既存在可以量化的因素,又有不能量化的因素,一般可适当建立模型,但无法确定最优方案。

sensitive area in marine ecological environment 海洋生态环境敏感区

It refers to the area that has a high target for marine eco-environmental functions and can hardly recover its function after being damaged, including spawning ground of marine fishery resources, major fishery waters, culture coastal areas of sea water, littoral wetlands, marine natural reserves, reserves for rare and endangered marine organisms, typical marine eco-systems (such as coral reefs, mangrove, estuaries), etc.

海洋生态环境功能目标很高,且遭受损害后很难恢复其功能的海域,包括海洋渔业资源产卵场、重要渔场水域、海水增养殖区、滨海湿地、海洋自然保护区、珍稀濒危海洋生物保护区,典型海洋生态系统(如珊瑚礁、红树林、河口)等。

series charts 系列海图

A number of nautical charts with different topics, they adopt the unified mathematical foundation and the same subdivision range.

表示不同主题内容的若干幅海图。采用统一的数学基础和相同的分幅范围。

services of marine aquatic improved varieties 海洋水产良种服务

They refer to services of providing improved varieties for mariculture.

为海水养殖提供优良品种的服务活动。

services of marine data processing 海洋数据处理服务

They refer to services of providing data logging, processing, storing, etc., for marine production and management, such as collecting, processing, filing and supplying of real-time and non-real-time data for a marine environment.

为海洋生产、管理提供数据录入、加工、存贮等方面的服务。如海洋环境实时、非实时资料的收集、处理、存档和供应的服务活动。

services of offshore oil and gas gathering and transportation system 海上油气集输系统服务

They refer to the services provided for an offshore oil and gas gathering and transportation system.

为海上油气集输系统提供的服务活动。

services of offshore oil and gas reservoir system 海上油气储油系统服务

They refer to the services provided for an offshore oil and gas reservoir system.

为海上油气储油系统提供的服务活动。

shallow sea culture 浅海养殖

It refers to the cultivation of marine economic flora and fauna in a shallow sea area below a low water line.

利用低潮线以下的浅海水域养殖海水经济动植物的生产活动。

shallow zone 浅海带

It refers to a coastal zone with shallow sea water.

海岸带深度较小的区域。

Shandong Peninsula Blue Economic Zone 山东半岛蓝色经济区

It refers to a geographic area, which is mainly supported by the marine industry and developed on the basis of marine resources and regional division of labour, and a multifunction area covering numerous elements of natural ecology, social economy, science and technology cultures, etc., whose main features are with dependence on oceans and coordination between sea and land, as the main feature.

依托海洋资源,以劳动地域分工为基础形成的、以海洋产业为主要支撑的地理区域,是涵盖了自然生态、社会经济、科技文化诸多因素的复合功能区。基本特征是:依托海洋,海陆统筹。

shelf ecosystem 陆架生态系统

It is a unity formed by material exchange, energy transmission and flows between marine communities in benthic and pelagic divisions and their surroundings within a continental shelf.

大陆架内海底区和水层区所有海洋生物群落与其周围环境进行物质交换、能量传递和流动所形成的统一整体。

shelf sea 陆架海

It refers to a sea area which occupies a continental shelf.

又称"陆棚海"。指占据大陆架的海域。

shell ridge 贝壳堤

Also called *a wetland beach ridge*, it is mainly composed of shells and its fragments. It is generally distributed in the vicinity of the high tide line of mudflat receding as a result of erosion.

又称"湿地滩脊",主要由贝壳及其碎片组成的滩脊。一般分布在侵蚀后退的淤泥质潮滩高潮线附近。

shoal 沙洲

It refers to the large ground formed by

sediment in the middle of a river bed or along its two sides, on the coast or in a shallow sea.

在河床中部或两侧、海滨或浅海中，由泥沙堆积而成的大片地面。

shoe-leather cost 皮鞋成本

It refers to the cost of time and effort that people spend in purchasing goods or exchanging currency into a more stable one in order to counteract the effects of inflation and maintain more actual value of currency. Therefore, it refers to the cost that a consumer or an enterprise pays to hold less cash.

由于通货膨胀，人们必须耗费相当多的时间和精力去购买物品或兑换成稳定的货币，以便保持货币更多的实际价值。因此，皮鞋成本指消费者或企业为了减少对现金的持有而付出的成本。

shore 滨

Also called *seashore*, it refers to the most distant land upward from the low tide line that can be influenced by waves or the area between the maximum upward flows. The backshore is a part normally above the high water mark but still influenced by the sea. The foreshore covers the area between high and low tide marks and is exposed at low tide. The nearshore is seaward of the foreshore and ends at the breaking point of the waves. The offshore, in coastal geomorphology, is the zone seaward of the breakers but in which material is moved by the waves.

又称"海滨"，自低潮线向上直至波浪所能作用到的陆上最远处或最大上冲流之间的地带，称为滨。海滨包括前滨与后滨。

shore ice 岸冰

It refers to sea ice that is frozen along river banks. It is divided into five types of shore ice, which are newborn, fixed, alluvial, renewable and residual.

沿河流岸边冻结的冰带。分为初生岸冰、固定岸冰、冲积岸冰、再生岸冰和残余岸冰5种。

shore station observation 岸站观测

It refers to the observation of tides, sea waves, sea water temperature and salinity, sea ice, marine meteorological factors and marine environmental quality conducted at a shore station of coastal waters, islands and platform.

在近岸、海岛和平台等岸站对潮汐、海浪、海水温盐、海冰等，以及海洋气象要素和海洋环境质量等进行的观测活动。

shoreface 滨面

It refers to a portion of narrow, steeply sloping zones between a seaward limit of the shore at a low water and nearly horizontal offshore zone. The sedimentation of this zone is mainly controlled by waves with active sand and mud movements.

位于低潮位以下，始终被海水覆盖的、狭窄的海滩部分，相当于内滨地带。其上界为低潮滨线，下界是较陡的滨面与平缓的滨外表面之间地形突变处。该带沉积作用主要受波浪控制，泥沙运动十分活跃。

shoreline 滨线

It refers to a boundary line between

the sea surface and beach. The boundary line between the sea surface of low tide and the beach is called low-tide shoreline, while the boundary line between the sea face of high tide and beach is called high-tide shoreline.

海面与海滩的交界线。低潮位海面与海滩的交界线称为低潮滨线,高潮位海面与海滩的交界线称为高潮滨线。

short-range urban planning 城市近期建设规划

It refers to the planning for significant items such as the important urban infrastructure, the public service facilities, housing construction for middle and low income residents and the ecological environmental protection, as well as the formulation of short-range urban construction in time sequence, development orientation and spatial arrangement.

是对城市的重要基础设施、公共服务设施和中低收入居民住房建设以及生态环境保护等重点内容作出的安排,以及对近期城市建设的时序、发展方向和空间布局的确定。

Silent Spring《寂静的春天》

It is a popular book on environment science written by American biologist, Rachel Carson and published by Houghton Mifflin in 1962. The book is widely credited with helping launch the contemporary American environmental movement. The book describes a scene about severe harm brought by pollution of pesticide, and migration and conversion of pollutants to expound close relationship between human beings and atmosphere, oceans, rivers, soil and animals and plants, preliminarily revealing effects of pollution on ecosystem. The publishing of the book raises widespread environment concerns, which helps fuel development of modern environmental science.

1962年在美国波士顿出版的一本环境科学普及读物,作者是美国生物学家R.Carson。书中描述了杀虫剂污染带来严重危害的景象,并通过对污染物迁移、转化的描述,阐明了人类同大气、海洋、河流、土壤、动植物之间的密切关系,初步揭示了污染对生态系统的影响。该书的出版引起人们对环境问题的普遍关注,对现代环境科学的发展起了积极的推动作用。

sill 海槛

It refers to the saddle part of a topographic rise which separates the sea basin.

分隔海盆的隆起地形的鞍部。

simple system 简单系统

It refers to a system in which the number of subsystems within the system is small and the relationship among them is simple. It is characterized by the homogeneity of elements within the system, large number of elements of the same type, and same structure and function of the elements.

组成系统的子系统数量较少,它们之间的关系比较简单。其特点是在系统中的元素往往是同质的,同一类的元素很多,它们的结构、功能都是一样的。

simultaneous measurement of groundwater level 地下水位统测

It refers to the measurement of wellhole in the studied area at the sametime, to find out simultaneously the distribution of the groundwater level, and to compile the contour map of groundwater, and the map of the depth of groundwater level for the moment.

对研究区内的井孔在同一时间进行水位测量,以便查明地下水水位的分布状况,编制此一时刻的地下水等水位线图和地下水埋藏深度图等。

sinking coast 下沉海岸

It is a coast formed by the subsidence of land, or rise of sea level, or the fact that the amount of land uplift is less than that of a sea level rise. It is often featured by a relatively tortuous coastline, lots of capes, peninsulas and islands, as well as deep waterways and good harbours.

因陆地下沉或海面上升或因陆地上升小于海面上升量而形成的海岸。一般具有较曲折的海岸线,多岬角、半岛和岛屿,并有深水道和良港。

six degrees of separation 六度分离假设

It is a theory put forward in 1967 by Stanley Milgram, a psychologist of Harvard University, which supposes that any two person in the world can connect to each other through no more than an average of 6 friends (acquaintances).

1967 年,美国哈佛大学心理学 Stanley Milgram 提出的,该假设是指世界上的任意两个人,可以通过平均不超过 6 个朋友(熟人)关系联系起来。

slope 斜坡

It is a single inclined surface with an inclined landform on a continental shelf, continental slope, island shelf or island slope; the slope surface is relatively wide and continuous. The average slope of the continental or island shelf is generally less than 1° and that of the continental or island slope is generally 3°—8°.

大陆架、大陆坡、岛架或岛坡上地形倾斜的单斜面,坡面相对宽阔连续,陆架、岛架斜坡平均坡度一般小于 1°,陆坡、岛坡斜坡坡度一般在 3°~8°之间。

small world networks 小世界网络

They describe in a simple language the fact that although most networks in reality have very large scales, there is a relatively short path between any two nodes.

以简单的措辞描述了现实中的大多数网络尽管规模很大,但是任意两个节点间却有一条相当短的路径的事实。

smooth-scale distribution urban system 顺序-规模分布型城镇体系

Also called *pyramid distribution urban system*, it refers to the fact that the number of cities and towns decreases with the increase in the size of urban systems.

又称"金字塔形分布的城镇体系",指城镇的数量随着城镇规模的增加而减少。

social economic attribute of the ocean 海洋社会经济属性

It refers to attributes with which the ocean is endowed through social and economic activities of human beings.

通过人类的社会经济活动赋予海洋的特性。

social feature of resources 资源社会特征

It is a social phenomenon in which the improvement of the level of social economy and human life is based on a large amount of investment and consumption of resources.

社会经济与人类生活水平的提高建立在资源的大量投入和消耗的基础上的所构成的社会现象。

social resources 社会资源

They collectively include the material and spiritual wealth provided by human labour in the development and exploitation of natural resources under certain spatial-temporal conditions. They cover a wide range of resources, in the current technical and economic conditions they mainly refer to the non-physical resources that constitute elements of social productivity such as labour resources, education resources, capital resources and science and technology resources, etc.

一定时空条件下，人类通过自身劳动在开发利用自然资源过程中所提供的物质和精神财富的统称。社会资源包括的范围十分广泛，在当前的技术经济条件下，主要是指构成社会生产力要素的劳动力资源、教育资源、资本资源、科技资源等非实物形态的资源。

soda manufacturing 纯碱制造

It refers to an activity of using sea salt or brine as raw materials to produce sodium carbonate.

以海盐或海盐卤水为原料生产纯碱的活动。

soft systems 软系统

Also called *ill-structured systems*, soft systems are the ones with unclear mechanisms and are hard to be described in a specific mathematical model, such as social system and biological system.

又称"不良结构系统"，是指机理不清，很难用明确的数学模型描述的系统，如社会系统和生物系统。

Songhua River 松花江

As the largest tributary of the Heilong River, the Songhua River is one of the seven major rivers in China, flowing about 1,927 kilometres through Jilin and Heilongjiang Provinces. The drainage area is about 545,000 square kilometres, accounting for 60% of the total area in Northeast China. Its main tributaries include Nen River, Hulan River, Mudan River, Tangwang River, etc.

黑龙江水系的最大支流，是中国七大江河之一。全长 1 927 千米，流域面积约为 54.5 万平方千米，占东北地区总面积的 60%，地跨吉林、黑龙江两省。其主要支流有嫩江、呼兰河、牡丹江、汤旺河等。

sounding 水深

It is the vertical distance from depth datum to water bottom.

自深度基准至水底的垂直距离。

sounding datum 深度基准

It refers to the starting surface of the water depth annotated in a nautical chart. "The lowest tide level in theory" is generally adopted as the sounding datum in coastal areas of China.

海图所注水深的深度起算面。中国沿海地区一般采用"理论最低潮面"作为深度基准。

South China Sea 南海

It is the Pacific marginal sea located between the south of Chinese mainland and the Philippines, Kalimantan Island, Sumatra, the Malay Peninsula and the Indo China Peninsula.

位于中国大陆南部与菲律宾群岛、加里曼丹岛、苏门答腊岛、马来半岛和中南半岛之间的太平洋边缘海。

Southern Ocean 南大洋

It refers to a special sea area which circulates the Antarctica and stops at subtropical convergence at the north border without a land boundary.

环绕南极大陆,北边无陆界,而以副热带辐合带为其北界的独特水域。

spatial pattern of community 群落空间格局

It refers to the distribution pattern formed by the changes in community structures of marine organisms along certain environmental gradients (such as latitudinal gradient, water depth, temperature gradient, salinity gradient, nutrient salinity gradient, sediment type, etc.).

沿一定的环境梯度(如纬度梯度、水深、温度梯度、盐度梯度、营养盐梯度、底质类型等)海洋生物群落结构发生相应改变而形成的分布型。

spawning ground 产卵场

It refers to a place where fish cluster to release the eggs, and has physicochemical and biological condition needed for spawning. It may contain many spawning sites.

鱼类集群产卵的场所,具有鱼类产卵所需要的理化和生物条件。产卵场内可能包含许多产卵地。

special sea-land protected area 海岛特别保护区

It refers to islands which are specially managed through effective protective measures and scientific development methods approved by the competent authorities. These islands, together with their surrounding sea areas, are normally characterised by special geographical conditions with landscape, humanity, ecology, living and non-living resources for the special needs of development and utilisation.

对具有特殊地理条件、景观、人文、生态、生物与非生物资源开发利用特殊需要的岛屿及其周边海域,经主管部门批准,采取有效的保护措施和科学的开发方式进行特殊管理的海岛。

species 物种

As basic units of classification, they refer to a group of organisms with the same gene bank, which can breed with each other and isolate from other reproduction.

基本的分类单元。指能相互繁殖、享有一个共同基因库的一群个体,并和其他种生殖隔离。

species diversity 物种多样性

It refers to species richness and individual number distribution in a biological community.

生物群落中物种的丰富度及其个体数量分布。

specific gravity of seawater 海水比重

It refers to the ratio of the seawater density at a certain temperature under atmospheric pressure to the density of distilled water at 4℃. The density value of the 4℃ distilled water approximates 1, and the numerical value of the specific gravity of seawater and its density are roughly the same in this case.

大气压力下具有某温度的海水密度与4℃蒸馏水密度的比值。4℃的蒸馏水密度值很接近于1,在这种情况下海水的比重与密度就具有大致相等的数值。

specific volume in situ 现场比容

It refers to the volume of unit mass of water at a marine specific point. It is the reciprocal of density.

海洋特定点单位质量海水的体积,为密度的倒数。

specific volume of water 海水比容

It refers to the volume of seawater per unit mass. The legal unit of measurement is Cubic meter per kilogram (m^3/kg) and its symbol is $\alpha = 1/\rho$ (ρ refers to the density of seawater).

单位质量的海水体积。法定计量单位为立方米每千克(m^3/kg),符号为:$\alpha = 1/\rho$(ρ为海水的密度)。

spillover effect 外部效应

They refer to externalities of economic activities or processes that affect those who are not directly involved. For example, the behavior of a person or business affects the welfare of other people or companies, but that person or business fails to consider this effect on others when making decisions due to a lack of incentive mechanisms.

一个人或企业的行为影响了其他人或企业的福利,但是还没有激励机制使得产生影响的人或企业在决策时考虑这种对别人的影响。

spit 沙嘴

It is a flat and narrow accumulation landform found off coasts, seaward like a sickle with its base attached to the land and fronted extending to the sea. It is formed when sediments are transported through the water by longshore currents and accumulated at cape and estuary.

向海突出的一种低平狭隘的海岸堆积地貌,略似镰刀形,基部依附陆地,前端突出海中。主要由泥沙经沿岸流搬运堆积而成,常见于海湾岬角处和河口附近。

spring tides 大潮

They refer to the tides just after a new or full moon, when there is the greatest difference between high and low water.

海洋的水面升降幅度最大的潮汐现象。

stagflation 滞胀

In economics, it refers to a situation in which the inflation rate is high, the economic growth rate slows, and unemployment remains steadily high. It raises a dilemma for economic policy, since actions designed to lower inflation may exacerbate unemployment, and vice versa.

全称停滞性通货膨胀,是在经济生活中,通货膨胀与经济停滞不前或低速增长交织并存的状态。

stakeholders 利益相关者

They refer to developers and users who have direct or indirect relationships with projects on sea use or who are affected by projects on sea use.

与项目用海有直接或间接连带关系或者受到项目用海影响的开发、利用者。

stand of tide 停潮

It occurs when the water level remains the same in a short period during high tide and low tide in the process of tidal fluctuation.

潮汐涨落过程中,高潮和低潮时出现的水位短时间不动现象。

standard sea water 标准海水

It refers to ocean water whose chlorine degree is adjusted to around 19.38 through a strict filtering process.

经过严格的过滤处理,调整其氯度为19.38左右的大洋海水。

standing stock 蕴藏量

It is the total amount of catchable and available aquatic economic animals and plants contained in the waters.

水域中蕴藏的可供采捕和利用的水产经济动、植物总量。

start point of the sea boundaries 海域界线起始点

It usually refers to coast boundary point. When boundary point of the administrative area of land at country level closest to the coast lies in estuary, the start point of the sea boundaries is the crossing point where oceanward boundary point of the administrative area of land at country level closest to the coast along channel center line or principal waterway center line connects with the most outstanding point of estuary banks.

海域界线起始点一般为海岸界点。最靠近海岸的陆域县级行政区域界点为河口时,海域界线起始点应为自陆域县级行政区域界线向海一侧最靠近海岸的界点沿河道中心线或主航道中心线与河口两岸最突出点连线的交点。

statistical general indicator of marine economy 海洋经济统计通用指标

It is a general indicator that various marine industries and enterprises commonly have extracted aiming at the common characteristics of all kinds of industries.

针对各类产业的普遍共性提取的各类涉海企业和单位共同拥有的指标。

statistical operational indicator of marine economy 海洋经济统计业务指标

It is an indicator designed on the basis of operating characteristics of various marine industries and their respective operational characteristics.

针对海洋产业的各自业务特点,按照各海洋产业业务经营特性而设计的指标。

steady-state 稳定状态

It occurs when capital stock and output, in which investment equals depreciation, remain stable (neither increasing nor decreasing) over time.

投资量等于折旧量的资本存量和产出随时间的推进一直是稳定的(既不增加也不减少)状态。

still tide 平潮

It refers to the stable condition of the tide level when the sea level reaches the highest position lasting for tens of minutes. The middle period is called *high tide time* and the height of still tide is called *height of high tide*.

海洋水位上涨到最高的位置一段时间内,潮位不升不降时的状态,时间一般为几十分钟。它的中间时刻称为"高潮时",而平潮时所达到的高度称为"高潮高"。

stock enhancement 资源增殖

Stock enhancement refers to the measures of artificially increasing the population and enriching marine economic animal resources such as fish and shrimp.

人为补充群体数量、丰富海上鱼虾类等经济动物资源的措施。

stock resources 存量资源

They refer to resources with storage features. They are the opposite of *running resources*.

"流量资源"的对称。系指具有储存特点的资源。

storm surge 风暴增水

Also called *storm surge*, *meteorological tsunami*, it is an anomalous rise and fall in water level caused by a low pressure weather system, typically tropical cyclones, extratropical cyclones, high winds as a result of the cold front passing across the ocean's surface and a sudden change in atmospheric pressure, etc.

又称"风暴潮"、"气象海啸",热带气旋、温带气旋和冷锋过境的强风作用以及气压骤变等天气系统,引起海面升降的异常现象。

storm tsunami 风暴潮

Also called *meteorological tsunami*, *storm surge*, storm tsunami is an anomalous rise and fall in water elevations caused by a low pressure weather system, typically tropical cyclones, extratropical cyclones, high winds pushing on the ocean surface and a sudden change in atmospheric pressure, etc.

又称"气象海啸"、"风暴增水"。热带气旋、温带气旋和冷锋过境的强风作用以及气压骤变等天气系统,引起海面升降的异常现象。

strait 海峡

A strait is a relatively narrow water channel connecting two seas or oceans between continents.

两块陆地之间连通两个海或洋的宽度较狭窄的水道。

Strait of Gibraltar 直布罗陀海峡

Lying between the southern tip of the Iberian Peninsula in Europe and the northwest corner of the African continent, the Strait of Gibraltar is the only channel connecting the Atlantic Ocean with the Mediterranean Sea.

位于欧洲伊比利亚半岛南端与非洲大陆西北角之间,是沟通大西洋与地中海的唯一水道。

Strait of Malacca 马六甲海峡

Located between the peninsular of Malaysia and Sumatra, it is a significant international shipping waterway connecting the Pacific Ocean and the Indian Ocean.

位于马来半岛与苏门答腊岛之间,沟通太平洋与印度洋的重要国际航运水道。

strait of the internal sea 内海海峡

It refers to a strait within a territorial sea baseline of coastal states and has the same legal status as inner waters under the exclusive sovereignty of coastal states. Coastal states regulate its closure or opening to all foreign ship or planes by enacting laws, management systems and requirements.

沿海国领海基线以内的海峡。与内水的法律地位相同,属于沿海国所有,对其享有完全的排他性主权。沿海国既可以规定内海海峡,不对一切外国船舶或飞机开放,也可以颁布法律规定外国船舶或飞机,通过内海海峡的管理制度和要求。

strait of the territorial sea 领海海峡

Also called *territorial strait*, it refers to a channel whose breadth is less than that of the territorial sea on both sides. It has the same legal status as the territorial sea of coastal states and allows innocent passage.

又称"领峡",指海峡宽度小于两岸领海宽度的海峡,具有与沿海国领海相同的法律地位,实行无害通过制度。

strange attractor 奇怪吸引子

It refers to an attractor in phase space of a dissipative system where the points never repeat themselves, and orbits never intersect, but they stay within the same region of phase space. Strange attractors are non-periodic, generally have a fractal dimension. They are a picture of non-linear, chaotic system.

一个耗散系统的相空间当时间趋于无穷大时,如果收缩到一个非整数维的点集,这就是一个奇怪吸引子。

strategy for economic development 经济发展战略

It is an overall plan for economic development, a global and long-term and fundamental decision, and basic fondation for guiding national economic development over a longer period.

为经济发展拟定的总体谋划,是带有全局性、长期性和根本性的决策,是较长时期内指导国民经济发展的基本依据。

stratified ocean 层化海洋

It refers to an ocean where its physical, chemical and biological properties, especially its temperature, salinity and density are in a vertical stratification structure.

海水的物理、化学和生物等特性,尤指温度、盐度和密度具有垂向分层机构的海洋。

stratum 地层

It is the generic term for the layer formed due to the sedimentation and the eruption of magma in a certain geologic age.

在某一地质年代因沉积作用以及岩浆喷出活动形成的地层的总称。

structural basin 构造盆地

It refers to a basin formed by geological structure action, which includes the basins of approximate circle or oval shapes formed by the leaning of rock strata towards the centre, and the basins formed by geological structure action such as geological depressions and fault depressions.

由地质构造作用形成的盆地。包括由岩层倾向中心而形成的近似圆形或椭圆形的盆地和地壳构造运动例如凹陷或断陷作用形成的盆地。

structural embeddedness 结构性嵌入性

It refers to a state of being located or secured within a larger entity. Organizations within a group, which have bilateral ties with each other and a third party, can be connected through the third party and form a systematic correlative structure. Namely, the relational network constructed by individuals or organizations is embedded in its social network and is influenced or decided by cultural and value elements from the social structure on a more macro level.

群体内组织之间不仅具有双边关系,而且与第三方有同样的关系,使得群体间通过第三方进行连接,并形成以系统为特征的关联结构,即在更宏观的层面上,由个人或组织所构成的关系网络是嵌入由其构成的社会网络结构之中的,并受到来自社会结构的文化、价值因素的影响或决定。

structural landform 构造地貌

It refers to a landform formed by a geologic structure and geotectonic.

地质构造和地壳构造运动所形成的地貌。

structural unemployment 结构性失业

It is a type of unemployment where the supply of labour quantity in labor markets surpasses the demand of labour quantity.

由于某些劳动市场上可提供的工作岗位数量不足以为每个想工作的人提供工作而引起的失业。

structured question 结构化问题

Also called *well-defined question* or *well-structured question*, it is a question which can be described and solved with a formalized (or formula) method.

又称"定义完善的问题"、"良构问题",能够通过形式化(或公式化)方法描述和求解的问题。

subaqueous delta 水下三角洲

It is a fan-shaped sediment accumulation mass formed on a continental shelf outside the estuary, which dips towards the sea in topography and whose surface has alternating banks and gullies.

河口外的大陆架上形成的扇状泥沙堆积体,地形向海倾斜,表面滩、沟相间。

subaqueous slope of coast 水下岸坡

It is a zone between low tide line and wave action base plane. The standard specifies its lower bound shall not exceed a nautical chart of 20 metres deep.

低潮线至波浪作用基面之间的地带。本标准规定,水下岸坡下界不超过海图 20 米水深。

subduction zone 俯冲带

It refers to a zone where the oceanic lithosphere of one plate sliding beneath the continental lithosphere due to the higher density of the oceanic lithosphere.

大洋板块与大陆板块相汇时,密度

较大的大洋板块岩石,俯冲于大陆板块之下的部分。

submarine breeding 海底养殖

It refers to the algae culture method of using submarine reef or rocks of shallow sea as the algae growth media for their breeding and reproduction.

又称"海底栽培"。利用浅海海底岩礁或石块作为海藻生长基进行生长繁殖的海藻养殖方法。

submarine cables 海底电缆

They refer to the (waterproof) metal wires or fiber optic cables to transmit electric currents or messages which are laid on or in the seabed.

铺设在海床上或海床内,用于传输电流或信息的(防水)金属线或光纤。

submarine canyon 海底峡谷

It refers to a steep-sided long and narrow valley under the sea. It is divided into a relatively narrow and deep submarine canyon which is located on a continental slope and a relatively broad and shallow submarine canyon which is located on an ocean basin.

两侧坡壁陡峭、狭长的谷地。可分为两种:一种位于大陆坡上相对较窄深的海底峡谷;另一种位于大洋盆地上相对较宽浅的海底峡谷。

submarine chimney 海底烟囱

It refers to a black or white chimney-like accumulation body formed by a ultra high temperature hydrothermal solution erupted from a sea floor.

由海底喷出的超高温的热水溶液所形成的呈黑色或白色的烟囱状堆积体。

submarine communication services 海底通信服务

They refer to communication services via submarine cables and optical fibers.

利用海底电缆、光缆进行的通信服务活动。

submarine control network 海底控制网

It refers to a marine geodetic control network on the seafloor, which can be used for marine geodetic surveys or local high-accuracy surveys.

海底的海洋大地测量控制网,用于海洋大地测量或局部高精度测量。

submarine control point 海底控制点

It refers to a control point installed on the seafloor to set up a marine geodetic control network.

为建立海洋大地测量控制网而设在海底的控制点。

submarine engineering operation services 海底工程作业服务

They refer to the services provided for submarine engineering operations.

对海底工程作业的服务活动。

submarine freshwater pipeline transportation 海底淡水管道运输

It refers to the transportation activities of freshwater through submarine pipelines.

通过海底管道对淡水的运输活动。

submarine hydrothermal deposit 海底热液矿床

It is the sulphide deposit and oxide deposit formed by a submarine hydrothermal process. It can be divided into the submarine multimetal mud and submarine sulphide deposit according to its shape.

由海底热液作用形成的硫化物和氧化物矿床。按其形态分为海底多金属软泥和海底硫化矿床两种。

submarine hydrothermal solutions 海底热液

They refer to the high-temperature hot water rich in various metal elements (mainly heavy metal elements), which is generated in mid-ocean ridges, the spreading axes of marginal seas, etc.

大洋中脊和边缘海扩张轴等处产生的富含多种金属元素(以重金属元素为主)的一种高温热水。

submarine levee 深海堤

It refers to the levee deposits formed naturally in a submarine canyon, submarine valley and trough edges.

位于海底峡谷、海底谷、海槽边沿处自然沉积形成的堤状堆积体。

submarine mineral resources 海洋矿产资源

They refer to various mineral deposits stored in surface sediments and rock strata of the seafloor, including mineral resources in seashores, neritic areas, abyssal areas, ocean basins and mid-ocean ridges.

又名"海底矿产资源"。赋存于海底表层沉积物和海底岩层中的各类矿藏。包括海滨、浅海、深海、大洋盆地和洋中脊底部的各类矿产资源。

submarine mining 海底采矿

It refers to the overall process of obtaining mineral resources from sediments on surface layer and rock stratum of the sea floor.

从海底表层的沉积物和海底岩层中获取矿产资源的全过程。

submarine oceanic ridges 海底洋脊

They refer to the extending submarine elevations with steep sides which are irregular or relatively flat in topography. The distance between the outer limit of the continental shelf and the baseline of territorial sea shall not exceed 350 nautical miles. This provision does not apply to submarine elevations that are natural components of the continental margins of coastal states.

地形不规则或比较平坦的海底延伸的高地。两侧陡峭，其上的大陆架外部界限，自领海基线起不应超过350海里。这一规定，不适用于作为沿海国大陆边缘自然构成部分的海底高地的情况。

submarine pipe construction 海底管道铺设

It refers to the construction of a pipeline used for the transport of petroleum, natural gas, water and others.

用于输油、输气、供水及输送其他物品的管道施工。

submarine pipelines 海底管道

They refer to the pipelines installed on a seabed, or buried deep under a seabed, or erected at a certain height above a seabed. They can be used for the transportation of water, natural gas and petroleum.

铺设在海床上或深埋于海床内，或架设在海床之上某一高度处的管线，用于输送水、天然气和石油。

submarine plateau 海底高原

It is a large or extensive submarine region that rises well above the level of the ambient seabed on the slope extending towards the ocean, having a steep edge and relatively flat summit. In parts of regions, the seamounts alternate with cloughs and the relative elevation difference is generally greater than 200 metres.

大陆向海延伸的坡地上范围广阔的地形隆起实体,边缘地形陡峭,顶部地形相对平坦,部分地区海山与沟谷相间,相对高差一般大于 200 米。

submarine platform 海台

It is a large, relatively flat submarine region rising sharply above the seafloor, with rather steep surrounding slopes. Its relative height difference is generally beyond 200 metres.

顶部宽阔平坦,周边斜坡比较陡峭的海底高地,相对高差一般大于 200 米。

submarine relief 海底地势

It refers to the overall situation of seabed terrain fluctuation. The basic configuration of the submarine landform is composed of the large-scale landform, such as the height, direction, distribution, location, etc.

海底地形起伏的总体态势。大型地貌组成的海底地貌基本形状,如高低、走向、分布和位置等。

submarine spring 海底泉

It refers to a natural outcrop of underground water in seabed strata.

海底地层中地下水的天然露头。

submarine sulphur mine 海底硫矿

It refers to a deposit which contains sulphur in the submarine rock strata.

海底岩层中含有硫的矿产。

submarine sulphur mining 海底采硫

It refers to the total process of obtaining sulphur deposits from the sea floor by adopting a downhole heating and extraction method.

采用井下加热熔融提取法,获取海底硫磺矿的全过程。

submarine topography 海底地形学

It is a discipline which studies submarine relief, morphological characteristics, the rule of the development and change, and the interaction between human activities and submarine environment.

研究海底表面起伏情况、形态特征、发展变化规律以及人类活动与海底环境相互作用的学科。

submarine topography 海底地貌

It refers to a form, pattern and structure of the surface of the seabed. Based on the location and basic characteristics, it can be divided into three basic geomorphic units: continental margin, ocean basin, and mid-ocean ridge.

海底表面的形态、样式和结构。按所处位置和基本特征,可分为大陆边缘、大洋盆地和大洋中脊 3 个基本地貌单元。

submarine tunnel construction 海底隧道工程建筑

It refers to the construction of underground buildings passable for pedestrian and vehicles under the sea.

建在海底之下供行人、车辆通行的地下建筑物的施工。

submarine valley 海底谷地

Also called *submarine canyon*, it refers to a steep-sided, long and narrow valley under the sea. It is divided into a relatively narrow and deep submarine valley located on a continental slope and a relatively broad and shallow submarine valley located on an ocean basin.

又称"海底峡谷"、"水下峡谷"。为横切大陆坡和大陆架的海底谷地。主要由混浊流侵蚀作用和海底滑塌作用形成。

submarine volcano 海底火山

It refers to a conical highland formed at the surface of a sea floor when underwater magma erupts from the fissure of the earth's crust.

地下岩浆沿地壳裂隙口喷出的海底表面所形成的圆锥状高地。

submarine warehouse construction 海底仓库建筑

It refers to the construction of a warehouse established on the sea floor to store manufacturing materials, the products of the submarine factories, military equipment, etc.

设置在海底供存放生产器材、海底工厂产品和军事装备等物资场所的施工。

submerged coast 下沉海岸

It is a coast formed by the subsidence of land, or rise of sea level, or the fact that the amount of land uplift is less than that of a sea level rise. It is often featured by a relatively tortuous coastline, lots of capes, peninsulas and islands, as well as deep waterways and good harbours.

因陆地下沉或海面上升或因陆地上升小于海面上升量而形成的海岸。一般具有较曲折的海岸线,多岬角、半岛和岛屿,并有深水道和良港。

submerged reef 暗礁

Also called *hidden dune*, it refers to the underwater reef which is not exposed even at low tide. It can be formed by organisms (such as a coral reef) or the extension parts of volcanic rocks or continental rocks under water. Normally 10 metres lower than the sea surface, a submerged reef is often isolated in the sea or close to the coast, which is unsafe for navigation.

又称"暗沙",低潮时仍不露出海面的水下礁石。可以由生物体组成(如珊瑚礁),也可以由火山岩礁或大陆岩石在水下的延伸部分组成。暗礁离水面一般不到10米,往往孤立海中或靠近海岸,对安全航行不利。

Submerged reefs 暗礁群

They refer to a mass ridge of rocks, or rock-forming organisms lying beneath the surface of the water, distributed widely in patches.

由众多暗礁组成、成片分布的地理实体。

subtidal zone 潮下带

It refers to the tidal flat areas between the lowest mark of ordinary tides to the end of the continental shelf.

平均大潮低潮位以下向海延伸的潮滩地区。

succession 演替

[*marine science and technology*] It is a development process in which a biological community is replaced by another biological community.

[*ecology*] It is a sequential replacement process in which the community at one particular location naturally evolves from one type to another type.

[海洋科技]某一生物群落被另一生物群落所替代的发展过程。

[生态学]某一地段上群落由一种类型自然演变为另一类型的有顺序的更替过程。

supervision and management of marine transportation 海上运输监察管理

It refers to the supervision and management conducted to ensure the transportation safety at sea.

为保证海上运输安全所进行的监察管理活动。

supply 供给

It refers to the quantity of a commodity that producers are willing and able to offer for sale at various possible prices in a given period of time.

生产者在一定时期内在各种可能的价格下愿意并且能够提供出售的该种商品的数量。

supply schedule 供给表

It refers to a numerical sequence table showing the relationship between the various prices indicating a certain commodity and the quantity supplied by the corresponding commodity.

表示某种商品的各种价格和与各种价格相对应的该商品的供给数量之间关系的数字序列表。

supply shock 供给冲击

It refers to an event that directly changes the cost and price of a commodity or service of an enterprise, which shifts the aggregate supply curve and the Phillips curve in turn.

直接改变企业的成本和价格,使经济中的总供给曲线移动,进而使菲利普斯曲线移动的事件。

supralittoral zone 潮上带

It is an area above a spring high tide line, on coastlines and estuaries, that is regularly splashed, but not submerged by ocean water. Seawater penetrates these elevated areas only during storms with high tides.

平均高潮位与较大潮或风暴潮时海浪所能作用到的陆上最远处之间的地带。

surf 拍岸浪

It is a phenomenon in which breaking waves are formed when waves move from deep water to shallow water, and as a result the wave velocity decreases, the length shortens, the height increases, and the crest inclines upon the shore.

又称"击岸波"。当波浪由深水向浅水传播时,由于海水变浅,波速减小,波长缩短,波高增大,波峰向海岸方向倾倒,形成波浪破碎的现象。

surf beat 矶波

It is a phenomenon of short-cycle rise and fall of coastal water level superposing above tidal water level.

叠加于潮汐水位之上的海岸水位短周期内的升降现象。

surface water 地表水

It is a generic term for the surface water in rivers, lakes, wetlands, glaciers, ice sheets, etc.

存在于河流、湖泊、沼泽、冰川和冰盖等地表水体中的水的总称。

surface water recharge 地表水补给

It refers to the process of the natural infiltration of the surface water to recharge the ground water due to the water head difference between the surface water (such as reservoirs, rivers, lakes, ponds) and the ground water.

因地表水(水库、河流、湖泊、坑塘等)和地下水之间的天然水头差,使地表水自然入渗补给地下水的过程。

Surveying and Mapping Law of the People's Republic of China《中华人民共和国测绘法》

As a basic law on surveying and mapping, it serves as the basic principle and basis for activities and management of surveying and mapping in China. On August 29, 2002, *the 29th Meeting of the Standing Committee of the Ninth National People's Congress* adopted the revised version. It came into effect as of December 1, 2002.

中国关于测绘的基本法律,是从事测绘活动和进行测绘管理的基本准则和基本依据。2002 年 8 月 29 日第九届全国人民代表大会常务委员会第二十九次会议修订通过,2002 年 12 月 1 日起施行。

sustainable consumption 可持续消费

It is a consumption mode which satisfies human needs by offering services or related products to improve life quality and reducing the use of environment-unfriendly materials without threatening future generations.

提供服务以及相关产品以满足人类的需求,提高生活质量,同时尽量减少对环境不利的材料的使用,从而不危及后代需求的消费模式。

sustainable development 可持续发展

It is a development mode which can satisfy the requirements of contemporaries without bringing harm to future generations.

满足当代人的需求,又不损害子孙后代满足其需求的能力。

sustainable ecological system 可持续生态系统

It is a perpetual ecological system which coordinates both economic development and environmental protection to satisfy the requirements of contemporaries without threatening the development of future generations.

将经济发展与环境保护协调一致,使之既满足当代人的需求,又不对后代人需求的发展构成危害的永续的生态系统。

sustainable management 可持续管理

It is a way of managing resources considering both short-term benefits and long-term interests.

对资源管理方式不仅满足短期利益,更重要着眼于长远的利益。

sustainable output theory 可持续产出论

It is a theory believing that the output capacity can be maintained limitlessly by keeping a balance between the annual output and annual net gain of population.

能无限地维持产出的水平,它可以通过使每年的产出等于每年的人口净增长来获得。

sustainable use 可持续利用

It refers to the scientific and proper use of renewable resources on the premise that environment or resource are not degrad.

对可更新资源以不导致环境及资源退化为前提,进行科学、适当地利用。

sustained utilization of plant resources 植物资源永续利用

It is a development and utilization mode in which resources are exploited in a planned and systematic manner according to the amount (abundance) and category of regional resources and the catch rotation method is used to synchronise growth and exploitation amounts.

根据地区的资源量(丰度)、种类有计划有步骤地进行资源开发,采取轮采方式,使生长量与开发量同步的开发利用方式。

swell 涌浪

It is a wave with a period of several seconds, which continues to propagate in the original wind action direction after the wind waves leave the wind area or after sudden changes in factors such as wind speed, and wind direction, etc., occur.

风浪离开风区后或风速风向等风要素突变后,继续按原风力作用方向传播的波浪。

synergetics 协同学

It was founded in 1973 by the famous German physicist, Hermann Haken, who notes that the mechanism of a system developing from a disordered structure to an ordered structure lies in a kind of adjustment capability and synergism, or self-organizing capability inherent in the system itself, which is the motive power for the system itself to exist and develop.

又称"协同论",德国著名物理学家赫尔曼·哈肯(Hermann Haken)于1973年创立。他指出,系统从无序形成有序结构的机理在于事物系统本身所固有的一种调节能力和协同作用,或者说自组织能力,它是系统自身存在和发展的动力。

system emergence 制度涌现

It is a new system structure arrangement formed in the process of interaction between system subjects under the condition of a certain institutional environment.

制度主体之间在一定的制度环境条件下,在相互作用的过程中所形成的新的制度结构安排。

system isomorphism 系统同构性

It is the consistency in existence modes and movement modes, exhibited by systems with different properties, i.e. a law commonly observed by all of these systems.

各个不同性质的系统之间所表现出来的存在方式和运动方式上的一致性,即所有系统共同遵守的规律。

system simulation 系统仿真

It is an abstract, intrinsical description and simulation of an actual system. It consists of three basic elements, namely system objects, system models, and computing tools. It is completed by three kinds of activities: system modelling, model sequencing, and analysis of simulation experiments.

对实际系统的一种抽象的、本质的描述及模拟活动。包括3个基本要素，即系统对象、系统模型、计算工具，并由3种活动来完成，即系统建模、模型程序化以及仿真实验分析。

system theory 系统论

It is the study of using the concept of system to master a study object, always regarding the object as a whole, and emphasizing the structure and functions of the system, as well as the study on the interrelations and the variation laws of the three aspects, i.e. system, factor and environment. In the way of thinking, it integrates analysis with synthesis dialectically, thus forming the following model of system approach: first, conduct system synthesis in terms of the whole to obtain various possible system solutions; second, systematically analyse various factors and their relations to establish mathematical models; finally, optimize the mathematical models and re-synthesise them into a whole.

用系统概念来把握研究对象，始终把对象作为一个整体来看待，并强调系统结构与功能的研究，以及系统、要素、环境三者的相互关系和变动的规律性研究。在思维方式上把分析和综合辩证地结合起来，使系统方法形成了如下模式：首先，从整体出发进行系统综合，得到各种可能的系统方案；其次，系统地分析各个要素及其关系，建立数学模型；最后，对数学模型进行优化选择并重新综合成整体。

systems of large integrated system 大型集成系统的体系

They refer to the large-scale systems engineering integrated via information interation which mainly relies on a series of criteria of organizations and agreements. It has a broad geographical distribution, without any fixed system structure and system boundary. Its differences with other systems are as follows: (1) each system of the integrated system is available independently; (2) each system can be managed independently; (3) each system is situated in a different geographic location; (4) it can adapt, evolve, and develop actively; (5) it can be elevated and enhanced integrally.

一个地域分布广泛，没有固定的系统组织结构和系统边界，主要依靠一系列组织和协议标准，通过信息交互、互动而集成的大型系统工程，和其他系统的区别是：(1)组成体系中的各系统独立可用；(2)各系统管理独立；(3)各系统分布不同的地理位置；(4)主动适应和演化发展；(5)涌现行为，整体提升。

systems of resource property right 资源产权制度

It is a generic term for a series of regulations to specify the emergence, definition, exercise, transaction and protection of resource property rights. It is a part of the legal management of resources.

资源产权的产生、界定、行使、交易和保护等进行规定的一系列制度的总称。它是资源法律管理的一部分。

T

Taiwan Strait 台湾海峡

Located between Fujian Province and Taiwan Province, it is the only channel to connect the South China Sea and the East China Sea.

位于中国福建省和台湾省之间,沟通东海和南海的唯一通道。

tax multiplier 税收乘数

It is the ratio of income change to tax change causing such an income change.

收入变动对引起这种变动的税收变动的比率。

tax revenue 税收收入

It covers revenue from Value Added Tax (VAT), business tax, corporate income tax, individual income tax, resource tax, fixed assets investment adjustment tax, urban maintenance and construction tax, property tax, stamp tax, urban land use tax, land value-added tax, vehicle and vessel tax, farmland occupation tax, deed tax, tobacco leaf tax, and other taxes. The unit of measurement is one hundred million yuan.

增值税、营业税、企业所得税、个人所得税、资源税、固定资产投资方向调节税、城市维护建设税、房产税、印花税、城镇土地使用税、土地增值税、车船税、耕地占用税、契税、烟叶税以及其他税收收入。计量单位:亿元。

technology of sea water resources exploitation 海水资源开发技术

It is the technology of extracting dissolved salt and other chemicals from seawater, desalinating seawater into fresh water, utilizing seawater directly, etc.

由海水中提取溶存的食盐和其他化学物质,将海水脱盐得到淡水,以及直接利用海水等的技术。

technological research for marine energy development 海洋能源开发技术研究

It refers to the research and experimental development for marine energy engineering technology.

对海洋能源工程技术进行的研究与试验发展活动。

technology of marine mineral resources exploitation 海洋矿产资源开发技术

It includes methods, equipment and facilities for the exploitation of petroleum, natural gas and other mineral resources stored in the sea floor.

开发蕴藏在海底的石油、天然气及其他矿产资源所使用的方法、装备和设施。

technology of ocean energy exploitation 海洋能开发技术

It is the technology of converting renewable energy in the ocean into electricity as well as other energy convenient for utilisation and transmission.

将蕴藏于海洋中的可再生能源转换成电能及其他便于利用与传输的能量的技术。

technology transfer rate 科技成果转化率

It is an index indicating the transferring of technological fruits to practical productivity, which refers to the percentage of the total number of technological fruits transferred to practical productivity to the total number of technological fruits within a certain period. The formula can be expressed as: technology transfer rate = (total number of transferred technological fruits/ total technological fruits) × 100%.

科技成果向现实生产力转化的指标。指一段时间里被转化为现实生产力的科技成果总数占科技成果总数的百分比。公式表达为:科技成果转化率 = (被转化的科技成果总数/科技成果总数) × 100%。

technology-intensive industry 技术密集型产业

Also called *knowledge-intensive industry in comparison to labor-intensive industry*, it refers to production and service sectors with advanced and sophisticated scientific technologies as the means of work. The technological density is proportional to mechanization and automation level in industries, departments or enterprises while it is inversely proportional to the number of manual persons in industries, departments or enterprises. It is characterized by low consumption of resources with equipment and production technology based on advanced scientific technology; large proportion of scientific and technological personnel among employees and high productive efficiency; complexity technological performance and rapid upgrading of products.

又称"知识密集型产业",与"劳动密集型产业"相对。以先进、尖端科学技术作为工作手段的生产部门和服务部门。其技术密集程度,往往同各行业、部门或企业的机械化、自动化程度成正比,而同各行业、部门或企业所用手工操作人数成反比。特点有:设备、生产工艺建立在先进的科学技术基础上,资源消耗低;科技人员在职工中所占比重较大,劳动生产率高;产品技术性能复杂,更新换代迅速。

temperature-difference energy 温差能

It is energy flow produced by a temperature gradient between the media.

在介质与介质之间由温度梯度所产生的能流通量。

temporal change of community 群落演变

It refers to the changes in the structure of biological community over time.

生物群落的结构随时间而发生的变化。

temporary utilization of uninhabited offshore islands 临时性利用无居民海岛

It refers to short-term landing or docking at uninhabited offshore islands for conducting business, teaching, scientific investigations, taking shelter and providing disaster relief.

因公务、教学、科学调查、救灾避险等需要而短期登临、停靠无居民海岛的行为。

terms of trade 贸易条件

They refer to the ratio of export prices to import prices in a country or area, which is generally represented as price index of imports and exports.

一个国家或地区输出商品价格与输入商品价格的比率,一般用进出口商品的价格指数表示。

terrace 海底阶地

It refers to a ladder-like geographic entity which is continuously distributed on the continental shelf and continental slope. It is formed by a flat step plane alternating with the steep slope in between.

大陆架、大陆坡上呈阶梯状连续分布的地理实体,由地形平坦的阶梯面及其之间的陡坡相间排列构成。

terrain 地势

It is the vertical and horizontal dimension of land surface. It also reflects geographic features.

地表高低起伏的状态或格局,也指地理上的形势。

terrestrial pollution source 陆地污染源

Also called *land-sourced pollutants*, it refers to places or facilities where pollutants are emitted from land to sea, resulting in or potentially causing marine environmental pollution.

又称"陆源污染源",从陆地向海域排放污染物,造成或者可能造成海洋环境污染的场所、设施等。

terrigenous organic matter 陆源有机物

It refers to organic matter which flows from land into sea by runoff, atmospheric transport or other means.

通过径流、大气输送和其他途径由陆地进入海洋的有机物质。

territorial sea 领海

It includes waters under the sovereign jurisdiction of a coastal nation or state, including both marginal seas and inland waters. State sovereignty includes the airspace over and the seabed below. According to the *United Nations Convention on the Law of the Sea*, a territorial sea extends at most 12 nautical miles from the baseline of a coastal state.

沿海国根据其主权划定的,邻接其陆地领土及内水以外,或群岛水域以外的一定范围的海域。国家对领海及其上空和海底行使主权。《联合国海洋法公约》规定领海的宽度为12海里。

territorial sky 领空

It includes all air space above the territory and territorial sea of a state, which is a component of state territory.

一国领土和领海范围内的全部上空,是一国领土的组成部分。

territory 领土

It is a spatial category which covers all the area under the jurisdiction of a sovereign state, including inner waters such as land, rivers, lakes and their subsurface layers, as well as ports connected to land, inland gulfs, territorial sky and sea.

主权国家管辖下的全部疆域,属于空间的范畴。包括陆地和河流、湖泊等内陆水域及其底下层,以及与陆地相连的海港、内陆湾、领空和领海。

theory of liquidity preference 流动性偏好理论

It refers to a psychological tendency indicating that people would rather keep cash at the expense of interest rather than hold bonds such as stocks and other less liquid bonds in general as well as government bonds to stay wealthy due to the flexibility of cash.

由于货币具有使用上的灵活性,人们宁愿以牺牲利息收入而储存不生息的货币来保持财富的心理倾向。

theory of Total Allowable Catch (TAC) 限额捕捞制度

TAC, namely, Total Allowable Catch is a management system, which is put forward based on the concept of maximum sustainable catch. To determine the TAC, the Maximum Sustainable Yield (MSY) of the object fish species needs to be evaluated first, and the social and economic factors have to be considered. When the total catch reaches the TAC target, further catching shall be totally banned at the same time.

TAC 管理制度,即总允许渔获量是根据最大持续渔获量的概念提出来的。首先需评估对象鱼种之最大持续生产量(MSY),并考虑社会经济要素决定 TAC。当渔获量累计达到 TAC 目标时,应全面同时禁止捕捞。

thermal pollution 热污染

It refers to the environmental pollution caused by the discharge of waste heat from modern industrial production and life.

现代工业生产和生活中排放的废热所造成的环境污染。

thermohaline convection 热盐对流

It refers to the vertical bidirectional movement of sea water due to the significant differences in temperature and salinity.

海水在垂直方向上由于温度与盐度的显著差异而引起的双向运动。

third law of thermodynamics 热力学第三定律

This law claims that it is impossible for any to be cooled to absolute-zero of temperature through a finite number of methods and procedure.

不可能用有限个手段和程序使一个物体冷却到绝对温度零度。

throughput of the international standard containers 国际标准集装箱吞吐量

It refers to the quantity of containers which pass in and out of the coastal harbour area by waterways after loading and unloading. The unit of measurement is ten thousand TEU, ten thousand tons. (TEU is the English abbreviation for Twenty Foot Equivalent Unit)

经由水路进、出沿海港区范围并经过装卸的集装箱数量。计量单位:万 TEU、万吨。("TEU"是"折合20英尺标准箱"的英文缩写语,下同。)

tidal age 潮龄

It refers to a time period between the first day and the 15th day of a lunar month and the spring tide, which lasts one to three days.

朔、望日到大潮来临的时段,约1~3天。

tidal bore 涌潮

It is a tidal phenomenon in which the leading edge of an incoming tide forms a wave (or waves) of water that travels up a river or narrow bay against the direction of the river or bay's current.

潮波在河口传播过程中产生的波陡趋于极限而破碎的潮水暴涨现象。

tidal channel 潮汐通道

It refers to a major channel followed by tidal currents, extending from the ocean into a tidal marsh or tidal flat.

由海伸向湿地或潮滩的潮流通道。

tidal creek 潮沟

It refers to a tidal flat-gully system caused by the action of tidal erosion.

潮流侵蚀作用形成的潮滩沟谷系统。

tidal current 潮流

It refers to the alternating horizontal flow of water associated with the rise and fall of the tide caused by astronomical tide-producing forces.

海水在潮位升降时发生的沿水平方向的流动。

tidal current energy 潮流能

It refers to the kinetic energy formed by the periodic reciprocating horizontal movement of seawater generated under the action of the tidal force of the moon and the sun, mainly concentrating on the shore, the waterway between islands and the bay mouth.

月球和太阳的引潮力使海水产生周期性的往复水平运动时形成的动能,集中在岸边、岛屿之间的水道或湾口。

tidal current energy power generation 海洋潮流能发电

It refers to a production activities of converting tidal current energy into electricity.

将海洋潮流能转化成电能的生产活动。

tidal current limit 潮流界

The upper bound of flood currents and ebb currents in a downstream of rivers flowing into a sea influenced by ocean tides is called a tidal current limit. When tidal waves in an open sea trace back along an estuary, flood currents and ebb currents will alternate in a certain section. The further the upstream is, the shorter the flood current lasts, the longer the ebb current lasts. A tidal current limit is where the flood current time is zero. A tidal current limit has a certain range of fluctuation, and generally, the average tidal current limit of spring tides in dry seasons for many years is regarded as the tidal current limit.

入海河流下游受海洋潮汐影响而出现往复涨、落潮流的上界,称为潮流界。外海潮波沿河口上溯时,在一定区段内存在涨落潮流的交替。愈往上游,涨潮流历时越短,落潮流历时越长。潮流界即是涨潮流历时为零的位置。潮流界的位置有一定的变动范围,一般将多年平均枯季大潮的潮流界作为河口潮流界。

tidal cycle 潮汐循环

It refers to the changing process of the rise and fall, ebb and flow of tides. Basically divided into semi-diurnal and diurnal, the tidal cycle reflects a long-term fluctuation of tides. The vertical direction shows the rise and fall of a tidal level and the horizontal direction shows the ebb and flow of tidal currents.

潮汐升降、涨落的变化过程。反映潮汐的长周期波动现象,通常分为半日或全日。垂直方向上表现为潮位的升降,水平方向上则表现为潮流的涨落。

tidal delta 潮流三角洲

It refers to a delta formed by sediment silt carried by a flood current and an ebb current inside and outside of a tidal channel. The accumulating pile formed by the sediment deposition of a flood current inside of a channel is called a flood current delta; the accumulating pile formed by the sediment deposition of an ebb current outside of a channel is called an ebb current delta.

又称"潮汐三角洲"。在潮汐通道内外由于涨、落潮流携带泥沙落淤而形成的三角洲。在通道内侧由涨潮流携带泥沙沉积而形成的堆积体,称为涨潮三角洲;在通道外侧由落潮流携带泥沙沉积而成的堆积体,称为落潮三角洲。

tidal energy 潮汐能

It refers to the energy formed as a result of periodic ebb and flow of the seawater under the action of tidal force of the sun and the moon for the earth.

在太阳、月亮对地球的引潮力的作用下,使海水周期性地涨落所形成的能量。

tidal flat 潮滩

It refers to a nearly flat coastal area between a seawall and low tide level. Alternately covered and exposed by tides, it consists of unconsolidated sediments of mud and silt.

海堤与低潮位之间的区域。由潮汐作用形成的平缓宽坦的淤泥粉砂质堆积体。

tidal mixing 潮混合

It refers to a phenomenon showing that the density of different water mixtures is greater than the average density of water involved in mixing.

不同水体混合后的密度大于参与混合水体的平均密度的现象。

tidal observation 潮汐观测

It refers to a process of using a tide gauge or a water gauge to test tide height.

采用验潮仪或水尺测量潮高的过程。

tidal phenomenon 潮汐现象

It refers to a periodic movement of seawater generated under the action of a tidal force of celestial bodies (mainly the moon and the sun).

海水在天体(主要是月球和太阳)引潮力作用下所产生的周期性运动。

tidal range 潮差

It is the vertical difference between the high tide and the succeeding low tide within a tidal cycle.

又称"潮幅"。指在一个潮汐周期内,相邻高潮位与低潮位间的差值。

tidal theories 潮汐理论

They refer to theories of studying the formation, development and change of tidal waves by applying the principles and methods of fluid dynamics.

运用流体力学的原理与方法,研究海洋潮波的形成、发展和变化的理论。

tidal water level 潮汐水位

It refers to the rise and fall of the water level of an ocean, harbour and waterway influenced by tides. In a tidal cycle, the water level when the water surface rises to the highest point is called the highest tidal level, while the water level when the water surface falls to the lowest point is called the lowest tidal level.

海洋或港湾、河道受潮汐影响而随时涨落的水位。在潮汐的一个涨落周期内,水面上升到最高点时的水位,称为最高潮水位;水面下降到最低点时的水位,称为最低潮水位。

tidal wave 潮波

It refers to a fluctuation of sea waves caused by a tidal force, namely the long term fluctuation based on semidiurnal or diurnal cycle.

引潮力引起的海水波动现象,即海洋中以半日或全日为周期的长周期波动。

tide 潮汐

It is the periodic rise and fall of sea levels caused by the combined effects of gravitational forces exerted by the Moon and the Sun and the rotation of the Earth.

在天体引潮力作用下产生的海面周期性涨落现象。

tide dominated estuaries 潮控河口湾

They are funnel-shaped estuaries dominated by tidal currents. Such estuaries are usually located in the centre of estuaries, from the mouth of estuaries to the top of the estuaries. With landforms shrinking and water ways getting thinner, distributed in a proper order are the tidal current ridges and the upper flow regime sand flats formed mainly under a tidal action, a meander channel segment under the interaction of runoffs and tidal currents and a straight channel segment under the action of rivers. While on both sides of estuaries, muddy tidal banks and salt marshes are widely developed due to the weak hydrodynamic forces.

以潮流作用为主的漏斗状河口湾。此类河口湾通常在河口湾中心,自湾口向湾顶。随着地形的收缩和水道变窄,依次分布着以潮流作用为主塑造的潮流脊和高流态沙坪、径流与潮流相互作用形成的曲流河段和以河流作用为主的顺直河段;而在河口湾的两侧,因水动力较弱,广泛发育淤泥质潮滩和盐沼。

tide-generating force 引潮力

It is the difference between the force of gravity exerted by the moon or the sun on unit mass object on the earth and that exerted on unit mass object at the centre of the earth, or the total force of the inertial centrifugal force generated when the earth make mass-centre motion around the earth-moon (sun) and the gravitational force of the moon(sun).

月球、太阳或其他天体对地球上单位质量物体的引力和对地心单位质量物体的引力之差,或地球绕地—月(日)质心运动所产生的惯性离心力与月(日)引力的合力。

tide level 潮位

It refers to a vertical height of a point of a tidal sea against a datum level.

某点潮汐海面相对于某一基准面的铅直高度。

tide rise 潮升

It refers to an average height of high tide, divided into rise of spring tides and rise of neap tides.

高潮位的平均高度,分为大潮升和小潮升。

tidology 潮汐学

It is a discipline that studies the tidal phenomenon and its process, including its formation causes, its variation rules and predictions for it.

研究潮汐现象及其过程的形成原因、变化规律和对其进行预报的学科。

tied island 陆地半岛

It refers to an island whose sandbar connects to a continent.

又称"陆连岛",指岛的沙坝与大陆相连的岛屿。

times of exhibition held by marine museum 海洋博物馆举办展览次数

Exhibition refers to various forms of exhibitions set up within this institution, designed and laid out by this museum. Travelling exhibitions of the same content are considered as one time exhibition. Exhibitions cohosted with institutions outside the system are counted by this museum while exhibitions cohosted with institutions within the system are counted by the host museum. Displays as long-term exhibitions are included in the index.

展览指在本机构内设置,由本馆设计布陈,形式比较多样的展出。同一内容的巡回展览,均按一个计算。与系统外机构合办的展览,由本馆统计;与系统内机构合办的,由主办馆统计。陈列作为长期展览,并入此指标项中。计量单位:次。

to tie islands by reclamation 填海连岛

It is to connect an island with land or island and island through methods such as marine reclamation lands.

通过填海造地等方式将海岛与陆地或者海岛与海岛连接起来的行为。

Tokyo Declaration 东京宣言

It was a declaration delivered by the *World Commission on the Environment and the Development of the UN* in Tokyo on February 27th, 1987. It formally raised the concept of sustainable development which can represent the collective intelligence of humanity to curb and reduce the rate of man made environmental pollution.

1987年2月27日,联合国环境与发展委员会在东京发表的宣言。其正式提出了代表全人类集体智慧的"可持续发展"概念,旨在控制和减缓人类污染环境的速度。

tombolo 连岛沙坝

Also called *tied bar*, it is a sandbar

connecting adjoining islands or the mainland to neighboring island. Composed of gravel, sand or shell scrap, a tombolo is formed when sediments are moving along the continent and island and gradually accumulate at wave shadow zone, which is between a continent and adjacent islands.

又称"连岛沙洲"。连接相邻岛屿或连接大陆与邻近岛屿的沙坝称为连岛沙坝。由砾石、沙及贝壳碎屑组成。沿大陆和岛屿沿岸运动的泥沙逐渐堆积于大陆与邻近岛屿之间的波影区便会形成连岛沙坝。

topographic feature 地貌

It refers to the various forms of the earth surface (including the seabed), caused by the interaction of the endogenic forces and exogenic forces.

地球表面(包括海底)的各种形态。由内营力和外营力相互作用而形成。

topography 地形

It refers to the shape of the surface relief. Generally the terrain can be plain, mountain, hill, basin, plateau, etc.

地面起伏的形状。地形一般有平原、山地、丘陵、盆地、高原等。

topological properties 拓扑性质

A "measurement" property of a complex network, which can be exhibited independent of the specific positions of its nodes and the specific morphology of its edges, is called a topological property of a network. The corresponding structure is called a topological structure of a complex network.

把复杂网络不依赖于节点的具体位置和边的具体形态就能表现出来的"度量"性质叫做网络的拓扑性质。相应的结构叫做复杂网络的拓扑结构。

toppling 倾倒

It refers to the disposal of waste and other harmful substances into the ocean by means of ship, aircraft, platform or other carrying tools, including the discarding of ship, aircraft, platforms and their auxiliary facilities and other floating tools.

通过船舶、航空器、平台或者其他载运工具,向海洋处置废弃物和其他有害物质的行为,包括弃置船舶、航空器、平台及其辅助设施和其他浮动工具的行为。

total assets 资产总计

They are economic resources, which are owned or controlled by an enterprise, and can be measured by money value including all property, obligatory rights, and other rights. Classified by the degree of liquidity (i.e. asset cashability and solvency), total assets include circulating assets, long-term investment, fixed assets, intangible assets, deferred assets, and other assets. The unit of measurement is ten thousand yuan.

企业拥有或控制的能以货币计量的经济资源,包括各种财产、债权和其他权利。资产按其流动性(即资产的变现能力和支付能力)划分为:流动资产、长期投资、固定资产、无形资产、递延资产和其他资产。计量单位:万元。

total current assets 流动资产合计

They refer to assets which can either be converted to cash or consumed within one year or a business production cycle of longer than 12 months, including cash, bank savings, short-term investments, receivables, prepayments, inventory, etc. The unit of measurement is ten thousand yuan.

企业可以在一年内或超过一年的一个生产周期内变现或者耗用的资产,包括现金及各种存款、短期投资、应收及预付款项、存货等。计量单位:万元。

total current liabilities 流动负债合计

They refer to the sum of money which is owed by a company and due within one year or an operating cycle of more than 12 months, including short-term loans, notes payable, account payable, deposit received, payroll payable, tax payable, profit payable, accrued expenses, etc. The unit of measurement is ten thousand yuan.

企业在一年内或超过一年的一个营业周期内需要偿还的债务,包括短期借款、应付票据、应付账款、预收账款、应付工资、应交税金、应付利润、预提费用等。计量单位:万元。

total energy consumption 能源消费总量

It refers to the total sum of energies consumed by all industries and residents in a given period, including raw coal, crude oil and their related products, natural gas, power, exclusive of the utilization of low calorific value fuel, biomass energy and solar energy. It consists of three parts: terminal energy consumption, loss during the process of energy conversion and energy loss. The unit of measurement is tons of standard coal equivalent.

一定时期内各行业和居民生活消费的各种能源的总和。能源消费总量包括原煤和原油及其制品、天然气、电力,不包括低热值燃料、生物质能和太阳能等的利用。能源消费总量分为终端能源消费量、能源加工转换损失量和能源损失量三部分。计量单位:吨标准煤。

total energy production 能源生产总量

It refers to the total sum of primary energy output in a given period, including generating capacity of raw coal, crude oil, natural gas, hydropower, nuclear energy and other kinetic forces (such as wind energy, geothermal energy, etc.), exclusive of the utilization of low calorific value fuel production, biomass energy, solar energy etc. and the secondary energy production transformed from primary energy. The unit of measurement is tons of standard coal equivalent.

一定时期内一次能源生产量的总和。一次能源生产量包括原煤、原油、天然气、水电、核能及其他动力能(如风能、地热能等)发电量,不包括低热值燃料生产量、生物质能、太阳能等的利用和由一次能源加工转换而成的二次能源产量。计量单位:吨标准煤。

total length of island coastline 岛屿岸线总长度

It refers to the length of the trace line of the sea-land demarcation when the islands are at the mean high water level. The unit of measurement is kilometre.

岛屿平均大潮高潮位时海陆分界的

痕迹线的长度。计量单位:千米。

total liability 负债合计

It refers to the aggregate of all debts or obligations a company is liable for, which can be measured in currency. Liabilities are settled through the transfer of assets or provision of service, including money, assets or services. A company's total liabilities can be split up into current liabilities and long-term liabilities based on the length of the payback period. The unit of measurement is ten thousand yuan.

企业所承担的能以货币计量,将以资产或劳务偿付的债务,偿还形式包括货币、资产或提供劳务。负债一般按偿还期长短分为流动负债和长期负债。计量单位:万元。

total profits 利润总额

They are surplus gained after the deduction of various costs from total earnings in the process of production and management, which reflects the total amount of profit and loss realised in a reporting period, including operating profit, subsidy income, net investment income and net non-operating income. The unit of measurement is ten thousand yuan.

企业在生产经营过程中各种收入扣除各种耗费后的盈余,反映企业在报告期内实现的亏盈总额,包括营业利润、补贴收入、投资净收益和营业外收支净额。计量单位:万元。

total volume of imports and exports 进出口总额

It refers to the gross amount of goods actually passing in and out of the Chinese border, including cargoes imported and exported in foreign trade, processed and assembled import and export goods, free-aid goods and gifts from the UN, international organizations and other countries, donations by overseas Chinese, compatriots from Hong Kong, Marco, Taiwan and Chinese of foreign nationality, leased goods which belong to tenants at the expiration of contract, processed import and export goods in local border trade and small-scale border trade, cargoes imported and exported and advertisement articles of Sino-foreign joint ventures, cooperative ventures and wholly foreign-owned enterprises, import goods which are picked up from bonded warehouses and sold in China and other import and export goods. According to Chinese law, export and import goods are calculated at free-on-board price and CIF price respectively. The unit of measurement is one hundred million yuan.

实际进出我国国境的货物总金额。包括对外贸易实际进出口货物、来料加工装配进出口货物,国家间、联合国及国际组织无偿援助物资和赠送品,华侨、港澳台同胞和外籍华人捐赠品,租赁期满归承租人所有的租赁货物,进料加工进出口货物,边境地方贸易及边境地区小额贸易进出口货物,中外合资企业、中外合作经营企业、外商独资企业进出口货物和广告品,从保税仓库提取在中国境内销售的进口货物,以及其他进出口货物。我国规定出口货物按离岸价格统计,进口货物按到岸价格统计。计量单位:亿元。

total water resources 水资源总量

It is the total amount of surface and underground water formed by precipitation in an area under evaluation; that is, the sum of the surface water amount and the amount of groundwater replenished by infiltrated rainfall, excluding the amount of passing—by water. The unit of measurement is cubic metre.

评价区内降水形成的地表和地下产水总量,即地表产流量与降水入渗补给地下水量之和,不包括过境水量。计量单位:立方米。

tourism area 旅游区

It refers to a sea area delimited for the exploitation of coastal and marine tourism resources and the development of tourism industries, including scenic spots, holiday tourism areas, etc.

为开发利用滨海和海上旅游资源,发展旅游业需要划定的海域,包括风景旅游区和度假旅游区等。

tourism resources 旅游资源

They refer to all things and elements from nature and human society, which can attract tourists, promote development of tourism industries and produce economic, social and environmental benefits.

自然界和人类社会凡能对旅游者产生吸引力,可以为旅游业开发利用,并可产生经济效益、社会效益和环境效益的各种事物和因素。

tourism-based city 旅游型城市

It refers to a city which prospers by relying on the development of tourism resources and scenic spots.

在开发旅游资源和景点的基础上发展起来的城市。

town and country planning 城乡规划

It is an overall arrangement made and the measures taken by a government for construction layout, land utilization and matters related to the economic and social development of a city, town and village over a certain period of time, which is also one basic means taken by the government to guide and control a town and country construction and development.

政府对一定时期内城市、镇、乡、村庄的建设布局、土地利用以及经济和社会发展有关事项的总体安排和实施措施,是政府指导和调控城乡建设和发展的基本手段之一。

town and country planning management 城乡规划管理

It refers to the organization, formulation and approval of urban and rural planning, as well as the implementation of planning control, guidance, supervision and inspection for land utilization and construction in a city, town and village.

组织编制和审批城乡规划,并对城市、镇、乡、村庄的土地使用和各项建设的安排实施规划控制、指导和监督检查。

town construction detailed plan 镇的修建性详细规划

Based on the overall planning and regulatory detailed planning of a town, it is planning and design to guide facilities and construction of various architecture and their projects within the town.

以镇的总体规划和控制性详细规划

为依据,制定的用以指导镇内各项建筑及其工程设施和施工的规划设计。

town detailed plan 镇的详细规划

Based on the overall planning of a town, it is specific arrangement and design in land use, spatial layout and construction land for some areas of the town over a period of time.

以镇的总体规划为依据,对一定时期内镇的局部地区的土地利用、空间布局和建设用地所作的具体安排和设计。

town master plan 镇的总体规划

It is the overall deployment, specific arrangements and implementing measures regarding a town's nature, development goals, development scale, land use, spatial layout, and all constructions over a period of time.

对一定时期内镇的性质、发展目标、发展规模、土地利用、空间布局以及各项建设的综合部署、具体安排和实施措施。

town regulatory detailed plan 镇的控制性详细规划

Based on the overall planning of a town, it determines the planning requirements for construction areas within the town, including the control indexes of land use nature and intensity, the controlling positions of roads and engineering pipelines, as well as the spatial environmental control.

以镇的总体规划为依据,确定镇内建设地区的土地使用性质和使用强度的控制指标、道路和工程管线控制性位置以及空间环境控制的规划要求。

township planning 乡规划

It is the overall deployment, specific arrangements and implementing measures regarding rural economic and social development, land use, spatial layout, and all constructions over a period of time.

一定时期内乡的经济和社会发展、土地利用、空间布局以及各项建设的综合部署、具体安排和实施措施。

toxic pollutant 有毒污染物

It refers to pollutant ingested directly or indirectly by an organism, causing morbidity, abnormal behaviour, genetic mutation, physiological functional disturbance, organism deformity or death of the organism and its offspring.

那些直接或者间接为生物摄入体内后,导致该生物或者其后代发病、行为反常、遗传变异、生理机能失常、机体变形或者死亡的污染物。

trade benefit 贸易利益

It describes a way of obtaining mutual benefits through the cooperation of two regions even if one possesses absolute higher efficiency than the other in all industries. The terms of trade are as follows: when relative efficiency difference exists between two regions in manufacturing different products, they should specialise in the production of goods in view of their respective local advantages and then exchange to obtain trade benefits.

即使两个区域中的一个在每一种行业上都比另一个具有较高的绝对效率,两个区域之间的贸易同样对双方有利,贸易条件是:在生产不同的产品上两个区域之间存在着相对的效率差异,这时,每个区域都专业化于本区域具有相对有利条件的商品,并用该商品去换取另一区域具有相对有利条件的商品,从而产生贸易利益。

trade deficit 贸易赤字

Also called unfavorable balance of trade, trade deficit refers to a negative balance of trade, where a country's total imports exceed its total exports over a certain period, indicating the country's disadvantageous position in foreign trade for that year.

又称"贸易逆差",是指一国在一定时期内出口贸易总额小于进口贸易总额,表示该国当年对外贸易处于不利地位。

trade policy 贸易政策

It refers to government regulations which directly influence the amounts of imports, exports and labour services of a country.

直接影响一国进口或出口的物品与劳务数量的政府政策。

trade surplus 贸易盈余

Also called *favorable balance of trade*, it refers to a positive balance of trade, where a country's total exports exceed its total imports over a certain period, indicating the country's favourable position in foreign trade for that year.

又称"贸易顺差",是指一国在一定时期内出口贸易总额大于进口贸易总额,表示该国当年对外贸易处于有利地位。

trade wind drift 信风漂流

Also called *trade wind current*, it refers to a current generated by low-latitude ocean surfaces under the action of a trade wind. The trade wind zone is near the north and south of the equator, known as the northeast trade wind in the northern hemisphere, and the southeast trade wind in the southern hemisphere.

又称"信风海流",指信风作用下,低纬度的洋面产生的海流。地球信风带位于赤道南北附近区域,北半球称为东北信风,南半球称为东南信风。

traditional marine industry 传统海洋产业

It refers to traditional production and service industries composed of marine fishing, sea salt and marine transportation industries.

由海洋捕捞业、海盐业和海洋运输业等组成的古老的生产和服务行业。

transgression 海侵

Also called *advance of sea*, it refers to the process during which seawater slowly transgresses into land from the coast due to the rising of sea level and sink of crustal structures.

又称"海进",由海平面上升或地壳构造下沉等引起的海水缓慢地从海岸入侵陆地的过程。

transportation capacity of submarine pipeline 海底管道输送能力

It refers to the maximum volume of petroleum (or natural gas), water, etc. which can be transported through a submarine pipeline in a reporting period. Generally it can be grouped according to the different calibers of the pipeline. The unit of measurement is cubic meter per year.

报告期内海底运输管道最大可能通过油(或气)、水等输送物品的数量。一般按管道的不同口径分组。计算单位：米³/年。

transportation medium 搬运介质

It refers to the substances acting as media in the process of ground transportation. Common transportation media include running water, air, glaciers, underground water, waves, tidal currents, etc.

地面物质搬运过程中起媒介作用的物质称为搬运介质。常见的搬运介质主要有流水、空气、冰川、地下水、波浪和潮流等。

transportation volume of submarine pipeline 海底管道输送量

It refers to the volume of petroleum (or natural gas), water, etc., transported through a submarine pipeline in a reporting period. Classified statistics can be conducted based on the oil pipeline, gas pipeline, freshwater pipeline and others. The unit of measurement is cubic metre.

报告期内海底运输管道通过油(或气)、水等输送物品的数量。按照输油管道、输气管道、输送淡水管道及其他分类统计。计量单位：立方米。

transversal coast 横向海岸

Also called *Atlantic-type coast*, it refers to the coast where the overall stretching direction of the coastline almost intersects vertically with the tectonic line.

又称"大西洋型海岸"，海岸线延伸的总方向与地质构造线走向近似垂直相交的海岸。

trench-arc-basin system 沟弧盆地

As the abbreviation of trench-arc-back-arc basin system, it refers to the tectonic geomorphology system which has genetic connections like trenches, island arc or back arc basins caused by the subduction of the oceanic plates towards the continental plates, such as the west Pacific Ocean near the Asian continental margin zone.

海沟-岛弧-弧后盆地系统的简称，由大洋板块向大陆板块俯冲形成的海沟、岛弧和弧后盆地等具有生成联系的构造地貌体系，如西太平洋近亚洲大陆边缘带。

tributary 支流

It is a river that flows into a main stream directly or indirectly.

直接或间接流入干流的河流。

trophic level 营养级

It is the nutrition level of producers and consumers at all levels in the process of energy flow in an ecosystem.

生态系统的能量流动过程中生产者和各级消费者的营养水平等级。

trophic structure 营养结构

A trophic structure of organisms in a community (or an ecosystem) includes producers, consumers and decomposers. It is a process of transferring energy and materials from plant to herbivores, and then to carnivores.

一个群落(或生态系统)的生物，营养结构可分为生产者、消费者和分解者，能量、物质从植物转到植食者，再转到肉食者的过程。

Tropical Oceans Global Atmosphere Project(TOGA) 热带海洋与全球大气实验

It is an international cooperation program which studies inter-annual variations of tropical oceans and global atmosphere. The purpose is to investigate the annual changes of tropical oceans with 20 degrees north and south latitude span and global atmospheric climate, thus determining a changing mechanism and a change predictive mechanism. Starting from January 1985, this project lasted for 10 years observing and recording real-time sea levels, sea surface roughness, surface currents, etc. In view of the close unity of ocean and atmosphere, TOGA will help to improve the accuracy of medium-and-long term weather forecasts.

研究热带海洋及全球大气年际变动的国际合作计划。目的是研究南北纬20度跨距的热带海洋和全球大气气候逐年变动,从而确定变动机制及变化预测性机制。该项计划从 1985 年 1 月开始,为期 10 年,10 年观测、实时记录海面水位、海面粗糙度、表层海流等。鉴于海洋和大气是紧密耦合的统一系统,热带海洋地球大气计划将有助于提高中、长期天气预报的准确性。

trough 海槽

It refers to an oblong deep sea depression with a U-shaped profile. Generally, a trough is shallower and smaller than a trench and the slopes of the two sides are gentle.

一般比海沟浅而小,两侧边坡较缓,剖面呈 U 字形的长椭圆状深海凹地。

tsunami 海啸

It is a series of ocean waves of long-period fluctuation caused by submarine earthquakes, volcanic eruptions or giant rock collapses, landslides, etc., which can lead to violent fluctuations on the coastal sea surface.

由海底地震、火山爆发或巨大岩体塌陷和滑坡等导致的海水长周期波动,能造成近海海面大幅涨落。

tsunami disaster 海啸灾害

It refers to the damage to a coastal area that is attacked by an enormous oceanic long wave (tsunami) triggered by marine earthquakes, collapses, landslides, volcanic eruptions, etc.

海洋地震、塌陷、滑坡、火山喷发等引起的特大海洋长波(海啸)袭击海岸地带造成的灾害。

Tsushima Channel 对马海峡

Located between Tsushima island and Kyushu, Honshu of Japan, it is an important channel connecting the Japan Sea and the Yellow Sea.

位于日本对马岛与九州、本州岛之间沟通日本海和黄海的重要通道。

turn of tide 转流

It refers to a tide which shifts the direction in ocean currents.

海流中,流向转变的潮流。

turnover of passenger traffic 旅客周转量

It refers to the sum of products of the total number of passengers carried by boat and by their transportation distance. The unit of measurement is ten thousand people per kilometre.

港口船舶实际运送的每位旅客与该旅客运送距离的乘积之和。计量单位：万人/千米。

two-center urban system 双中心型城镇体系

It is defined as the urban system in some areas in which the largest city is on equal terms with the second one, either in size or in socioeconomic status.

在一些地区的城镇体系中，最大的城市无论在规模上，还是在经济社会地位上都与第二位城市不相上下，这样的城镇体系被称为双中心型城镇体系。

typhoon 台风

It is a tropical cyclone with a maximum wind speed (near the centre) of at least 32.7m/s (on a scale of 12 and above).

最大风速(中心附近)不小于32.7米/秒(12级及以上)的热带气旋。

typhoon disaster 台风灾害

It is a typhoon-induced disaster, such as a powerful wind, rough sea, storm surge, flood, etc. A typhoon is a tropical cyclone that occurs in the western Pacific and South China Sea and whose central maximum wind speed reaches a scale of 12 and above.

台风造成的灾害。台风所到之处引发的强风、巨浪、风暴潮和洪涝等灾害。台风为发生在西太平洋和南海，中心最大风力达12级以上的热带气旋。

U

ultimate productivity 终级生产力

It refers to the production capacity of carnivorous fish and other marine carnivorous creatures.

肉食性鱼类和其他海洋肉食性生物的生产能力。

unbalanced development strategy 非均衡发展战略

It refers to the strategy of investing limited resources in regions or industries which can bring higher economic benefits to obtain rapid growth in the regional economy, thus stimulating the development of other industries, based on the differences in resource endowment and configuration.

立足于资源禀赋与配置的差异性，通过将有限的资源首先投向效益较高的区域和产业，以获得区域经济的高速增长，并带动其他区域、其他产业的发展的战略。

unbalanced regional development 区域发展不平衡

It refers to an unbalanced economic development in geographic between various regions due to the differences of location, natural environment, economic development, policy support, degree of openness, etc. It is a significant factor influencing the harmony of society.

因地理位置、自然环境、经济发展、政策倾斜、开放程度等条件的差异而形成的不同地区之间经济发展水平的不均衡。是影响社会和谐的一个重要原因。

underground brine 地下卤水

It refers to the groundwater whose total mineralization is more than 50.0 grams per litre.

总矿化度大于 50.0 克/升的地下水。

undersea coal mine 海底煤矿

It refers to a deposit which contains coal in the submarine rock strata.

又称"海底煤田"。海底岩层中含有煤的矿床。

undersea feature 海底地理实体

It refers to a geomorphic unit that can measure and demarcate the boundaries on the sea floor.

海底可以测量并可划分界限的地貌单元。

undersea feature names 海底地名

They refer to the names given to undersea features. Broadly, undersea feature names can be divided into two categories: geographical proper names and geographical general names.

人们赋予海底地理实体的名称，广义上包括地名专名和地名通名两部分。

undersea iron mine 海底铁矿

It refers to a deposit which contains iron in the submarine rock strata.

海底岩层中含铁的矿床。

undersea technology 海洋水下技术

It is about the research and development of engineering technology within an undersea environment, including diving technology, undersea construction, diving instrument development, salvage technology, etc.

研究和发展在海洋水下环境条件下的工程技术，包括潜水技术、水下作业施工、潜水器开发、打捞技术等。

undersea tin mine 海底锡矿

It refers to a deposit which contains tin lode in the submarine rock strata.

海底岩层中含有锡矿脉的矿床。

undersea well logging 海底测井

It refers to the means to apply various geophysical methods, such as sound, light, electricity, magnetism and radioactive well logging to a geological profile to study the properties of the submarine strata and prospect for oil, gas and other mineral resources.

将各种地球物理方法，如声、光、电、磁、放射性测井等应用到井下地质剖面中，以研究海底地层的性质，寻找油气及其他矿产资源的方法。

union 工会

It refers to an employee association established to promote employee rights and work-related welfare, such as negotiating with employers to promote employee wages, allowances and working conditions.

与雇主就工资、津贴和工作条件进行谈判的工人协会。

unit of account 计价单位

It is the standard by which people express price and record debt.

人们用来表示价格和记录债务的标准。

United Nations Convention on the Law of the Sea (UNCLOS)《联合国海洋法公约》

It is an international public law which standardizes the legal status of all waters inside and outside national jurisdiction, and adjusts the relationship in marine aspects between countries, countries and international organizations with the approval of the UN conferences. The Law came into force on 16th, November 1994 and was adopted by the Standing Committee of the National People's Congress of China on 15th, May 1996.

由联合国召开有关会议通过的规范各国管辖范围内外各种水域的法律地位，调整国家之间、国家与国际组织之间在海洋方面关系的国际公法。1994年11月16日生效。我国全国人民代表大会常务委员会于1996年5月15日通过该公约。

United Nations Environment Programme (UNEP) 联合国环境规划署

As an international institution that coordinates United Nations environmental activities, UNEP was founded formally in January 1973 and has its headquarters in Nairobi of Kenya. It is a business subsidiary institution and reports annual performance through the *United Nations Economic and Social Council* with the following missions: promote international cooperation on environmental issues and offer policy advice; illustrate a general principle of guiding and coordinating environmental planning within the UN system and review planned periodic reports; examine global environmental conditions to arouse the attention of all governments on environment issues which have international influence; review regularly the impacts of national and international environmental policies on developing countries by and measures and the issue of increasing costs incurred; enhance the acquisition of environmental knowledge and exchange of information.

作为联合国统筹全世界环保工作的组织,联合国环境规划署于1973年1月正式成立,总部现设在肯尼亚首都内罗毕。环境规划署是一个业务性的辅助机构,它每年通过联合国经济和社会理事会向大会报告自己的活动,其宗旨是促进环境领域内的国际合作,并提出政策建议;在联合国系统内提供指导和协调环境规划总政策,并审查规划的定期报告;审查世界环境状况,以确保可能出现的具有广泛国际影响的环境问题得到各国政府的适当考虑;经常审查国家和国际环境政策和措施对发展中国家带来的影响和费用增加的问题;促进环境知识的取得和情报的交流。

United Nations Framework Convention on Climate Change (UNFCCC) 《联合国气候变化框架公约》

It is an international environmental treaty approved at the *United Nations Conference* on 4th, June 1992, which regulates that as contracting nations, developed countries should take measures to limit greenhouse gas emissions, provide new additional funds for developing countries to pay increased expenses needed as a result of fulfiling the obligations under the convention and promote and facilitate technology transfer by adopting all feasible means.

联合国大会于1992年6月4日通过的一项公约。《公约》规定发达国家为缔约方,应采取措施限制温室气体排放,同时要向发展中国家提供新的额外资金以支付发展中国家履行《公约》所需增加的费用,并采取一切可行的措施促进和方便有关技术转让的进行。

unstructured question 非结构化问题

Also called *imperfectly defined question*, or *ill-structured question*, it refers to the question that is hard to depict in a decisive manner. People solve the question mainly based on their experience since there are no fixed laws as to the solution procedures and solution methods.

又称"定义不完善的问题"、"劣构问题",难以用确定的形式来描述,其求解过程和求解方法没有固定的规律可以遵循,主要根据人的经验求解的问题。

upper layer 上层

It refers to the aquifer which is about 100 metres thick in the ocean with a high temperature, low density and even vertical mixing under solar radiation.

大洋中被太阳辐射加热的温度较高、密度较小、垂向混合较均匀、厚约100米的水层。

upper water 上层水

It is a generic term for surface water and subsurface water in the ocean.

海洋的表层水和次表层水的总称。

upwelling 上升流

It refers to a phenomenon or process involving the ascending motion of water in the ocean from the ocean depth to the surface.

从深层向上涌升的海水流动。

urban detailed planning 城市详细规划

It is the detailed arrangement and design made for land utilisation, spatial arrangements and construction lands in a local urban area over a certain period of time, based on comprehensive urban planning.

以城市的总体规划为依据,对一定时期内城市的局部地区的土地利用、空间布局和建设用地所作的具体安排和设计。

urban disease 城市病

They refer to those problems resulting from urban development process, including traffic congestion, housing shortage, inadequate water and energy supply, environmental pollution, disorders, unbalance between input and output of material and energy flows, and growing demand and supply conflicts. These problems reduce the urban construction and development to unbalanced situations and disorders, resulting in a huge waste of resources, a decrease of living standards of residents and an increase of economic development costs, hindering sustainable urban development to a certain degree.

城市在发展过程中出现的交通拥挤、住房紧张、供水不足、能源紧缺、环境污染、秩序混乱,以及物质流、能量流的输入、输出失去平衡,需求矛盾加剧等问题。这些问题使城市建设与城市发展处于失衡和无序状态,造成资源的巨大浪费、居民生活质量下降和经济发展成本提高,在一定程度上阻碍了城市的可持续发展。

urban ecology 城市生态学

It is a discipline that studies the relations between human activities and their physical and living environments in an urban or urbanised context. It studies the ecological relations inside or between different levels including molecule, cell, individual, community, city group and urbanised region. It mainly studies the basic rules of the structure, function, evolution and process of the complex ecological systems dominated by human activities, the mechanism of the ecological service and the systematic method of planning, construction and management.

研究城市或城市化环境下人类活动与其物理和生命环境关系的学科。其研究层次可以包括从分子、细胞、个体、社区到城市、城市群乃至城市化区域不同尺度内部和之间的生态关系。主要研究以人类活动为主导的复合生态系统的结构、功能、演化、过程的基本规律、生态服务的机制和规划、建设、管理的系统方法。

urban economic ecology 城市经济生态学

It is a discipline which focuses on the study of the conversion and utilisation efficiency of material flow, energy flow and information flow in the process of urban metabolism from an economic perspective.

从经济学角度重点研究城市代谢过程的物流、能流和信息流的转化、利用效率等问题的学科。

urban landscape ecology 城市景观生态学

It is a discipline which studies the conversion and utilisation efficiency of material flow, energy flow and information flow in the process of urban metabolism between different ecological systems from a landscape perspective.

从景观尺度研究城市不同生态系统之间代谢过程的物流、能流和信息流的转化、利用效率等问题的学科。

urban master planning 城市总体规划

It refers to the comprehensive deployment, specific arrangement and implementation measures regarding the nature, development goal and scale, land use, spatial layout, and all construction of a city over a certain period of time. It also serves as an important basis and means to guide and control urban construction, as well as protect and manage urban space resources.

对一定时期内城市的性质、发展目标、发展规模、土地利用、空间布局以及各项建设的综合部署、具体安排和实施措施,是引导和控制城市建设,保护和管理城市空间资源的重要依据和手段。

urban natural ecology 城市自然生态学

It is a discipline that studies the positive and negative impact of urban human activities on its natural ecological system and also the impact of regional natural factors on human activities, in other words, it is a discipline that studies the mutual relation between urban human activities and regional natural ecological system factors.

研究城市的人类活动对所在地域自然生态系统的积极和消极影响,以及地域自然要素对人类活动的影响,即人的城市活动与地域的自然生态系统要素之间相互关系的学科。

urban planning 城市规划

It refers to a comprehensive deployment, specific arrangement and implementation measures of economic and social development, land use, spatial layout and various construction projects of a city over a certain period of time.

一定时期内城市的经济和社会发展、土地利用、空间布局以及各项建设的综合部署、具体安排和实施措施。

urban socioecology 城市社会生态学

It is a discipline which studies the impact of the urban environment on people's physiology and mentality, and the problems such as population, transportation and energy that people meet in the process of urban construction and nature transformation.

研究城市环境对人的生理和心理的影响、效应及人在建设城市、改造自然的过程中所遇到的人口、交通、能源等问题的学科。

urban system 城镇体系

It is urban groups of different functions, scales and ranks having close economic and social relationships with each other in a certain region.

在一个区域内经济社会联系密切的、具有不同职能、不同规模、不同等级的城镇群体。

urban system planning 城镇体系规划

It is the distribution and development planning for urban areas with different population sizes, ranks and functional divisions in a certain region based on its rational distribution of regional productivity and division of urban functions.

对一定地域范围内,以区域生产力合理布局和城镇职能分工为依据,确定不同人口规模等级和职能分工的城镇的分布和发展规划。

urban-rural dualistic structure 城乡二元结构

It is a system structure formed by a series of system arrangements which maintains the dualistic economic form of modern industry in urban areas and traditional agriculture in rural areas, and the dualistic social form of urban society and rural society isolated from each other. It contains an urban-rural dualistic economic structure and an urban-rural dualistic social structure.

维持城市现代工业和农村传统农业二元经济形态,以及城市社会和农村社会相互分割的二元社会形态的一系列制度安排所形成的制度结构。包括城乡二元经济结构和城乡二元社会结构。

usability of natural resources 自然资源可用性

Differing from natural conditions, it is the function and performance of natural resources available to human beings under certain technical and economic conditions.

在一定技术、经济条件下,自然资源可被人类利用的功效和性能。即自然资源可用性,亦以此与自然条件相区别。

use of sea area 海域使用

It refers to any exclusive and continuous use of a certain sea area over three months within the inland waters and territory seas, including sea area used for fishery, transportation, industry and mining tourism entertainment, sea floor project, pollution discharge and dumping, reclamation, special purpose, ect.

在内水、领海持续使用特定海域3个月以上的排他性用海活动。包括渔业用海、交通运输用海、工矿用海、旅游娱乐用海、海底工程用海、排污倾倒用海、围海造地用海、特殊用海及其他用海。

using quantity of seawater by circulating and cooling 海水循环冷却利用量

It refers to actual seawater using quantity when seawater, as a cooling medium, is cooled and circulated within a reporting period. The unit of measurement is ten thousand tons.

报告期内以海水为冷却介质进行循环冷却的实际海水利用量。计量单位:万吨。

usufruct of immovable property 用益物权

It is the right to possess, use and benefit from, within a certain range, the things owned by others.

用益物权是对他人所有的物,在一定范围内进行占有、使用和收益的权利。

urbanization 城市化

It refers to a natural historical process from a traditional agricultural society to a modern urban society, which is manifested as the migration of rural populations into towns and cities, an increase in the number of cities, an expansion of city scales, and an improvement of urban modernisation levels. It reflects tremendous development space due to fundamental changes in social and economic structures.

由传统的农业社会向现代城市社会发展的自然历史过程。它表现为人口向城市的集中、城市数量的增加、规模的扩大以及城市现代化水平的提高,是社会经济结构发生根本性变革并获得巨大发展空间的表现。

urbanization level 城市化水平

Also called *urbanisation rate*, it is a quantitative index to measure the development degree of urbanization, which is generally indicated by the proportion of urban population in a total population in certain geographical areas.

又称"城市化率",是衡量城市化发展程度的数量指标,一般用一定地域内城市人口占总人口比例来表示。

urbanization rate 城市化率

Also called *urbanization level*, it is a quantitative index to measure the development degree of urbanisation, which is generally indicated by the proportion of urban population in a total population in certain geographical areas.

又称"城市化水平",是衡量城市化发展程度的数量指标,一般用一定地域内城市人口占总人口比例来表示。

urbanization velocity 城市化速度

It is closely related to urbanization level, which is measured by the number of percentage points of annual average increase in urbanization level.

城市化速度与城市化水平密切相关,它以城市化水平年平均提高的百分点数来衡量。

utility feature of resources 资源效用特征

It is the performance and characteristics of resource products meeting certain needs of the user or consumer.

资源产品满足用户或消费者某种需要的性能及特点。

utilization of ocean energy 海洋能利用

It refers to a process of converting ocean energy into electricity and other available energy with the application of scientific principles, technical measures, equipment and devices.

应用科学原理、技术措施和设备装置将海洋能转换成电能或其他形式的可用能的过程。

V

value of marginal product(VMP) 边际产品价值

It refers to the increased value of output resulting from employing one more unit of a particular input, which is equal to the product of marginal physical product multiplied by the unit price of the output. VMP is a measure of a firm's revenue contributed by the last unit of productive factor employed.

增加一个单位生产要素所增加的产量的价值等于边际物质产品与价格的乘积。

velocity of money 货币流通速度

It is the average frequency with which a unit of money is spent on new goods and services produced domestically over a certain period of time. If the frequency is high, the velocity is increasing, and vice versa. Velocity of money has to do with the transactions of commodities. If more transactions are occurring frequently, the velocity is increasing, and vice versa.

同一单位的货币在一定时期内平均周转的次数,次数越多表明流通速度越快,次数越少表明流通速度越慢。货币流通速度是由商品流通速度决定的,商品流通越快,货币流通也就越快,反之亦然。

vertical stability 垂直稳定度

It indicates the stability of the aqueous layers in the ocean. It is the ratio of the density difference to the vertical distance between two adjacent aqueous layers in the ocean.

表示海洋水层稳定程度的量,为相邻两层海水的密度差与该两层海水之间的垂直距离之比值。

village planning 村庄规划

It refers to the comprehensive deployment, detailed arrangement and implementation measures taken for the economic and social development, land use, special layout and all construction projects of the village over a certain period of time.

对一定时期村庄的经济和社会发展、土地利用、空间布局以及各项建设的综合部署、具体安排和实施措施。

volcanic coast 火山海岸

Accompanying modern volcanic activity, it is formed by substances from volcanic eruption or outpouring. It is usually an island or peninsula coast with volcanic structural features. With tongue-shaped or ring-shaped shoreline, volcanic coast can be primarily divided into tongue-shaped (ring-shaped) coast and caldera island coast.

与现代火山活动相半生,由火山喷出或溢出物质构成的海岸称为火山海岸。通常是具火山构造特征的岛屿或半岛海岸。火山海岸的滨线常呈舌状或环状,可分为舌状(环状)海岸和破火山口岛海岸两种主要类型。

volcanic island 火山岛

It is formed in the sea under volcanism. For example, the Hawaiian island is a volcanic island created by eruption of undersea volcanoes.

由火山作用在海中形成的岛屿称为火山岛。如夏威夷岛即是由海底火山喷发形成的火山岛。

volume of the international standard containers 国际标准集装箱运量

It refers to the quantity of containers which is actually shipped by port vessels. The unit of measurement is ten thousand TEU, ten thousand tons.

港口船舶实际运送的集装箱数量。计量单位:万 TEU、万吨。

W

warm water tongue 暖水舌

It refers to warm water whose isotherm is distributed from high to low in the shape of a tongue on a map of sea water temperature distribution.

在海洋水温分布图上,等温线从高到低呈舌状分布的暖水。

waste water and other pollutants discharged from ocean engineering buildings 废水及其他污染物排海工程建筑

It refers to the construction of urban sewage discharge projects at an estuary and other projects for discharging pollutants to other sea areas.

入海口处的城市污水排海工程和其他向海域排放污染物的建设工程的施工活动。

water and soil loss 水土流失

It refers to a phenomenon where the soil lacking in effective protection can not effectively keep soil moisture, thus resulting in water loss. Meanwhile, due to water loss, soil erosion and scouring occur, leading to soil loss.

缺少有效保护的土壤不能有效地将水分保持在土壤中而造成水分流失的现象,同时伴随水的流失,产生对土壤的侵蚀和冲刷,也使土壤流失。

water collecting area 汇水面积

Water collecting area is the area surrounded by a drainage divide. In other words, waters falling on the area converge to a watercourse along the slope at the exit of the basin where waters join another water body. Size and shape of drainage area directly affect the formation of river runoffs.

又称"受水面积"、"流域面积",指流域分水线所包围的面积,即降落在流域面积上的降水都沿着地面斜坡汇入河道,经流域出口断面流出。流域面积的大小和形状,直接影响河流径流形成的过程。

water consumption 水消费量

It is the enterprise's actual consumption quantity of all kinds of water during the reporting period, including surface water, groundwater, tap water, the water from pipeline supply without standard treatment, the recycled water after the treatment from an urban sewage plant, and seawater which is composed of hot water and geothermal water. Water consumption = The amount of water consumption/The unit price of water. The unit of measurement is cubic metre (ton).

报告期间企业实际消费各种水的数量,包括地表水、地下水、自来水、由管道供应的未经达标处理的水、经城市污水处理厂处理后回用的中水、海水,包括热水、地热水。水消费量=水的消费金额/水的单价。计量单位:立方米(吨)。

water cycle 水循环

It is a continuous cycle in which the

moisture in the lithosphere, hydrosphere, and biosphere enters the atmosphere through evaporation, and then enters the lithosphere, hydrosphere, and biosphere through precipitation under appropriate conditions.

又称"水文循环"。岩石圈、水圈、生物圈中的水分通过蒸发蒸腾作用进入大气圈,在适当条件下又以降水形式进入岩石圈、水圈、生物圈的不断循环过程。

water environment 水环境

As an important component of the natural environment, it refers to the conditions of various water bodies in a natural system.

水环境是自然环境的一个重要组成部分,指自然界各类水体在系统中所处的状况。

water environmental capacity 水环境容量

It studies the upper limit for pollutants that waters can hold under certain natural conditions and goals of social needs, namely the maximum number of pollutants that water can hold under the condition that water environment functions are not destroyed.

研究水域在一定的自然条件和社会需求目标下,所允许容纳的污染物上限,即水环境功能在不受破坏的条件下,受纳水体能够接受污染物的最大数量。

water environmental factor 水环境要素

It refers to the independent, distinct, but interrelated basic material components which reflect the conditions of a water environment.

反映水环境状态的各个独立的、性质不同而又相互联系的基本物质组分。

water mass 水团

It is an enormous and identifiable body of water with a common cradle and formation history. It has relatively even physical, chemical and biological features with basically the same change trend, which is distinct from its surrounding water.

源地和形成机制相近,具有比较均匀的物理、化学和生物特征及大体一致的变化趋势,而与周围海水存在明显差异的宏大水体。

water pollution 水污染

It is a phenomenon where the chemical, physical, biological, or radioactive properties, etc., of a water body are changed due to the intervention of a substance, thus influencing the effective utilization of water, endangering human health or destroying the ecological environment, causing water degradation.

水体因某种物质的介入,而导致其化学、物理、生物或者放射性等方面特性的改变,从而影响水的效有利用,危害人体健康或者破坏生态环境,造成水质恶化的现象。

water quality standard 水质标准

It is the requirement specified by the state on all kinds of water and discharge water in terms of physical, chemical and biological properties.

国家规定的各种用水和排放水在物理性质、化学性质和生物性质方面的要求。

water resources supporting capacity 水资源承载力

[*geography*] It refers to the largest population and economic scale (or gross volume) that can be maintained and supported by the local water resources of an area (or a basin) under certain social, economic and technical conditions, and under the precondition of sustainable utilization of water resources.

[*resources technology*] It refers to the capacity of available water resources within a certain range to sustain the survival and development of human society and a natural environment.

[地理学]在一定的社会经济和技术条件下,在水资源可持续利用前提下,某一区域(流域)当地水资源能够维系和支撑的最大人口和经济规模(或总量)。

[资源科学]一定范围内,可利用水资源能够维护和支撑人类社会和自然环境生存与发展的能力。

water-salt balance 水盐均衡

It refers to the correlation between the amount of ground water and the input and output items salt, and it is one of the research contents in hydrogeological study for soil improvement.

地下水的水量和盐分的收入项和支出项的对比关系,是土壤改良水文地质的研究内容之一。

water spout 海龙卷

It refers to a tornado occurring at sea. The storm is of a small diameter and it rotates violently in a strong cumulonimbus cloud, which approximates a funnel-shaped cloud at the base of the cloud.

海上发生的龙卷风。强积雨云中,小直径剧烈旋转的风暴,近似云底下垂的漏斗云。

wave age 波龄

It refers to the physical quantity to show the growing condition of wind waves. It is the ratio between wind wave propagation speed and wind speed causing waves.

表征风浪成长情况的物理量,用风浪传播速度与引起风浪的风速之比表示。

wave characteristics 波浪要素

It refers to the main characteristic value (eigenvalue) showing wave status. The commonly used characteristic values are wave mode, wave direction, wave height and periodicity.

表征海浪状态的主要特征值。常用的特征值为波型、波向、波高和周期。

wave climate 波候

It refers to long-term statistical properties of wave conditions of a sea area, such as its mean value, variance, extreme probability, etc.

某一海域的波浪状况的长期统计特征。如平均值、方差、极值概率等。

wave cutting 浪蚀

It refers to the wave erosion of lakes, coasts and bottoms by waves, which is presented as the grinding to a coast and water bottom by wave impact and scraps carried in wave.

波浪对湖、海岸和底部的侵蚀作用,

表现为波浪的冲击及所挟带碎屑物质对岸边和水底的磨削作用。

wave diffraction 波浪绕射

When waves encounter obstacles such as breakwaters or islands in the process of transmission, some of them are blocked and reflect back, while others can pass around the obstacles to get to their geometric shelter area. This phenomenon of horizontal wave transmission is called wave diffraction. At that time, the wave can spread along a wave crest line towards the shelter area, with the gradual attenuation of wave height.

又称"波衍射",波浪在传播过程中遇到防波堤或岛屿等障碍物时,一部分受阻反射,另一部分可以绕过障碍物而传至其几何掩蔽区域内,这种波能横向传递的现象称为波浪绕射。此时波能沿波峰线向掩蔽区域扩散,波高也逐渐衰减。

wave direction 波向

It refers to the direction of waves.

波浪传来的方向。

wave disaster 海浪灾害

It refers to a disaster caused by wild winds and huge waves at sea or along the coast, such as the economic loss and casualties caused by a typhoon (hurricane) or the rough sea generated by a cold air gale towards sailing ships, facilities like marine oil production platform and so on, marine fishery fishing ships, coastal and offshore aquaculture farms, harbour wharf, coastal dyke and other coastal construction.

海上和海岸的大风巨浪造成的灾害。如台风(飓风)和冷空气大风引起的凶猛海浪,对航行船舶、海洋石油产生平台等设施、海上渔业捕捞船只,沿岸及近海水产养殖场、港口码头、防潮堤等海岸工程造成的经济损失和人员伤亡。

wave dominated delta 波浪型三角洲

It is formed mainly under wave action. Such a delta is usually formed in an area with less river sand transport, steeper front slopes of deltas and strong alongshore currents. The wave action can reassign the sediment of most rivers flowing into the sea and the sedimentation of delta happens in the delta front, hence the formation of linear and sheeted sand bodies parallel to the water front such as the Gerry Galva Delta in the Gulf of Mexico.

又称"浪控型"三角洲,以波浪作用为主形成的三角洲。此类三角洲一般形成于河流输沙少、三角洲前坡较陡和沿岸流强的地区。波浪作用能够对绝大多数河流入海的沉积物进行再分配,三角洲的沉积作用沿整个三角洲前缘发生,因而形成与岸线平行的线状或席状沙体,如墨西哥湾的格里加尔瓦三角洲。

wave energy 波浪能

It refers to the energy deposited in a wave movement which is generated by wind and other natural forces passing over the ocean surface.

风和其他自然力作用在海面上产生的波浪运动中所存储的能量。

wave forecasting diagram 海浪预报图

It refers to a pictorial diagram describing the spatial distribution characteristics of wave elements of the future ocean.

描述未来洋面上海浪要素的空间分布特征的直观图。

wave height 波高

It refers to a vertical distance between an adjacent wave crest and wave trough on the wave profile.

波剖面上相邻的波峰与波谷间的垂直距离。

wave induced current 波流

It refers to aperiodic horizontal movement or current movement along a certain direction caused by wave motion, which includes littoral currents caused by oblique waves along a coast in breaking wave areas; onshore currents caused by the waves whose movement of propagation is perpendicular to a coast in breaking wave areas; and water displacement caused by a closed movement of fluctuation in water points.

波动引起的非周期性水平运动或波动引起的流体沿一定方向的运动，包括沿海岸斜向传播的波浪引起的破波带内的沿岸水流；垂直于海岸传播的波浪引起的破波带内的向岸流；波动水质点的封闭运动引起的水体位移等。

wave length 波长

It refers to the distance between two successive wave crests on the wave profile.

波剖面上相继两波峰间的距离。

wave period 波周期

It refers to time intervals when two successive wave crests pass through a point on the wave profile.

波剖面上相继两波峰通过某一点的时间间隔。

wave profile 波剖面

It refers to a vertical profile that is perpendicular to a wave crest line or that cuts waves along a wave direction line.

垂直于波峰线或沿波向线切割波浪的铅直剖面。

wave refraction 波浪折射

It refers to a phenomenon that occurs when there is a change of wave direction lines and wave crest lines due to the change of water depth as waves approach the shore.

波浪从外海向近岸传播时，因水深变化而发生波向线和波峰线转折的现象。

wave steepness 波陡

It refers to a physical quantity indicating wave form by the ratio between wave height and wave length.

用波高与波长之比表示波形的物理量。

wave-cut platform 波蚀台

It refers to a narrow flat area outside a sea cliff mainly in an intertidal region, which is composed of bedrocks. Due to the erosion of waves, the sea cliff steps back, the platform inclines seaward slightly and a wave-cut platform takes shape with minor fluctuations including wave-cut grooves, wave-cut columns, eroded bulging, etc. Sand and gravel sediment in some areas develop clearly in bedrock coastlines under strong wave impacts.

又称"浪蚀台"，指海蚀崖外侧的平坦台地，主要分布于潮间带，多由基岩组成。因波浪冲蚀，海蚀崖不断后退所成，台面向海微倾，上有浪蚀沟、浪蚀柱以及溶蚀的小凸地等微小起伏形态，局部见有沙砾沉积物，在波浪作用较强的基岩海岸，发育明显。

weak mineralized groundwater 地下微咸水

It refers to groundwater whose total mineralization is between 1.0 and 3.0 grams per litre.

总矿化度在1.0~3.0克/升之间的地下水。

weathering 风化作用

It refers to the process of physical disintegration and chemical decomposition of rocks, minerals at or near the Earth's surface by the action of solar radiation, atmosphere, water, and living things. During the process, the rocks and materials break down into grains or soils, and their mineral compositions change into new substances.

地表岩石与矿物在太阳辐射、大气、水和生物参与下理化性质发生变化，颗粒细化，矿物成分改变，从而形成新物质的过程。

wetland ecology 湿地生态学

It is a discipline that studies the community structures, functions, ecological processes and evolution laws of various types of swamp wetland ecosystems, as well as the interaction with physical and chemical factors, and biological components.

研究各种类型沼泽湿地生态系统的群落结构、功能、生态过程和演化规律及其与理化因子、生物组分之间相互作用机制的学科。

wharf 码头

It is a structure where ships may dock to load and unload cargo and passengers. In a broad sense, it also includes supporting warehouses, storage yards, roads, railways and other facilities.

供船舶停靠并装卸货物和上下旅客的建筑物。广义还包括与之配套的仓库、堆场、道路、铁路和其他设施。

wholesale of marine aquatic products 海洋水产品批发

It refers to the wholesale as well as the import and export of marine aquatic products and processed aquatic products. Aquatic products include fresh and frozen aquatic products while aquatic processed products include dried fish floss, fish fillet, fish ball, fish sauce, etc.

海洋水产品和水产加工品的批发和进出口，水产品包括新鲜水产品和冷冻水产品，水产加工品包括鱼松、鱼片、鱼丸、鱼酱等。

wholesale of marine biological medicines 海洋生物药品批发

It refers to the wholesale as well as the import and export of marine bio-chemical medicine, traditional Chinese medicinal materials and Chinese patent medicine.

海洋生物化学药品、中药材及中成药的批发和进出口。

wholesale of marine chemical fertilizers 海洋化肥批发

It refers to the wholesale as well as the import and export of marine chemical fertilizers, such as potassium fertilizer, magnesium fertilizer, and compound fertilizer.

海洋化肥的批发和进出口,如钾肥、镁肥、复混肥料。

wholesale of marine fishery machineries 海洋渔业机械批发

It refers to the wholesale as well as the import and export of machinery and equipment for marine fishery and related service industries.

海洋渔业及相关服务业的机械设备的批发和进出口。

wholesale of marine fishery machinery 海洋渔业管理

It refers to the management of marine fishery transactions by government departments at all levels.

各级政府部门对海洋渔业事务的管理活动。

wholesale of marine health products 海洋保健品批发

It refers to the wholesale as well as the import and export of marine health products.

海洋保健食品的批发和进出口。

wholesale of marine petrochemical products 海洋石化产品批发

It refers to the wholesale as well as the import and export of marine petrochemical products.

海洋石油化工产品的批发和进出口。

wholesale of marine petroleum and products 海洋石油及制品批发

It refers to the wholesale as well as the import and export of marine petroleum, natural gas and marine petroleum products.

海洋石油和天然气、海洋石油制品的批发和进出口。

wholesale of seawater desalination products 海水淡化产品批发

It refers to the wholesale, the import and export of barreled water and bottled water produced after seawater desalination.

海水淡化后的桶装及瓶装水的批发和进出口。

width of the territorial sea 领海宽度

It refers to a vertical distance from a baseline to the outer limit of a territorial sea of coastal states. According to the *United Nations Convention on the Law of the Sea*, "Every State has the right to establish the breadth of its territorial sea up to a limit not exceeding 12 nautical miles, measured from baselines determined in accordance with this Convention." (Article 3 of the *Convention*) It is reaffirmed in the *Law of the People's Republic of China Concerning Territorial Waters and Contiguous Zone* in Feb, 1992 that the width of the territorial sea of China is 12 nautical miles.

沿海国领海基线至领海外部界限的垂直距离。《联合国海洋法公约》规定："每一国家有权确定其领海的宽度,直至从按照本公约确定的基线量起不超过十二海里的界限为止"(《联合国海洋法公约》第三条)。1992年2月《中华人民共和国领海及毗连区法》重申,中国领海的宽度为12海里。

wind duration 风时

It refers to the time of duration during which wind of the same state acts on the sea surface continuously.

状态相同的风持续作用于海面的时间。

wind energy 风能

It is the kinetic energy resulting from the airflow on the surface of the earth and it is a conversion form of solar energy. The higher the wind speed is, the more powerful its energy will be.

地球表面空气流动所形成的动能。风能是太阳能的一种转化形式,风速愈大,它具有的能量愈大。

wind erosion 风蚀

It refers to the deflation resulting from the wind's removal of sand and loose topsoil (and the abrasion resulting from the wearing down of surfaces by the grinding action and sandblasting of wind-borne particles). It is an important form of soil erosion.

风对沙、尘的吹扬造成的吹蚀作用,以及风吹沙尘对地面产生的磨蚀作用。是土壤侵蚀的一种重要形式。

wind wave 风浪

It refers to a wave motion resulting from the direct action of wind on a water surface.

海面在风力直接作用下产生的波动现象。

wind-generated noise 风生海洋噪声

It refers to the noise generated due to the action of the wind on the sea surface.

由于风对海面作用产生的噪声。

World Climate Conference in Copenhagen 哥本哈根世界气候大会

It is the 15th session of the conference for contracting parties of the *United Nations Framework Convention on Climate Change* (UNFCCC) and the 5th session of the conference for contracting parties of the *Kyoto Protocol*, which was held in Copenhagen, capital of Denmark, from December 7th to 18th, 2009. Negotiators from 192 countries attended the summit, and discussed the follow-up program when the first commitment period of the *Kyoto Protocol* expires, namely the agreement to curb the global greenhouse gas emissions from 2012 to 2020.

哥本哈根世界气候大会全称《联合国气候变化框架公约》第 15 次缔约方会议暨《京都议定书》第 5 次缔约方会议,于 2009 年 12 月 7—18 日在丹麦首都哥本哈根召开,来自 192 个国家的谈判代表召开峰会,共同商讨《京都议定书》一期承诺到期后的后续方案,即 2012—2020 年的全球减排协议。

World Climate Research Programme (WCRP) 世界气候研究计划

It was jointly sponsored by the *World Meteorological Organisation and International Council for Science*. The objectives of the Programme are to study the Earth's physical climate system and climate processes needed to determine to what extent the climate can be predicted and the influence of human activities on the climate. The Programme encompasses studies of the global atmosphere, oceans, cryosphere, the land, as well as the interaction and feedback among them. Launched in the 1970s and performed in the 1980s, WCRP is one of the earlier programs to develop in global change research.

世界气候研究计划(WCRP)由世界气象组织与国际科学联合会联合主持,以气候的可预报程度和人类活动对气候的影响为目标,主要研究地球系统中有关气候的物理过程,涉及整个气候系统。其主要部分是大气、海洋、低温层(冰雪圈)和陆地以及这些组成部分之间的相互作用和反馈。此计划在 20 世纪 70 年代开始酝酿,80 年代开始执行,是全球变化研究中开展得较早的一个计划。

World Commission on Environment and Development 世界环境与发展委员会

It was formally established in May 1984 by a resolution of the *38th United Nations (UN) General Assembly* in 1983. Mrs. Brundtland, the leader of the *Norwegian Labor Party* was the chairman of the Commission under the nomination of the *UN Secretary General*. The Commission's mission is to review the key issues in world environment and development, to creatively put forward the practical proposals to address problems, and to improve the awareness of individuals, groups, business circles, research institutions and national governments on environment and development. The influential report known as *Our Common Future* was released in 1987 at a Tokyo environment special conference.

1983 年第 38 届联合国大会通过成立的决议,1984 年 5 月正式成立,并由联合国秘书长提名挪威工党当时领袖布伦特兰夫人(Brundtland)任委员会主席。委员会的主要任务是:审查世界环境和发展的关键问题,创造性地提出解决这些问题的现实行动建议,提高个人、团体、企业界、研究机构和各国政府对环境与发展的认识水平。在 1987 年于东京召开的环境特别会议上提出极具影响力的报告《我们共同的未来》。

World Meteorological Organization (WMO) 世界气象组织

Originated from the *International Meteorological Organization* (IMO), it was founded in Vienna in 1873. As a specialised agency of the United Nations (UN), WMO fosters cooperation on meteorological services and sciences between governments around the world.

世界气象组织是联合国的专门机构之一。其前身为国际气象组织(International Meteorological Organization, IMO),于 1873 年在维也纳成立。是世界各国政府间开展气象业务和气象科学合作活动的国际机构。

World Ocean Circulation Experiment (WOCE) 世界海洋环流试验

Initiated by the *Intergovernmental Oceanographic Commission*, it is a large international cooperation plan implemented between 1990 to 2002 for the observation and research of ocean circulation.

由政府间海洋学委员会发起的,在 1990—2002 年期间实施的大型国际合作海洋环流观测和研究计划。

World Wide Fund For Nature(WWF) 世界自然基金会

Established in 1961 and headquartered in Switzerland, it is the world's largest independent non-governmental conservation organization working on the protection of the environment. It is a prestigious organization with about 5.2 million supporters worldwide and a network that is active in more than 100 countries. The WWF is committed to stopping the degradation of the planet's natural environment and protecting the world's biodiversity; ensuring the sustainable utilization of renewable natural resources; and promoting the reduction of pollution and wasteful consumption to build a future in which humans live in harmony with nature.

该组织成立于1961年,总部设于瑞士,是在全球享有盛誉的、最大的独立性非政府环境保护组织之一,致力于环保事业,在全世界拥有将近520万支持者和一个在100多个国家活跃着的网络。WWF致力于遏止地球自然环境的恶化,保护世界生物多样性;确保可再生自然资源的可持续利用;推动降低污染和减少浪费性消费的行动,创造人类与自然和谐相处的美好未来。

X

Xisha Islands 西沙群岛

Located in the northwest of the South China Sea, and southeast of Hainan Island, they are one of the four largest groups of islands in the South China Sea. Composed of Yongle and Xuande Islands, the Xisha Islands have a total of 22 islands, 7 shoals, and over 10 submerged reefs and hidden shoals. Situated in the central tropics, they belong to a tropical monsoon climate, hot and humid, but without intense heat.

位于南海的西北部,海南岛东南方,中国南海四大群岛之一,由永乐群岛和宣德群岛组成,共有22个岛屿,7个沙洲,另有10多个暗礁暗滩。地处热带中部,属热带季风气候,炎热湿润,但无酷暑。

Y

Yangtze River 长江

It is the longest river in China and the third longest river in the world, just second to the Nile in Africa and the Amazon in South America. Originating in Geladaindong Peak the Tanggula Mountains on the Qinghai-Tibet Plateau, it flows southeast through 11 provinces and municipalities including Qinghai, Tibet, Sichuan, Yunnan, Chongqing, Hubei, Hunan, Jiangxi, Anhui, Jiangsu before emptying into the East China Sea at Shanghai. The total length is 6,397 kilometres and it is regarded as the mother river of China as is the Yellow River.

中国第一大河,世界第三长河。仅次于非洲的尼罗河与南美洲的亚马孙河。它发源于青藏高原唐古拉山主峰各拉丹冬雪山,流经青海、西藏、四川、云南、重庆、湖北、湖南、江西、安徽、江苏、上海等11个省(自治区、直辖市),最后在上海注入东海。全长6 397千米和黄河并称为中华民族的"母亲河"。

Yangtze River delta 长江三角洲

It refers to the alluvial plain formed by the Yangtze River flowing into the sea, generally comprising the triangular-shaped territory of Wu-speaking Shanghai, southern Jiangsu province and northern Zhejiang province of China. This area incorporates more than sixteen relatively developed cities in three provinces, specifically including Shanghai municipality, Nanjing, Suzhou, Wuxi, Changzhou, Zhenjiang, Nantong, Yangzhou and Taizhou in Jiangsu province, Hangzhou, Ningbo, Jiaxing, Huzhou, Shaoxing, Zhoushan and Taizhou in Zhejiang province. It covers a land area of 109,600 square kilometers.

长江入海而形成的冲积平原,包括上海市、江苏省和浙江省的部分地区。由沪、苏、浙三地16个地级以上城市组成的复合型区域,具体包括上海市、江苏省的南京、苏州、无锡、常州、镇江、南通、扬州和泰州,以及浙江省的杭州、宁波、嘉兴、湖州、绍兴、舟山和台州,其土地面积为10.96万平方千米。

Yangtze river delta economic zone 长江三角洲经济区

It is an economic region in the Yangtze River delta area that encompasses the sea and land area of Shanghai municipality, Zhejiang and Jiangsu province.

长江三角洲沿岸地区所组成的经济区域,主要包括江苏省、上海市和浙江省两省一市的海域与陆地。

Yellow River 黄河

It is the second longest river in China, with a total length of 5,464 km and a basin area of 752,400 square kilometers. Maqu (or Yueguzong Canal) originating in the Bayan Har Mountains in Qinghai province, and Kariqu originating in the Anemaqen (Amne Machin) Mountains are upper sources that join each other near E-ling, then flow eastward through Sichuan, Gansu, Ningxia, Inner Mongolia, Shaanxi, Shanxi, Henan and other provinces and finally empty into Bohai Sea in Northern Shandong.

我国第二大河。上源马曲(约古宗列渠)出自青海省巴颜喀拉山脉雅拉达泽山麓,卡日曲出自各姿各雅山麓,在鄂陵附近相汇,向东流经四川、甘肃、宁夏、内蒙古、陕西、山西、河南等省(自治区),在山东省北部入渤海。全长5 464千米,流域面积75.24万平方千米。

Yellow River Delta 黄河三角洲

In broad sense, it refers to the broad flood plain of Yellow River, stretching to the east of Gong County in Henan, to Tianjin in the north and to the abandoned Yellow River mouth (Huai River) in the south, covering an area of 240,000 square kilometers. It is the largest delta in the world. In a narrow sense, it is a land form with the axis point located in Kenli County in the city of Dongying in Shandong, which stretches to Taoer River mouth in the north, Zimai River mouth in the south, and disperses eastward in fan shape. Its altitude is lower than 15 meters. Its total basin area reaches 5,450 square kilometers and total population is about 9.85 million.

广义的黄河三角洲指河南巩县以东,北至天津,南到废黄河口(淮河)的广大黄河冲积平原,面积约24万平方千米,是世界上最大的三角洲。狭义的黄河三角洲指以山东省东营市垦利县宁海为轴点,北起套尔河口,南至滋脉河口,向东撒开的扇状地形,海拔高程低于15米,面积达5450平方千米,总人口约985万人。

Yellow Sea 黄海

It is a marginal sea of the Western Pacific Ocean between the Chinese mainland and the Korean Peninsula.

位于中国大陆与朝鲜半岛之间的西太平洋边缘海。

Z

Zero Energy Development (ZED) 零能源发展

It is a new mode of energy utilization, whose design philosophy is to maximise the use of natural resources, reduce environmental destruction and pollution, realise zero fossil energy use and establish a new residential mode by which energy demand and waste disposal can be recycled. Zero energy development provides a comprehensive solution for the sustainable development of urban residential housing.

能源利用方式的一种新形式。零能源发展系统的设计理念在于最大限度地利用自然资源、减少环境破坏与污染、实现零化石能源使用的目的,能源需求与废物处理实现基本循环利用的居住模式。零能源发展为城市住宅建筑实现可持续发展提供了一个综合性解决方案。

zero-energy coast 零能海岸

It refers to a coast where an average breaking wave is 3-4 centimetres high or less and the wave energy is close to zero. This kind of coast is generally indented in a leeward bay with wide and shallow offshore slopes. Tidal swamps and low moors are distributed in the nearshore zone.

平均破波高度3~4厘米或更小、波浪能量趋近于零的海岸,称为零能海岸。此类海岸一般位于背风的凹入海湾内,滨外斜坡宽浅,近滨地带分布潮汐沼泽和低位沼泽。

Zhongsha Islands 中沙群岛

About 200 kilometres west of Yongxing Island in the Xisha Islands, they are located in the centre of the four large Islands in South China Sea. Its main parts are composed of three underwater hidden dunes, shoals, reefs and islets. About 140 kilometres long (exclusive of Huangyan Island) and 60 kilometres wide, they have a slightly elliptical shape, extending from northeast to southwest.

中国的南海诸岛四大群岛中位置居中的群岛。西距西沙群岛的永兴岛约200千米。它的主要部分由隐没在水中的3座暗沙、滩、礁、岛组成。长约140千米(不包括黄岩岛),宽约60千米,从东北向西南延伸,略呈椭圆形。

References 参考文献

GB 12319-998. 1990. 中国海图图式.北京:中国标准出版社.
GB/12763.6-2007. 2007.海洋调查规范 第 6 部分:海洋生物调查.北京:中国标准出版社.
GB/T 12763.9—2007. 2007.海洋调查规范第 9 部分:海洋生态调查指南.北京:中国标准出版社.
GB/T 14157-93. 1993. 水文地质术语.北京:中国标准出版社.
GB/T 15918-1995. 1995. 海洋学综合术语.北京:中国标准出版社.
GB/T 18972-2003. 2003. 旅游资源分类、调查与评价.北京:中国标准出版社.
GB/T 19485-2004. 2004. 海洋工程环境影响评价技术导则.北京:中国标准出版社.
GB/T 19571-2004. 2004. 海洋自然保护区管理技术规范.北京:中国标准出版社.
GB/T 20794—2006. 2010. 海洋以及相关产业分类.北京:中国标准出版社.
GB/T 24050-2004/ISO 14050:2002. 2004.环境管理术语.北京:中国标准出版社.
HJ/T 19—1997. 1997. 环境影响评价技术导则 非污染生态影响.北京:中国环境出版社.
HY 070-2003. 2003. 海域使用面积测量规范.北京:中国标准出版社.
HY/T 118—2008. 2008. 海洋特别保护区功能分区和总体规划编制技术导则.北京:中国标准出版社.
HY/T124-2009. 2009. 海籍调查规范.北京:中国标准出版社.
安建.2009.中华人民共和国城乡规划法释义.北京:法律出版社.
蔡佳亮,殷贺,黄艺.2010.生态功能区划理论研究进展.生态学报.
陈同庆,王明新,李维博,等.2010.海洋腐蚀与防护辞典(修订版).北京:海洋出版社.
大气科学辞典编委会.1994.大气科学辞典.北京:气象出版社.
邓绶林.1992.地学辞典.石家庄:河北教育出版社.
范如国.2011.制度演化及其复杂性.北京:科学出版社.
封吉昌.2011.国土资源实用词典.武汉:中国地质大学出版社.
冯士筰,李凤岐,李少菁.1999.海洋科学导论.北京:高等教育出版社.
高鸿业.2007 西方经济学(宏观部分)第四版.北京:中国人民大学出版社.
国家海洋局.2008.海域使用分类体系.北京:国海管字〔2008〕273 号.
海洋大辞典编辑委员会.1998.海洋大辞典.沈阳:辽宁人民出版社.
韩立民,任新君.2009.海域承载力与海洋产业布局关系初探.太平洋学报.
韩双林.1993.证券投资大辞典.哈尔滨:黑龙江人民出版社.
韩增林,刘桂春.2007.人海关系地域系统探讨.地理科学.
韩战涛,刘树金.2001.关于实行捕捞限额制度的探讨.中国渔业经济.

侯纯扬.2002(4).海水冷却技术.海洋技术.
环境科学大辞典编委会.2008.环境科学大辞典.北京:中国环境科学出版社.
黄达.2009.金融学(第二版)精编版.北京:中国人民大学出版社.
解景林.1990.国际金融大辞典.哈尔滨:黑龙江人民出版社.
李宜良,王震,王晶.2011.海岛统计调查指标体系研究.中国渔业经济.
刘大海,纪瑞雪,关丽娟等.2012.海陆二元结构均衡模型的构建及其运行机制研究.海洋开发与管理,(7).
刘树成. 2005.现代经济词典. 南京:江苏人民出版社.
卢宁,韩立民.2008.海陆一体化的基本内涵及其实践意义.太平洋学报,(3).
罗肇鸿. 1995.资本主义大辞典. 北京:人民出版社.
马志荣. 2005.新世纪实施科技兴海战略的思考. 科技进步与对策.
曼昆.2009.经济学原理(宏观经济学分册)第5版.梁小民,梁砾译.北京:北京大学出版社.
农业大词典编辑委员会.1998.农业大词典.北京:中国农业出版社.
全国科学技术名词审定委员会. 1991.海洋科学名词. 北京:科学出版社.
全国科学技术名词审定委员会.1989.地理学名词(第二版,定义版).北京:科学出版社.
全国科学技术名词审定委员会.1991.古生物学名词. 北京:科学出版社.
全国科学技术名词审定委员会.1996.大气科学名词(第三版,定义版).北京:科学出版社.
全国科学技术名词审定委员会.1997.航海科技名词. 北京:科学出版社.
全国科学技术名词审定委员会.1998.水利科技名词(定义版).北京:科学出版社.
全国科学技术名词审定委员会.2002.测绘学名词(第三版,定义版).北京:科学出版社.
全国科学技术名词审定委员会.2002.水产名词(定义版).北京:科学出版社.
全国科学技术名词审定委员会.2007.海洋科技名词. 北京:科学出版社.
全国科学技术名词审定委员会.2008.资源科学技术名词. 北京:科学出版社.
全国科学技术名词审定委员会审定.1998.地理学名词. 北京:科学出版社.
全国科学技术名词审定委员会审定.2002.电力名词. 北京:科学出版社.
全国科学技术名词审定委员会审定.2007.生态学名词.北京:科学出版社.
全国人民代表大会常务委员会法制工作委员会.2010.中华人民共和国海岛保护法释义. 北京:法律出版社.
任超奇.2006.新华汉语词典. 武汉:崇文书局.
沈文周.2010.简明数字海洋科技文化词典.北京:海洋出版社.
孙久文,叶裕民.2010.区域经济学教程. 北京:中国人民大学出版社.
王倩,李彬.2011.关于"海陆统筹"的理论初探.中国渔业经济,(3).

王伟光,郑国光.2011.德班的困境与中国的战略选择. 北京:社会科学文献出版社.
魏振瀛.2007.民法(第三版). 北京:北京大学出版社.
吴金培,李学伟.2010.系统科学发展概论. 北京:清华大学出版社.
奚洁人.2007.科学发展观百科辞典.上海:上海辞书出版社.
辛仁臣,刘豪.2008.海洋资源.北京:中国石化出版社.
邢继俊,黄栋,赵刚.2010.低碳经济报告. 北京:电子工业出版社.
熊武一.1988.当代军人辞典. 北京:新华出版社.
许力以.2008.百科知识数据辞典. 青岛:青岛出版社.
杨青山,梅林.2001.人地关系、人地关系系统与人地关系地域系统. 经济地理.
殷克东.2010.海洋经济统计术语.北京:经济科学出版社.
张玉忠,彭晓敏.2004.浅谈海水循环冷却处理技术. 工业水处理,(8).
赵瑞林,陈公雨,王诗成.2004.山东省海域使用管理条例释义. 北京:海洋出版社.
中国农业百科全书总编辑委员会. 1994. 中国农业百科全书水产业卷(上). 北京:农业出版社.
中华法学大辞典(简明本). 2003.北京:中国检察出版社.
中华人民共和国国务院.国务院关于印发全国海洋经济发展规划纲要的通知. 北京:国发〔2003〕13 号.
周成虎.2006.地貌学词典. 北京:中国水利水电出版社.
朱贤姬,郝艳萍,梁熙掂.2010.关于"渤海碧海行动计划"的几个思考.海洋开发与管理.
朱晓东,李扬帆,吴小根,等.2005. 海洋资源概论. 北京:高等教育出版社.